Graph Coloring Problems

A complete list of titles in this series appears at the end of this volume

Graph Coloring Problems

TOMMY R. JENSEN
BJARNE TOFT
Odense University

A Wiley-Interscience Publication
JOHN WILEY & SONS
New York • Chichester • Brisbane • Toronto • Singapore

This text is printed on acid-free paper.

Library of Congress Cataloging in Publication Data:

Jensen, Tommy R.
Graph coloring problems / Tommy R. Jensen and Bjarne Toft.
p. cm. — (Wiley-Interscience series in discrete mathematics
and optimization)
"A Wiley-Interscience publication."
Includes bibliographical references and index.
ISBN 0-471-02865-7 (pbk. : acid-free)
1. Map-coloring problem. I. Toft, Bjarne. II. Title.
III. Series.
QA612.18.J46 1995
511′.5—dc20 94-11418

Printed in the United States of America

10 9 8 7 6 5 4 3 2

to Paul Erdős

Contents

Preface

WHY GRAPH COLORING?

Graph coloring problems? The four-color problem has been changed into the four-color theorem, so is there really much more to say or do about coloring? Yes there is, and for several reasons!

First, the last word on the four-color problem has not been said. The ingenious solution by K. Appel, W. Haken, and J. Koch [1, 2], based on the approach of H. Heesch, is a major achievement, but to some mathematicians the solution is unsatisfactory and raises new questions, both mathematical and philosophical.

Second, graph coloring theory has a central position in discrete mathematics. It appears in many places with seemingly no or little connection to coloring. A good example is the Erdős–Stone–Simonovits theorem [3] in extremal graph theory, showing that for a fixed graph G the behavior of the maximum number $f(n, G)$ of edges in a graph on n vertices not containing G as a subgraph depends on the chromatic number $\chi(G)$ of G:

$$\lim_{n \to \infty} \frac{f(n, G)}{n^2} = \frac{\chi(G) - 2}{2\chi(G) - 2}.$$

Third, graph coloring theory is of interest for its applications. Graph coloring deals with the fundamental problem of partitioning a set of objects into classes, according to certain rules. Time tabling, sequencing, and scheduling problems, in their many forms, are basically of this nature.

Fourth, graph coloring theory continually surprises by producing unexpected new answers. For example, the century old five-color theorem for planar graphs due to P.J. Heawood [4] has recently been furnished with a new proof by C. Thomassen [5], avoiding both the use of Euler's formula and the powerful recoloring technique invented by A.B. Kempe [6], thus making it conceptually simpler than any previous proof.

And finally, even if many deep and interesting results have been obtained during the 100 years of graph coloring, there are very many easily formulated, interesting problems left. This is the most important reason for us, and our book is an attempt to exemplify it. As far as we know it is the first book devoted to unsolved graph coloring problems, but a number of papers sharing the same topic have preceded it—for example, many of the "problems and results" papers by P. Erdős (referred to throughout the book), surveys by W. Klotz [7], Z. Tuza [8], and J. Kahn [9], problem sections in proceedings and newsletters (such as the column by D.B. West [10]), and lists of "problems from the world surrounding perfect graphs" by A. Gyárfás [11] and V. Chvátal [12]. A list of 50 carefully selected problems in graph theory is contained in the book by J.A. Bondy and U.S.R. Murty [13]. Finally, two interesting collections of geometry problems, by H.T. Croft, K.J. Falconer, and R.K. Guy [14], and by W. Moser (McGill University, Canada) and J. Pach [15], share some of our general ideas and contain some coloring problems.

In a delightful paper W.T. Tutte [16] described several difficult coloring conjectures, many of them generalizing the four-color theorem. The paper showed, in Tutte's words, that *"The Four Colour Theorem is the tip of the iceberg, the thin end of the wedge and the first cuckoo of spring."*

THE PROBLEMS

In selecting and presenting the more than 200 problems for this book we had four main objectives in mind:

1. Each problem should be simple to state and understand, and thus problems requiring several or complicated definitions are not included. Only a few of the problems have the character of a broad research program; most of them are specific questions. We have aimed to select for each problem its most attractive formulation, which may not always be the most general or the most specific. But very often we mention more general versions and/or special cases in the comments.
2. The list of problems should tell not only what is *not* known in graph coloring theory. The comments should also provide an exposition of the major known graph coloring results.
3. The history of the problems, and the credit for them and for the results presented, should be as accurate and complete as possible.
4. The list should not consist just of "impossible" problems, but also of questions where progress is definitely possible.

We did not intend to write a textbook to be read from beginning to end, but rather a catalog suitable for browsing. Chapter 1 contains a common basis of graph coloring terminology and a collection of important theorems. The remaining 16 chapters comprise the main body of the book, each containing a list of open problems within a separate area. The necessary background for understanding each problem and the information directly related to it appear together with the statement of the problem. Each chapter is intended to be self-contained and is closed by its own separate list of references. We have paid a price in terms of having to allow some redundancies, but we think that the level is tolerable even for the thorough reader. To make the presentation short and succinct, we have included very few proofs and pictures. Proofs, outlines of arguments, or figures have been added in a few cases when we did not have an appropriate source of reference.

There is one remark we should make concerning the organization of the references. When consulting any given one of the bibliographies it may seem strange that different papers by the same author(s) and published within the same year are not always listed in a consecutively numbered fashion. For example, there is a reference to a paper of Edmonds [1965b] in Chapter 2, but there is no reference to a paper of Edmonds [1965a] preceding it in the bibliography. The explanation is that we have chosen to maintain a consistent numbering of the references throughout the entire book. In other words, the numbering is exactly as it would have been, had the references all been put together into one big list. Thus the same paper is being referred to in the same manner throughout.

UPDATES

The present activity in discrete mathematics is so extensive that a work of this nature is outdated before it is written! Solutions, partial results, and new ideas appear all the time. And there will be interesting questions that we have overlooked, and also, solutions or partial solutions. In some cases we have probably not met objective 3. We apologize for all such cases, and we shall be grateful for corrections, comments, and information.

For easy access to any new and updated information, we have installed an `ftp`-archive at Odense University, Denmark. You can reach this facility via `ftp` using the address

`ftp.imada.ou.dk`

logging in as "`anonymous`" and giving your `e-mail` address as the password. The archive is located in a directory which can be reached by typing the command `cd pub/graphcol`, where a short `README` file is available for further information on how to proceed.

`World Wide Web` access to the archive is also available. You either need to locate the menu of `Danish Information Servers`, and then click successively on the menus for `IMADA`, listed under Odense University, `Research Activities`, and `Graph Theory`. Or you may use the address

`http://www.imada.ou.dk`

to directly reach the `IMADA` Home Page.

The contents of the `ftp`-archive will depend largely on new information (papers, abstracts, questions, solutions, etc.) sent by our readers. Contributions should be e-mailed to the address

graphcol@imada.ou.dk

to be considered for inclusion.

In addition to the ftp-archive, we shall consider writing updates from time to time in the form of articles. Such papers will be submitted to the *Journal of Graph Theory*.

ACKNOWLEDGMENTS

The senior author, Bjarne Toft, learned graph theory from G.A. Dirac, starting in 1966, at the University of Aarhus in Denmark. Dirac's lectures were captivating. He presented the subject as a general and serious mathematical theory, and he did so with rigor and care, to an extent which made some of his colleagues think that graph theory is just a collection of easy facts with trivial proofs. Dirac's thorough style, strongly influenced by the book of Dénes König [17], was, however, a delight for his students. Later, in 1968–1970 and in 1973, Toft spent semesters at the Hungarian Academy of Sciences in Budapest, Hungary, the University of London, England, and the University of Waterloo, Canada. Later, in 1985–1986, while visiting the University of Regina, Canada, and Vanderbilt University, United States, the first steps toward this book were taken.

Tommy Jensen first learned graph theory at Odense University, Denmark, in the first half of the 1980s under the supervision of Bjarne Toft, and later, while studying at the University of Waterloo, Canada, enjoyed the benefit of receiving supervision from D.H. Younger. He owes the beginning of his interest in mathematics to his father, Emil Jensen, who sadly did not get to see the finished version of this book.

We are glad for the opportunities given to us to learn from some of the greatest mathematicians in graph theory, such as P. Erdős, to whom we dedicate this book, and J. Edmonds, T. Gallai, and W.T. Tutte. We are most grateful for the helpfulness and generosity we have met from many sides. In connection with the present work we would like to thank a very large number of people who supplied information—their names may be found in appropriate places in this book. In particular, we wish to thank the following for showing their interest in the project by giving us highly qualified comments and suggestions at various stages: M.O. Albertson, N. Alon, O.V. Borodin, F. Jaeger, M.K. Goldberg, R. Häggkvist, D. Hanson, A.V. Kostochka, L. Lovász, P. Mihók, G. Sabidussi, M. Stiebitz, C. Thomassen, and two anonymous referees.

Of course we alone are responsible for all errors, inaccuracies, and omissions.

Finally, we would like to express our appreciation of the support received from Wiley-Interscience and their editorial staff, in particular Maria Allegra, Elizabeth Murphy, Kimi Sugeno, and Angela Volan, from the Danish Natural Science Research Council, the University of Regina, Canada, and Odense University, Denmark. We also thank Margit Christiansen of the Mathematical Library in Odense—may our readers be as adept in problem solving!

BIBLIOGRAPHY

[1] K. Appel and W. Haken. Every planar map is four colorable. Part I: Discharging. *Illinois J. Math.* **21**, 429–490, 1977.

[2] K. Appel, W. Haken, and J. Koch. Every planar map is four colorable. Part II: Reducibility. *Illinois J. Math.* **21**, 491–567, 1977.

[3] P. Erdős and M. Simonovits. A limit theorem in graph theory. *Studia Sci. Math. Hungar.* **1**, 51–57, 1966.

[4] P.J. Heawood. Map colour theorem. *Quart. J. Pure Appl. Math.* **24**, 332–338, 1890.

[5] C. Thomassen. Every planar graph is 5-choosable. *J. Combin. Theory Ser. B* **62**, 180–181, 1994.

[6] A.B. Kempe. On the geographical problem of four colours. *Amer. J. Math.* **2**, 193–200, 1879.

[7] W. Klotz. Clique covers and coloring problems of graphs. *J. Combin. Theory Ser. B* **46**, 338–345, 1989.

[8] Z. Tuza. Problems and results on graph and hypergraph colorings. In: M. Gionfriddo, editor, *Le Matematiche*, volume XLV, pages 219–238, 1990.

[9] J. Kahn. Recent results on some not-so-recent hypergraph matching and covering problems. In: P. Frankl, Z. Füredi, G.O.H. Katona, and D. Miklós, editors, *Extremal Problems for Finite Sets*, volume 3 of *Bolyai Society Mathematical Studies*. János Bolyai Mathematical Society, 1994.

[10] D.B. West. Open problems. A regular column in *SIAM J. Discrete Math. Newslett.*

[11] A. Gyárfás. Problems from the world surrounding perfect graphs. *Zastos. Mat.* **XIX**, 413–441, 1987.

[12] V. Chvátal. Problems concerning perfect graphs. Manuscript. Dept. Computer Science, Rutgers University, New Brunswick, NJ 08903, USA, 1993.

[13] J.A. Bondy and U.S.R. Murty. *Graph Theory with Applications*. Macmillan, 1976.

[14] H.T. Croft, K.J. Falconer, and R.K. Guy. *Unsolved Problems in Geometry*. Springer-Verlag, 1991.

[15] W. Moser and J. Pach. *Research Problems in Discrete Geometry*, 1994.

[16] W.T. Tutte. Colouring problems. *Math. Intelligencer* **1**, 72–75, 1978.

[17] D. König. *Theorie der endlichen und unendlichen Graphen*. Akademische Verlagsgesellschaft M.B.H. Leipzig, 1936. Reprinted by Chelsea 1950 and by B.G. Teubner 1986. English translation published by Birkhäuser 1990.

MATH. AND COMP. SCI. DEPT.
ODENSE UNIVERSITY
DK–5230 ODENSE M
DENMARK

MAY 1994

TOMMY R. JENSEN
BJARNE TOFT

Graph Coloring Problems

1

Introduction to Graph Coloring

1.1. BASIC DEFINITIONS

Partitioning a set of objects into classes according to certain rules is a fundamental process in mathematics. A conceptually simple set of rules tells us for each pair of objects whether or not they are allowed in the same class. The theory of graph coloring deals with exactly this situation. The objects form the set of vertices $V(G)$ of a graph G, two vertices being joined by an edge in G whenever they are not allowed in the same class. To distinguish the classes we use a set of colors C, and the division into classes is given by a *coloring* $\varphi : V(G) \to C$, where $\varphi(x) \neq \varphi(y)$ for all xy belonging to the set of edges $E(G)$ of G. If C has cardinality k, then φ is a *k-coloring*, and when k is finite, we usually assume that $C = \{1, 2, 3, \ldots, k\}$. For $i \in C$ the set $\varphi^{-1}(i)$ is the ith *color class*. Thus each color class forms an *independent* set of vertices; that is, no two of them are joined by an edge. The minimum cardinal k for which G has a k-coloring is the *chromatic number* $\chi(G)$ of G, and G is *$\chi(G)$-chromatic*. The existence of the chromatic number follows from the Well-Ordering Theorem of set theory, and conversely, considering cardinals as special ordinals, the existence of the chromatic number easily implies the Well-Ordering Theorem. However, even if it is not assumed that every set has a well-ordering, but maintaining the property that every set has a cardinality, then the statement "Any finite or infinite graph has a chromatic number" is equivalent to the Axiom of Choice, as proved by Galvin and Komjáth [1991].

If the condition $\varphi(x) \neq \varphi(y)$ for all $xy \in E(G)$ is dropped from the definition of coloring, then φ is called an *improper coloring* of G. Accordingly, the term *proper coloring* is sometimes used when we want to emphasize that this condition holds.

For a hypergraph H with vertex set $V(H)$ and edge set $E(H)$, a *coloring* $\varphi : V(H) \to C$ must assign at least two different colors to the vertices of every edge in H. That is, no edge is monochromatic. If the edges of H all have the same size r, we say that H is *r-uniform*. Thus the 2-uniform hypergraphs are exactly the graphs. We do not normally allow loops in graphs, nor edges of size at most 1 in hypergraphs; when we do, it will be stated explicitly. We do allow multiple edges. A graph or hypergraph without multiple edges is *simple*. The term *multigraph* is used when we explicitly want to say that multiple edges are allowed in a graph, and the *multiplicity*

$\mu(G)$ will denote the maximum number of edges joining the same pair of vertices in a multigraph G.

The theory of hypergraph coloring is extremely rich, and graph coloring is just one special case. Ramsey theory can be viewed naturally as another special case (see Graham, Rothschild, and Spencer [1990]).

A *homomorphism* of a graph G into a graph H is a mapping $f : V(G) \to V(H)$ such that $f(x)f(y)$ is an edge of H if xy is an edge of G. A k-coloring of G can then be thought of as a homomorphism of G into the complete k-graph K_k. In general, a homomorphism of G into a graph H is called an *H-coloring* of G.

An *edge coloring* of a hypergraph (or graph) H is a mapping $\varphi' : E(H) \to C$, where nondisjoint edges are mapped into distinct elements of the color set C. If C has k elements, then φ' is a *k-edge coloring*. The minimum cardinal k for which H has a k-edge coloring is the *edge-chromatic number* $\chi'(H)$, and H is said to be *$\chi'(H)$-edge-chromatic*.

A *face coloring* of a map M on a surface $\mathbf{S}$ (i.e., a bridge-less graph embedded on $\mathbf{S}$) with a set $F(M)$ of faces (or countries) consists of a mapping $\varphi : F(M) \to C$, where neighboring faces (those with a common borderline) are mapped into different elements of the color set C. This corresponds to a vertex coloring of the *dual graph* G, defined by having vertex set $V(G) = F(M)$ and an edge $xy \in E(G)$ for every edge of M on the common borderline of the faces x and y. When the map M is embedded on $\mathbf{S}$, its dual graph can also be embedded on $\mathbf{S}$ without crossing edges.

As with face coloring, both hypergraph coloring (with at least three colors) and edge coloring can be translated into vertex-coloring of graphs, as we shall see.

In the following we deal almost exclusively with graphs rather than with maps, even in cases where the results were initially obtained for face coloring. In the time before the papers of Whitney [1932b] and Brooks [1941], coloring theory dealt almost exclusively with maps, even though Kempe [1879] had drawn attention to vertex colorings of graphs: *"If we lay a sheet of tracing paper over a map and mark a point on it over each district and connect the points corresponding to districts which have a common boundary, we have on the tracing paper a diagram of a 'linkage,' and we have as the exact analogue of the question we have been considering, that of lettering the points of the linkage with as few letters as possible, so that no two directly connected points shall be lettered with the same letter. Following this up, we may ask what are the linkages which can be similarly lettered with no less than n letters? The classification of linkages according to the value of n is one of considerable importance."*

Vertex coloring of infinite graphs with a finite number of colors, or more generally H-coloring with a finite graph H, can always be reduced to finite instances. For vertex coloring, this is the content of the following theorem, which may be derived from a theorem of Rado [1949]. Gottschalk [1951] gave a short proof of Rado's theorem using compactness. A similar proof gives an extension of the theorem that includes H-coloring in general.

Theorem 1 (de Bruijn and Erdős [1951]). *If all finite subgraphs of an infinite graph G are k-colorable, where k is finite, then G is k-colorable.*

A short direct graph-theoretic proof of Theorem 1 was obtained by L. Pósa, and may be found for example in the book by Wagner [1970]. It was actually already contained in the Ph.D. thesis of G.A. Dirac at the University of London in 1951. However, as pointed out by G. Sabidussi [personal communication in 1993], this particular proof does not generalize to H-colorings as readily as the proof of Gottschalk. Because of Theorem 1 we shall only deal with finite graphs in the following, except when explicitly stating otherwise.

The reader looking for proofs of the theorems in this chapter may in many cases have to consult the references. However, a well-written general exposition of graph coloring theory, including proofs of several of the theorems we mention, can be found in the classical book on extremal graph theory by Bollobás [1978a]. Another good general source is the forthcoming *Handbook of Combinatorics*, edited by L. Lovász, R.L. Graham, and M. Grötschel, and published by North-Holland.

1.2. GRAPHS ON SURFACES

Many other areas of graph theory besides coloring theory originated from the *four-color problem* of Francis Guthrie: Is every planar graph 4-colorable? Well-written accounts of the problem are contained in the monographs by Ringel [1959], Ore [1967], Biggs, Lloyd, and Wilson [1976], Saaty and Kainen [1977], Barnette [1983], and Aigner [1984].

The four-color problem seems first to have been mentioned in writing in an 1852 letter from A. De Morgan to W.R. Hamilton, written on the same day as De Morgan first heard about the problem from his student Frederick Guthrie, Francis Guthrie's brother. It first appeared in print in an anonymous book review by De Morgan in 1860 (see Wilson [1976]), and later as an open problem raised by Cayley [1878] at a meeting in the London Mathematical Society and in a paper by Cayley [1879]. A proposed solution by Kempe [1879] stood for more than a decade until it was refuted by Heawood [1890] in his first paper. Heawood proved the five-color theorem for planar maps and the best possible twelve-color theorem for the case where each country consists of at most two connected parts. Moreover, he extended the problem to higher surfaces. Dirac [1963] gave an excellent survey of Heawood's achievements.

The higher surfaces (i.e., compact 2-dimensional manifolds) can be classified into three types as follows (see, e.g., Massey [1991]). The sphere with g handles attached is denoted by $\mathbf{S}_g$ (of Euler characteristic $\varepsilon = 2 - 2g$), the projective plane with g handles attached by $\mathbf{P}_g$ (of Euler characteristic $\varepsilon = 1 - 2g$), and the Klein bottle with g handles attached by $\mathbf{K}_g$ (of Euler characteristic $\varepsilon = -2g$). In each case g may assume the value zero. Note that the surfaces $\mathbf{S}_g$ are orientable, whereas $\mathbf{P}_g$ and $\mathbf{K}_g$ are nonorientable.

Theorem 2 (Heawood [1890]). *Let* $\mathbf{S}$ *be a surface of Euler characteristic* ε. *When* $\varepsilon < 2$, *every graph* G *on* $\mathbf{S}$ *can be colored using the Heawood number* $H(\varepsilon)$ *of colors,* *given by*

$$H(\varepsilon) = \left\lfloor \frac{7 + \sqrt{49 - 24\varepsilon}}{2} \right\rfloor.$$

A graph of seven mutually adjacent vertices, the complete 7-graph K_7, embeds on the torus $\mathbf{S}_1$, hence 7 (= $H(0)$) colors are both sufficient and necessary for toroidal graphs.

The topological prerequisite for Heawood's formula is Euler's formula, implying that every graph G embedded on a surface $\mathbf{S}$ of Euler characteristic ε has at most $3|V(G)| - 3\varepsilon$ edges. Since a minimal k-chromatic graph G has minimum degree $\delta(G) \geq k - 1$, such a graph G satisfies

$$(k - 1)|V(G)| \leq 2|E(G)| \leq 6|V(G)| - 6\varepsilon.$$

Since $|V(G)| \geq k$, it follows for $k \geq 7$ that $(k - 7)k + 6\varepsilon \leq 0$, which in turn implies that $k \leq H(\varepsilon)$.

For the Klein bottle $\mathbf{K}_0$ the Heawood formula gives a seven-color theorem. However, Franklin [1934] proved that six colors suffice to color any graph on the Klein bottle. This is the only case where the Heawood number is not the right answer to the coloring problem for higher surfaces.

Theorem 3 (Heffter [1891], Tietze [1910], Ringel [1954, 1959, 1974], Ringel and Youngs [1968]). *For a surface $\mathbf{S}$ of Euler characteristic $\varepsilon < 2$, where $\mathbf{S}$ is not the Klein bottle, the Heawood number $H(\varepsilon)$ is the maximum chromatic number of graphs embeddable on $\mathbf{S}$.*

The proof of this major result, completed in 1968, was obtained by embedding the complete $H(\varepsilon)$-graph $K_{H(\varepsilon)}$ on the surface with Euler characteristic ε. This is of course sufficient for a proof of Theorem 3. It is in fact also necessary.

Theorem 4 (P. Ungar and Dirac [1952b], Albertson and Hutchinson [1979]). *For a surface $\mathbf{S}$ of Euler characteristic $\varepsilon < 2$, and $\mathbf{S}$ different from the Klein bottle, any $H(\varepsilon)$-chromatic graph on $\mathbf{S}$ contains $K_{H(\varepsilon)}$ as a subgraph.*

Dirac's arithmetic did not cover the cases $\varepsilon = -1$ and 1, but these cases were later settled by Albertson and Hutchinson [1979]. The idea of the result of Theorem 4 and a proof in the case of the torus were first obtained by P. Ungar, as mentioned by Dirac [1952b].

After various attempts and the achieving of partial results on the four-color problem by many mathematicians, Appel and Haken [1976a] announced a complete proof. The four-color theorem for plane triangulations (i.e., plane graphs in which all faces are triangles), and hence for all planar graphs, follows immediately by induction from

Theorem 5 (Appel and Haken [1977a], Appel, Haken and Koch [1977]). *There exists a set U of 1482 configurations such that*

- **(a)** *Unavoidability: any plane triangulation contains an element of U, and*
- **(b)** *Reducibility: a 4-coloring of a plane triangulation containing an element of U can be obtained from 4-colorings of smaller plane triangulations .*

This is the same basic idea as in Kempe's proof, where U consisted of vertices of degree at most 5. Kempe's only mistake was in his argument for the reducibility of vertices of degree equal to 5. The detailed techniques of Appel, Haken, and Koch are further developments of methods of Heesch [1969], who was the first to emphasize strongly the possibility of a proof of the four-color theorem along these lines (see Bigalke [1988]). The proof of part (a) is based on Euler's formula and an elaborate "discharging procedure." Whereas this part of the proof can in principle be carried out by hand, Appel, Haken, and Koch had to use computer programs to verify that each member of their unavoidable set U of configurations submits to one of two types of reducibility that Heesch had named "C-reducibility" and "D-reducibility." Combining this fact with results of Bernhart [1947], they proved that U satisfies (b) of Theorem 5.

Several surveys of the proof of Theorem 5 exist: for example, Appel and Haken [1977b, 1978] and Woodall and Wilson [1978]. Due to its length, extensive use of verification by computer, some inaccuracies, and omissions of details, the proof of Theorem 5 has been surrounded by some controversy. Appel and Haken [1986, 1989] have themselves addressed the questions raised. Recent accounts of the situation have been given by F. Bernhart [*Math. Reviews* **91m**:05005] in an informative review of the book by Appel and Haken [1989], and by Kainen [1993].

Very recently, N. Robertson, D.P. Sanders, P.D. Seymour, and R. Thomas [personal communication from N. Robertson and P.D. Seymour in 1994] have obtained a new, improved proof of the four-color theorem by using the same general approach as that of Appel, Haken, and Koch. This proof has less than 700 configurations and is based on a simpler discharging procedure. In addition, the proof avoids some of the more problematic details of the proof by Appel, Haken, and Koch (we describe these in Problem 2.1). However, it still relies on extensive computer checking.

An early approach to coloring problems for plane maps and graphs concerned studying the number $P(G,k)$ of all possible different k-colorings of a graph G with colors $1, 2, \ldots, k$. Birkhoff [1912] noted that $P(G,k)$ as a function of k can be expressed as a polynomial, the so-called *chromatic polynomial* of G, $P(G,k) = a_1k^n + a_2k^{n-1} + \cdots + a_nk$ of degree $n = |V(G)|$. In particular, $\chi(G)$ is the smallest nonnegative integer that is not a zero of $P(G,k)$. Whitney [1932a, 1932b], Birkhoff and Lewis [1946], Tutte [1954, 1970b], and Read [1968] are some of the researchers who have developed the theory of chromatic polynomials. A well-written survey was given by Read and Tutte [1988].

One of Tutte's surprising and beautiful results is the following *golden identity.*

Theorem 6 (Tutte [1970b]). *Let M be a plane triangulation on n vertices. Then*

$$P(M, \tau + 2) = (\tau + 2) \cdot \tau^{3n-10} \cdot (P(M, \tau + 1))^2,$$

where τ is the golden ratio $\frac{1}{2}(1 + \sqrt{5})$, with $\tau + 1 = \tau^2$ and $\tau + 2 = \sqrt{5}\tau$.

Tutte [1970b] noted that $P(M, \tau + 1) \neq 0$. Hence we have the curious consequence that $P(M, \tau + 2)$ is positive, where $\tau + 2 \simeq 3.618\ldots$. Of course, the four-color theorem is equivalent to the statement that $P(M, 4)$ is positive.

As explained by Saaty [1972] and Saaty and Kainen [1977] the four-color theorem has many equivalent formulations. A particularly noteworthy result is

Theorem 7 (Wagner [1937]). *If all planar graphs are 4-colorable, then 4-colorability extends to the class $\mathcal{G}$ of all graphs from which a complete 5-graph K_5 cannot be obtained by deletions (of vertices and/or edges) and contractions of edges (removing possible loops that might arise).*

Thus the four-color theorem is equivalent to the case $k = 5$ of the famous

Hadwiger's Conjecture (Hadwiger [1943]). *Let $\mathcal{G}$ be a class of graphs closed under deletions (of edges and/or vertices) and contractions of edges (removing possible loops that might arise). Then the maximum chromatic number of the graphs in $\mathcal{G}$ equals the number of vertices $(k - 1)$ in a largest complete graph in $\mathcal{G}$.*

For $k = 4$ this was proved by Dirac [1952a]. Recently, Robertson, Seymour, and Thomas [1993a] gave a complete characterization of all 6-colorable graphs from which the complete graph K_6 cannot be obtained by deletions and contractions. As a corollary of the characterization, all such graphs are in fact 5-colorable, assuming the four-color theorem. This proves that Hadwiger's conjecture for $k = 6$ is also equivalent to the four-color theorem. Hadwiger's conjecture is true for $\mathcal{G}$ the class of all graphs embeddable on the same surface **S**. This follows from Theorems 3, 4, and 5 above, and from a paper by Albertson and Hutchinson [1980a] for the Klein bottle.

A deep extension of the five-color theorem for planar graphs was conjectured by Grünbaum [1973] and proved by Borodin [1979a]. The proof is reminiscent of the four-color proof by Appel, Haken, and Koch; it involves an unavoidable set of some 450 reducible configurations (but no computers).

Theorem 8 (Borodin [1979a]). *Every planar graph has an acyclic 5-coloring, that is, a 5-coloring in which each pair of color classes induces a subgraph without cycles.*

As for 3-colorings of planar graphs, the most important results are

Theorem 9 (Heawood [1898]). *A plane triangulation can be 3-colored if and only if all vertices have even degrees.*

Theorem 10 (Grötzsch [1959]). *Every planar graph G without triangles is 3-colorable. Moreover, any proper 3-coloring of a 4-cycle or a 5-cycle can be extended to a 3-coloring of all of G.*

Recently, Thomassen [1993c] gave a short proof of Grötzsch's theorem and proved that every graph on the torus without cycles of length three or four is 3-colorable. The induction proofs given by Grötzsch and Thomassen both depend on *coloring extension* arguments—prescribed colorings of special subgraphs, such as 4-cycles and 5-cycles in the case of Theorem 10, may be assumed extendable to all of G. Another new proof of the first, essential, part of Theorem 10, free of coloring extension arguments, was obtained by Borodin [1993a, 1994a].

Grünbaum [1963] made the assertion that if G is a planar graph with at most three triangles, then G is 3-colorable. To prove this extension of Grötzsch's theorem, Grünbaum extended the statement a bit further to make the induction run. However, a counterexample to Grünbaum's extended theorem was found by T. Gallai. The basic claim of Grünbaum (that if a planar graph G has at most three triangles, then G is 3-colorable) is true, as proved by Aksionov [1974], with a proof based on the ideas introduced by Grünbaum and Grötzsch.

An informative survey of 3-colorability in general was given by Steinberg [1993a], recently updated by Steinberg [1993b].

1.3. VERTEX DEGREES AND COLORINGS

Brooks [1941] observed that every graph G may be colored by $\Delta(G) + 1$ colors, where $\Delta(G)$ is the maximum degree of G, and he characterized the graphs for which $\Delta(G)$ colors are not enough.

Theorem 11 (Brooks [1941]). $\chi(G) \leq \Delta(G) + 1$ *holds for every graph G. Moreover,* $\chi(G) = \Delta(G) + 1$ *if and only if either* $\Delta(G) \neq 2$ *and G has a complete* $(\Delta(G) + 1)$*-graph* $K_{\Delta(G)+1}$ *as a connected component, or* $\Delta(G) = 2$ *and G has an odd cycle as a connected component.*

The *coloring number* $\text{col}(G)$ is defined as the smallest number d such that for some linear ordering $<$ of the vertex set, the "back-degree" $|\{y : y < x, xy \in E(G)\}|$ of every vertex x is strictly less than d. In other words, if the vertices of G are $x_1, x_2, \ldots, x_n$, then

$$\text{col}(G) = 1 + \min_p \max_i \{d(x_{p(i)}, G_{p(i)})\},$$

where the minimum is taken over all permutations p of $\{1, 2, \ldots, n\}$, and $G_{p(i)}$ is the subgraph of G induced by $x_{p(1)}, x_{p(2)}, \ldots, x_{p(i)}$, and where $d(x, H)$ denotes the degree of a vertex x in a graph H. It is clear that $\text{col}(G) \leq \Delta(G) + 1$, hence the inequality

$$\chi(G) \leq \text{col}(G),$$

obtained by sequentially coloring the vertices of G with col(G) colors, strengthens the inequality of Theorem 11.

A *smallest last order* $x_{p(1)}, x_{p(2)}, \ldots, x_{p(n)}$ of the vertices of G is obtained by letting $x_{p(i)}$ be a vertex of minimum degree $\delta(G_i)$ in the subgraph

$$G_i = G - \{x_{p(i+1)}, x_{p(i+2)}, \ldots, x_{p(n)}\},$$

for $i = n, n-1, \ldots, 1$. It follows that

$$\mathrm{col}(G) \leq 1 + \max_{1 \leq i \leq n} \delta(G_i) \leq 1 + \max_{G' \subseteq G} \delta(G').$$

In fact:

Theorem 12 (Halin [1967], Matula [1968], Finck and Sachs [1969], Lick and White [1970]).

$$\mathrm{col}(G) = 1 + \max_i \delta(G[x_{p(1)}, x_{p(2)}, \ldots, x_{p(i)}]) = 1 + \max_{G' \subseteq G} \delta(G'),$$

where $x_{p(1)}, x_{p(2)}, \ldots, x_{p(n)}$ *is a smallest last order.*

Theorem 12 implies that col(G) is the smallest number $k+1$ such that G is k-*degenerate*; that is, every nonempty subgraph G' of G has a vertex of degree at most k in G'. Moreover, the middle term in Theorem 12 shows that it is easy to calculate col(G): Among the $n!$ possible orders on the vertex set of G, the minimum in the definition of col(G) is attained by a smallest last order, and among the 2^n induced subgraphs of G it is possible to find a G' giving the maximum value of $\delta(G')$ among the only n easily obtained induced subgraphs $G[x_{p(1)}, x_{p(2)}, \ldots, x_{p(i)}]$.

The coloring number was introduced and studied by Erdős and Hajnal [1966], the so-called Szekeres–Wilf number $1 + \max_{G' \subseteq G} \delta(G')$ by Szekeres and Wilf [1968], and the property of being k-degenerate by Lick and White [1970]. The Szekeres–Wilf number was used implicitly by Vizing [1965a], who proved that the edge-chromatic number of a graph G equals the maximum degree Δ when $\Delta \geq 2k$ and $\max_{G' \subseteq G} \delta(G') \leq k$.

Another useful and deep generalization of the inequality $\chi(G) \leq \Delta(G) + 1$ is

Theorem 13 (Hajnal and Szemerédi [1970]). *A graph G may be colored by* $\Delta(G) + 1$ *colors such that for any two colors i and j, where* $1 \leq i < j \leq \Delta(G) + 1$, *the numbers of vertices of color i and color j differ by at most one (i.e., the* $\Delta(G) + 1$ *color classes are equal or almost equal in size).*

1.4. CRITICALITY AND COMPLEXITY

An element t of $V(G) \cup E(G)$ is *critical* if $\chi(G - t) < \chi(G)$. G is *critical*, or more precisely, $\chi(G)$-*critical*, if all edges and vertices of G are critical. The k-critical graphs are thus the minimal graphs with respect to subgraph inclusion in the class of all k-chromatic graphs. If the vertices of G are critical, but not necessarily the edges, we say that G is *$\chi(G)$-vertex-critical.* No k-critical graph can be infinite by Theorem 1. For $k = 1, 2$, and 3 they are K_1, K_2, and odd cycles, respectively.

Theorem 14 (König [1916, 1936]). *A graph is 2-colorable if and only if it does not contain an odd cycle.*

The importance of the notion of criticality is that problems for k-chromatic graphs in general may often be reduced to problems for k-critical graphs, and these are more restricted than k-chromatic graphs in general. Critical graphs were first defined and used by Dirac [1951, 1952a, 1952b, 1952c, 1953]. In terms of critical graphs Theorem 11 of Brooks may be reformulated as follows.

Theorem 15 (Brooks [1941]). *In a k-critical graph G all vertices have degree at least $k - 1$. Moreover, if all vertices have degree equal to $k - 1$, then either G is the complete k-graph K_k, or $k = 3$ and G is an odd cycle.*

A later generalization, the basis for a simple proof of Theorem 4 due to Dirac [1957b], is

Theorem 16 (Dirac [1957c]). *Let G be a k-critical graph. If $k \geq 4$ and $G \neq K_k$, then*

$$2|E(G)| \geq (k-1)|V(G)| + (k-3).$$

Dirac's original proof of Theorem 16 was rather complicated. Shorter and more elegant proofs were given by Kronk and Mitchem [1972a] and by Weinstein [1975]. Dirac [1974] characterized all k-critical graphs G for which equality holds in Theorem 16 (in each of these cases $|V(G)| = 2k - 1$ holds). A further very important and elegant generalization of Brooks' theorem is

Theorem 17 (Gallai [1963a]). *Let G be a k-critical graph with $k \geq 4$ and $G \neq K_k$. Then every block in the subgraph of G induced by the vertices of degree $k - 1$ in G is either complete or an odd cycle. Moreover,*

$$2|E(G)| \geq (k-1)|V(G)| + \frac{k-3}{k^2-3}|V(G)|.$$

Using matching theory, Gallai [1963b] also obtained deep structural results on k-critical graphs on at most $2k - 1$ vertices. All these results on k-critical graphs are positive in the sense that they restrict the behavior of critical graphs. In his doctoral dissertation Stiebitz [1985] proved that in general the structure of k-critical graphs is manageable so long as the subgraph induced by the vertices of degree at least k is $(k - 2)$-colorable or complete (this is also explained in the survey by Sachs and Stiebitz [1989]).

There are, however, also a large number of negative results showing that critical graphs may have a very unrestricted behavior. Thus k-critical graphs for $k \geq 4$ may have many edges (Dirac [1952a], Toft [1970a], V. Rödl and Stiebitz [1987b]), many independent vertices (Brown and Moon [1969], Simonovits [1972], Lovász [1973b]), and high minimum degree (Simonovits [1972], Toft [1972b]). This is in sharp contrast to the situation for $k = 3$, when the only critical graphs are the odd cycles. A further such observation, based on the existence of 4-critical graphs with many edges, was made by V. Rödl and published by Toft [1985]: The number of nonisomorphic 4-critical graphs on n vertices is larger than $c^{(n^2)}$ for some $c > 1$.

A constructive characterization of k-chromatic graphs was obtained by G. Hajós in his approach to the four-color problem: A graph is *Hajós–k-constructible* if it can be obtained from complete k-graphs K_k by repeated applications of the following two operations.

(a) *Hajós' construction*: Let G_1 and G_2 be already obtained disjoint graphs with edges x_1y_1 and x_2y_2. Remove x_1y_1 and x_2y_2, identify x_1 and x_2, and join y_1 and y_2 by a new edge.

(b) Identify independent vertices.

Theorem 18 (Hajós [1961]). *A graph has chromatic number at least k if and only if it contains a Hajós–k-constructible subgraph. Every k-critical graph is Hajós–k-constructible.*

Hajós' construction is extremely useful. However, despite its seemingly large generality and power, no interesting applications of Theorem 18 have been found. In particular, obtaining a proof of the nonexistence of 5-critical planar graphs, which was perhaps what Hajós hoped to achieve, seems far beyond its reach.

A much more restricted type of critical graphs was introduced by Dirac [1960, 1964a]. Let the notation $G_2 \leq G_1$ denote that G_2 can be obtained from G_1 by deletions of vertices and/or edges and contraction of edges, in any order. The terminology that G_2 is a *minor* or a *subcontraction* of G_1 is often used. The k-chromatic graphs minimal with respect to $\leq$ are called *k-contraction-critical*. For each value of k there is only one known k-contraction-critical graph, namely K_k. The conjecture that there is no other is equivalent to Hadwiger's conjecture, known to be true for $k \leq 6$, with the cases $k = 5$ and 6 depending on the four-color theorem (see Section 1.2).

Criticality also applies to hypergraphs, both uniform and nonuniform. If we say, based on the negative results stated above, that a rich and uncontrolled behavior of k-critical graphs begins with $k = 4$, then for hypergraphs a similar rich and uncontrolled behavior starts in the 3-critical 3-uniform case (Lovász [1973a], Toft [1975a], Müller, Rödl, and Turzík [1977]).

The difficulty in characterizing the 4-critical graphs or the 3-critical hypergraphs is reflected in the difficulty of obtaining polynomial algorithms for graph 3-coloring or hypergraph 2-coloring. Both types of problems are ***NP**-complete* (for definitions of ***NP***-complete and the classes ***NP*** and ***P***, see the authoritative book by Garey and Johnson [1979]).

Theorem 19 (Karp [1972], Lovász [1973a], Stockmeyer [1973], Garey, Johnson, and Stockmeyer [1976], Toft [1975a], see also Garey and Johnson [1979]). *The following problem types are* ***NP****-complete. In particular, there is a polynomial algorithm for one of them if and only if there are polynomial algorithms for all of them.*

1. *Hypergraph k-colorability.*
2. *Graph k-colorability.*
3. *Graph 3-colorability.*
4. *3-colorability for planar graphs of maximum degree 4.*
5. *Hypergraph 2-colorability.*
6. *2-colorability for 3-uniform hypergraphs.*

An interesting recent negative result on approximative algorithms, mentioned in the very informative survey by Johnson [1992], is the following.

Theorem 20 (Lund and Yannakakis [1993]). *There is a positive constant ε_0 such that if there is a polynomial graph coloring algorithm using, for every graph G, at most $\chi(G) \cdot |V(G)|^{\varepsilon_0}$ colors, then there is a polynomial algorithm to determine $\chi(G)$ for all G.*

At first sight the if part seems to ask for little, but the then part shows that it is in fact a lot, in particular ***NP*** would be equal to ***P*** and all six problem types in Theorem 19 would be polynomially solvable.

An important class of hypergraphs deserves mentioning since it allows a polynomial algorithm for the problem of k-colorability. If the edges of a hypergraph form the circuits of a matroid, then a coloring of the hypergraph corresponds to a partitioning of the elements of the matroid into classes that are independent in the matroid sense. This coloring problem has a very beautiful and satisfactory solution by Edmonds [1965a], both in terms of characterizing the chromatic number by a min-max theorem and in terms of a polynomial algorithm for determining its exact value. For definitions of matroids and a fine exposition of the basic theory, including the partitioning results of Edmonds, see the book by Welsh [1976], in particular Chapters 8 and 19.

In general, hypergraphs might be expected to behave much more chaotically than graphs. However, for $k \geq 4$ the k-critical hypergraphs seem no worse than the k-critical graphs, since k-critical hypergraphs can be transformed into k-critical graphs as follows. For an edge A of a hypergraph H take a $(k-1)$-critical graph G (disjoint from H) such that $|V(G)| \geq |A|$. Remove the edge A from H without removing any vertices, and join each vertex of G to exactly one vertex of A in the remaining hypergraph so that each vertex of A is joined to at least one vertex of G. If the hypergraph H' obtained is $(k-1)$-colorable, then H is $(k-1)$-colorable. In many special cases, for instance a graph consisting of an odd cycle completely joined to a K_{k-4}, the hypergraph H' is $(k-1)$-colorable if and only if H is $(k-1)$-colorable, and

H' is k-critical if and only if H is k-critical. In particular, hypergraph $(k-1)$-coloring can be reduced to graph $(k-1)$-coloring when $k \geq 4$.

The construction method described above was used by Toft [1974a, 1975a], but the basic idea is due to B. Descartes in the case where H is the k-critical r-uniform hypergraph consisting of all subsets of size r on $(k-1)(r-1)+1$ vertices. Thus constructively, the following theorem can be proved by induction over k.

Theorem 21 (Descartes [1947, 1948b, 1954], Kelly and Kelly [1954]). *For every $k \geq 4$ there exists a k-chromatic graph in which every cycle has length 6 or more.*

1.5. SPARSE GRAPHS AND RANDOM GRAPHS

The size of a largest complete subgraph, $\omega(G)$, is obviously a lower bound for the chromatic number $\chi(G)$, and $\omega(G) = 1$ implies that $\chi(G) = 1$. Zykov [1949] showed by examples that apart from these there are no relations between ω and χ in general. Theorem 21 shows this when $\omega(G) = 2$.

The constructions of both Descartes and Zykov give k-chromatic triangle-free graphs for all k, but the graphs obtained have vertex numbers exponential in k, as have the graphs of Mycielski [1955]. Erdős [1958] proved by a geometric construction the existence of k-chromatic triangle-free graphs with only a polynomial number (k^{50}) of vertices. Shortly thereafter Erdős obtained the following even more striking result with his elegant nonconstructive probabilistic method.

Theorem 22 (Erdős [1959]). *For all $g \geq 4$ and for sufficiently large k there exist k-chromatic graphs on at most k^{cg} vertices ($c > 0$ constant) in which all cycles have lengths at least g.*

The proof given by Erdős shows that the best possible constant c of Theorem 22 satisfies $c \leq 2$. A good exposition of the proof of a result very similar to Theorem 22, with the condition of k being "sufficiently large" removed at the cost of increasing the constant to $c = 4$, is given in the seminal book of Bollobás [1978a]. Constructive inductive proofs of the existence for all g and k (without the bound on the number of vertices) were obtained by Lovász [1968] and by Nešetřil and Rödl [1979]. These constructions also included r-uniform hypergraphs—this extension was in fact essential for making the induction work. More recently, Lubotzky, Phillips, and Sarnak [1988] showed that a class of Ramanujan graphs provides explicit examples of graphs satisfying Theorem 22 for all g and k with $c = 4$ (their constructions do not involve the consideration of hypergraphs).

The proof of Theorem 22 by Erdős was one of the first applications of the probabilistic method to graph coloring. The classical paper by Erdős and Rényi [1960] contained the definition of a *random graph*. For a positive integer n and real number p with $0 \leq p \leq 1$, a random graph $G(n,p)$ is a graph on n vertices so that for each unordered pair x, y of these, xy is an edge of $G(n,p)$ with probability p

independently of all other pairs. The fine book by Bollobás [1985] contains a vast collection of techniques and results in this area of graph theory.

One of the most studied problems on random graphs has been the determination of the chromatic number of random graphs $G(n, p)$, for p a fixed edge probability, as n approaches infinity. For many years it seemed possible that $\chi(G(n, p))$ might be distributed over a relatively wide range of values, even when n gets very large. However, that this is not the case was shown by Bollobás, who managed to prove

Theorem 23 (Bollobás [1988]). *For p fixed with $0 < p < 1$ a random graph, $G(n, p)$ satisfies*

$$\left(\frac{1}{2} - o(1)\right) \log(1/(1-p)) \frac{n}{\log n} \leq \chi(G(n, p)) \leq \left(\frac{1}{2} + o(1)\right) \log(1/(1-p)) \frac{n}{\log n}$$

with probability $\longrightarrow 1$ as $n \longrightarrow \infty$, where $o(1)$ denotes a function of n converging to zero as $n \longrightarrow \infty$.

It follows from Bollobás' theorem with $p = \frac{1}{2}$ that almost all graphs G satisfy

$$\chi(G) \sim \frac{|V(G)|}{2 \log_2 |V(G)|}.$$

Bollobás, Catlin, and Erdős [1980] and independently Kostochka [1982] proved the important probabilistic result that Hadwiger's conjecture (see Section 1.2) is true for almost all graphs. Alon and Spencer [1992] gave a well-written in-depth treatment of the probabilistic method, including outlines of proofs of Theorems 22 and 23.

1.6. PERFECT GRAPHS

The opposite extreme of sparse graphs are perfect graphs. A graph G is *perfect* if G and each of the induced subgraphs of G has chromatic number equal to the size of its largest complete subgraph. The idea of perfect graphs grew out of the work of T. Gallai, who was among the first to emphasize the importance of min-max theorems for graphs and to apply linear programming duality in combinatorics (see Gallai [1958, 1959]). Gallai [1958] published a result, due to himself and D. König dating back to 1932, that in today's language would be phrased "the complement of any bipartite graph is perfect." Other such early results were obtained by Hajnal and Surányi [1958], Berge [1960, 1961], Dirac [1961], and Gallai [1962]. The definition of perfect graphs appeared in Berge [1963a, 1966],[1] and this topic has been one of the

[1]We do not know if the term "perfect graph" was used or if the perfect graph conjecture was mentioned in any publication prior to Berge [1963a]. Two other papers by Berge [1961, 1963b], often mentioned as the original sources to "perfectness" and to the perfect graph conjecture, do not explicitly refer to perfect graphs. Many historic details are contained in a manuscript by C. Berge, "The history of the perfect graphs" EC 94/02, Institut Blaise Pascal, Université Paris VI, March 1994.

most fruitful in graph theory, relating also to linear programming and computational complexity. Grötschel, Lovász, and Schrijver [1981] proved, based on the existence of a polynomial algorithm for linear programming, that the chromatic number of a perfect graph can be computed in polynomial time (see also the recent interesting survey by Lovász [1994]). For a general graph the problem of deciding if it is k-colorable is ***NP***-complete for $k \geq 3$, as first proved by Karp [1972]. An excellent survey of perfect graphs was presented by Lovász [1983a] (see also the interesting Introduction written by Berge and Chvátal [1984] containing many historical details).

Many classes of graphs are perfect, among them bipartite graphs and their complements, triangulated graphs (where every cycle of length at least 4 has a diagonal, also called chordal or rigid circuit graphs) and their complements, comparability graphs (where the vertices are the elements of a partial order, and $xy \in E(G)$ if and only if x and y are comparable), interval graphs (the intersection graphs of finite sets of intervals on the real line), and the graphs of Meyniel [1976] (where every odd cycle of length at least 5 has at least two diagonals).

Inspired by a conjecture of Shannon [1956] on optimal codes, C. Berge and A. Ghouila-Houri studied the class of graphs containing neither odd cycles of length greater than or equal to 5 nor their complements as induced subgraphs. Such graphs are now commonly referred to as *Berge graphs*. Berge conjectured in the early 1960s that these graphs are precisely the perfect graphs. This implies the weaker, but also at the time unsolved conjecture that a graph is perfect if and only if its complement is perfect. In an attack on this weaker conjecture, Fulkerson [1971, 1972] considered a seemingly more restricted class of pluperfect graphs. Let G be a graph with vertices $x_1, x_2, \ldots, x_n$ having nonnegative integer weights $(w_1, w_2, \ldots, w_n) = \underline{w}$. Let $\chi(G, \underline{w})$ denote the minimum number of independent subsets of $V(G)$, where the same set may be counted more than once, such that x_i is in at least w_i of the sets for each i. Let $\omega(G, \underline{w})$ denote the maximum weight of a complete subgraph of G—thus $\chi(G, \underline{1}) = \chi(G)$ and $\omega(G, \underline{1}) = \omega(G)$, where $\underline{1} = (1, 1, \ldots, 1)$. It is easy to see that $\chi(G, \underline{w}) \geq \omega(G, \underline{w})$. A graph is *pluperfect* if $\chi(G, \underline{w}) = \omega(G, \underline{w})$ for all nonnegative integer weights $\underline{w}$. Note that a graph is perfect, by definition, if and only if $\chi(G, \underline{w}) = \omega(G, \underline{w})$ for all $(0, 1)$-valued weights $\underline{w}$.

Based on the duality theory of linear programming Fulkerson proved that the complement of a pluperfect graph is pluperfect. Lovász then proved the following replacement theorem: If G is a perfect graph and x is a vertex of G, then replacing x by a perfect graph K joined completely to all the neighbors of x in G, one again obtains a perfect graph. This theorem implies that the pluperfect and the perfect graphs are in fact the same. Thus Lovász obtained the first proof of Berge's weak perfect graph conjecture, changing it to the *perfect graph theorem*.

Theorem 24 (Lovász [1972a, 1972b]). *A graph is perfect if and only if its complement is perfect.*

Lovász gave two further proofs of the perfect graph theorem, the first of which was based on hypergraphs. In the second proof he extended Theorem 24 by characterizing the perfect graphs as follows.

Theorem 25 (Lovász [1972a]). *A graph G is perfect if and only if every induced subgraph H of G satisfies*

$$\alpha(H) \cdot \omega(H) \geq |V(H)|,$$

where $\alpha(H)$ is the size of a largest independent set of vertices in H, and $\omega(H)$ is the size of a largest complete subgraph of H.

The stronger conjecture by Berge, also discovered by P.C. Gilmore, remains a challenging unsolved problem. The bulk of evidence in support of the conjecture has been steadily increasing over the years, beginning with the pioneering work by Gallai, Berge, and others mentioned above.

Strong Perfect Graph Conjecture (Berge [1963a, 1966]). *The perfect graphs are precisely the Berge graphs. Equivalently, the nonperfect graphs for which all induced proper subgraphs are perfect are precisely the odd cycles of length at least 5 and their complements.*

Chvátal [1984b] formulated the following "semistrong perfect graph conjecture":

If G and H are graphs on the same vertex set, with the property that any four vertices induce a path in G if and only if they induce a path in H, then G is perfect if and only if H is perfect.

This statement implies the perfect graph theorem since the complement of a path on four vertices is again a path. It is also implied by the strong perfect graph conjecture. The semistrong perfect graph conjecture was proved by Reed [1987]. A probabilistic argument of Prömel and Steger [1992] implies that the strong perfect graph conjecture is "almost true." They proved that almost all Berge graphs are perfect.

1.7. EDGE-COLORING

Let us first note that the edge-chromatic number $\chi'(H)$ of a graph or hypergraph H equals $\chi(L(H))$, where $L(H)$ is the line graph of H, defined by $V(L(H)) = E(H)$ and

$$E(L(H)) = \{xy : x \text{ and } y \text{ are distinct nondisjoint edges of } H\}.$$

Thus edge coloring is equivalent to vertex coloring of a line graph.

Compared with vertex coloring the theory of edge coloring has received less attention until relatively recently. However, much studied areas, such as map coloring, matching theory, and Latin squares, have strong connections to edge coloring. The first important observation on edge coloring was Tait's theorem.

Theorem 26 (Tait [1880]). *The faces of a 3-regular plane graph G can be 4-colored if and only if G can be 3-edge-colored.*

The next important result was a theorem of D. König, obtained in connection with his study of factorizations of bipartite graphs.

Theorem 27 (König [1916]). *For a bipartite graph G (with multiple edges allowed) the edge-chromatic number* $\chi'(G)$ *equals the maximum degree* $\Delta(G)$.

An n-edge coloring of a complete bipartite graph $K_{n,n}$ corresponds to a Latin square. The following theorem answers a conjecture of T. Evans from 1960 on completing partial Latin squares and also characterizes the extremal cases.

Theorem 28 (Andersen and Hilton [1983]). *A partial edge coloring* φ *of at most* n *edges of* $K_{n,n}$ *can be extended to an* n*-edge coloring of* $K_{n,n}$ *except in the following two cases:*

- **(a)** *For some uncolored edge* xy *there are* n *colored edges of different colors, each one incident with* x *or* y.
- **(b)** *For some vertex* x *and some color* i*, the color* i *is not incident to* x*, but it is incident to all vertices* y *for which* xy *is uncolored.*

Following Tait and König, the next breakthrough was the theorem of Vizing, obtained independently by Gupta.

Theorem 29 (Vizing [1964, 1965b], Gupta [1966]). *Let G be a multigraph of multiplicity* $\mu(G)$. *Then* $\chi'(G) \leq \Delta(G) + \mu(G)$.

From this result one can prove the theorem of Shannon [1949] that $\chi'(G) \leq 3\Delta(G)/2$.

It follows immediately from Vizing's theorem that $\chi'(G)$ is equal to either $\Delta(G)$ or $\Delta(G) + 1$ when G is simple ($\mu(G) = 1$). For the line graph H of G we have $\chi(H) = \chi'(G)$ and $\Delta(G) = \omega(H)$ [provided that $\Delta(G) \geq 3$]. Therefore, Theorem 29 with $\mu(G) = 1$ may be formulated as follows. When H is a line graph of a simple graph, $\chi(H)$ equals $\omega(H)$ or $\omega(H) + 1$.

It was proved by Beineke [1970] that H is a line graph $L(G)$ of a simple graph G if and only if H does not contain any one of nine specified graphs as induced subgraphs. To obtain the same conclusion for $\chi(H)$ as above, all but two of these forbidden subgraphs can in fact be eliminated. Thus we have the following interesting generalization of Vizing's theorem for simple graphs.

Theorem 30 (Kierstead and Schmerl [1983]). *If H is a simple graph containing neither a complete 5-graph with an edge missing nor a complete bipartite graph* $K_{1,3}$ *as an induced subgraph, then* $\chi(H)$ *is equal to* $\omega(H)$ *or* $\omega(H) + 1$.

1.8. ORIENTATIONS AND INTEGER FLOWS

The chromatic number of a graph G has been characterized in terms of orientations of G in two different ways. The "only if" parts are easy to obtain by directing the edges from smaller to larger colors.

Theorem 31 (Minty [1962]). *A graph G is k-colorable if and only if G has an orientation in which the flow ratio of any cycle C (i.e., the maximum of m/n and n/m, where n is the number of edges of C pointing in one direction and m is the number of edges of C pointing in the opposite direction) is at most $k - 1$.*

Theorem 32 (Roy [1967], Gallai [1968]). *A graph G is k-colorable if and only if G has an orientation in which the length of every directed path is at most $k - 1$.*

Weaker versions of Theorem 32 were obtained earlier by Vitaver [1962] and Gupta [1963], and a stronger version is due to Bondy [1976].

Replacing "cycle" by "edge-cut" in the formulation of Minty's theorem, then instead of characterizing k-colorability, the new statement characterizes the property of having a k-flow. A (nowhere zero, integer) *k-flow* in an undirected graph is an assignment to each edge of a direction and a weight from $\{1, 2, \ldots, k - 1\}$ such that for each vertex v the sum of the weights on the edges directed into v equals the sum of the weights on the edges directed away from v. Since there is a duality between cycles and edge cuts in graphs, "k-coloring" and "k-flow" are in a sense dual concepts. For example, Theorem 14 (König's theorem) restricted to planar graphs can be formulated in the following dual version: A planar graph has a 2-flow if and only if all its vertices have even degrees. The fact that the word "planar" can be removed is essentially the classical theorem of Euler [1736] that a necessary and sufficient condition for a connected graph to have an Euler tour is that all degrees be even. Theorem 9 of Heawood and Theorem 26 of Tait can similarly be extended to nonplanar graphs by omitting the word "planar" in their integer flow reformulations: A 3-regular planar graph has a 3-flow if and only if it is bipartite, and it has a 4-flow if and only if it is 3-edge-colorable.

The following conjecture may be seen as a far-reaching generalization of the five-color theorem for planar graphs (note that "bridge" and "loop" are dual concepts).

5-Flow Conjecture (Tutte [1954]). *Every graph without bridges has a 5-flow.*

As opposed to the five-color theorem for planar graphs, the 5-flow conjecture, which includes nonplanar graphs, would be the best possible. The Petersen graph has a 5-flow but not a 4-flow. The best result in the direction of Tutte's conjecture, strengthening an earlier 8-flow theorem proved by Kilpatrick [1975] in his Ph.D. thesis and independently obtained by Jaeger [1976a], is

Theorem 33 (Seymour [1981a]). *Every graph without bridges has a 6-flow.*

In a k-flow the numbers $1, 2, \ldots, k-1$ may be thought of as elements of $\mathbf{Z}_k$, the integers modulo k, and the flow can then be described as a (nowhere zero) $\mathbf{Z}_k$-*flow*. For an Abelian group $\mathbf{A}$, a (nowhere zero) $\mathbf{A}$-*flow* in a graph G is an assignment of a direction and a nonzero value from $\mathbf{A}$ to every edge of G such that for every vertex $v \in V(G)$ the sum in $\mathbf{A}$ of the weights of edges directed into v equals the sum in $\mathbf{A}$ of the weights of edges directed away from v.

It is easy to see that the number of different (not necessarily nowhere zero) $\mathbf{A}$-flows in G is $|\mathbf{A}|^{|E(G)|-|E(F)|}$, where F is a spanning forest in G which satisfies $|E(F)| = |V(G)|-$(the number of connected components in G). Using the principle of inclusion and exclusion it follows that the number of different (nowhere zero) $\mathbf{A}$-flows in G is independent of the detailed group structure of $\mathbf{A}$ and depends only on the order $|\mathbf{A}|$. This observation is due to Tutte [1954]. Tutte also observed that a graph has a k-flow if and only if it has a $\mathbf{Z}_k$-flow, and hence if and only if it has an $\mathbf{A}$-flow for some Abelian group $\mathbf{A}$ of order k. Younger [1983] and Jaeger [1988] gave original well-written surveys of integer flows.

1.9. LIST COLORING

A surprising type of relation between colorings and orientations was obtained recently.

Theorem 34 (Alon and Tarsi [1992]). *Let $\vec{G}$ be an orientation of a graph G. If the number of spanning Eulerian subgraphs of $\vec{G}$ with an even number of edges differs from the number of spanning Eulerian subgraphs with an odd number of edges, then for any assignment of a set $\Lambda(v)$ of $d^+(v)+1$ colors to every vertex v of G, where $d^+(v)$ denotes the out-degree of v in $\vec{G}$, there exists a proper coloring φ of G such that $\varphi(v) \in \Lambda(v)$ for all $v \in V(G)$.*

A directed graph $\vec{H}$ is *Eulerian* if each vertex $v \in V(\vec{H})$ has out-degree $d^+(v)$ equal to its in-degree $d^-(v)$. The number of even spanning Eulerian subgraphs of a directed graph $\vec{G}$ is always at least one, since $\vec{H}$ with $V(\vec{H}) = V(\vec{G})$ and $E(\vec{H}) = \emptyset$ is such a subgraph. Thus in particular an orientation without odd directed cycles satisfies the hypothesis of Theorem 34. When applied to the special case of acyclic orientations, the theorem generalizes the inequality $\chi(G) \leq \mathrm{col}(G)$.

The proof of Theorem 34 is algebraic and based on a study of the properties of the *graph polynomial*

$$f_G(x_1, x_2, \ldots, x_n) = \prod_{\substack{v_i v_j \in E(G)\\ i<j}} (x_i - x_j)$$

in $n = |V(G)|$ variables $x_1, x_2, \ldots, x_n$, where the vertices of G have been arranged in some fixed order $v_1, v_2, \ldots, v_n$. This polynomial was introduced by Sylvester [1878] and studied also by Petersen [1891]. Interesting descriptions of graph colorability in terms of the graph polynomial were found by D.J. Kleitman and Lovász [1982, 1994] and by Alon and Tarsi [1992]. Kleitman and Lovász's result simply states that $\chi(G) \geq k$ if and only if f_G is the sum of graph polynomials f_H for a family of graphs H with $V(H) = V(G)$ and $K_k \subseteq H$. Alon and Tarsi's result is slightly more involved to formulate. Given n polynomials $Q_1(x), Q_2(x), \ldots, Q_n(x)$ without multiple zeros over the field of complex numbers $\mathbf{C}$, Alon and Tarsi proved that G has no vertex coloring $\varphi : V(G) \to \mathbf{C}$ such that $Q_i(\varphi(v_i)) = 0$ for all $i = 1, 2, \ldots, n$, if and only if there exist polynomials $g_i(x_1, x_2, \ldots, x_n)$ such that the graph polynomial for G can be expressed as

$$f_G(x_1, x_2, \ldots, x_n) = \sum_{1 \leq i \leq n} g_i(x_1, x_2, \ldots, x_n) \cdot Q_i(x_i).$$

The "if" part of this statement is the easy part. The proof of Theorem 34 is obtained by combining this result with results describing the connection between the coefficients of f_G and orientations of G.

The idea of associating with each vertex v of G a list $\Lambda(v)$ from which the color of v has to be chosen in a coloring of G is due independently to Vizing [1976] and to Erdős, Rubin, and Taylor [1979]. The *list-chromatic number* $\chi_\ell(G)$ of G is the smallest number k for which, for any assignment of a list $\Lambda(v)$ of size at least k to every vertex $v \in V(G)$, it is possible to color G so that every vertex gets a color from its list. G is said to be *k-choosable* if $\chi_\ell(G) \leq k$, and the list-chromatic number of G is also sometimes called the *choice number*, denoted by ch(G).

It is clear that $\chi(G) \leq \chi_\ell(G)$. But equality does not hold in general. Let H be a 3-chromatic k-uniform hypergraph with n edges. Then $\chi_\ell(K_{n,n}) \geq k + 1$, as can be seen from assigning lists corresponding to the edges of H to the vertices on both sides of $K_{n,n}$. Thus $H = K_3$ gives $\chi_\ell(K_{3,3}) \geq 3$. However, the following remarkable extension of Brooks' theorem to list colorings is true. It was obtained by Vizing and independently by Erdős, Rubin, and Taylor.

Theorem 35 (Vizing [1976], Erdős, Rubin, and Taylor [1979]). $\chi_\ell(G) \leq \Delta(G) + 1$ *holds for every graph G. Moreover,* $\chi_\ell(G) = \Delta(G) + 1$ *if and only if either* $\Delta(G) \neq 2$ *and G has a complete* $(\Delta(G) + 1)$*-graph* $K_{\Delta(G)+1}$ *as a connected component, or* $\Delta(G) = 2$ *and G has an odd cycle as a connected component.*

The list-coloring problem for planar graphs has been solved quite recently. A conjecture that Heawood's five-color theorem can be generalized to list coloring is due to Vizing in 1975 [personal communication from O.V. Borodin in 1985]. The conjecture was also posed independently by Erdős, Rubin, and Taylor [1979]. Voigt [1993] proved that unlike the situation for the usual coloring of planar graphs, "five" cannot be replaced by "four" in this conjecture, by giving an example of a planar graph with 238 vertices that has list-chromatic number five. The upper bound of five was proved by C. Thomassen with a very short and elegant induction argument.

Theorem 36 (Voigt [1993], Thomassen [1993e]). *The maximum list-chromatic number of planar graphs is five. Moreover, if each vertex on the boundary of the* *infinite face of a plane graph G has a list of three colors assigned to it, and if each remaining vertex has a list of five colors assigned to it, then there exists a coloring of G so that every vertex receives a color from its list.*

The proof by Thomassen of the last part of Theorem 36 does not make use of Euler's formula, and the procedure for finding a coloring of G does not involve re-coloring of previously colored vertices. Thomassen remarked that the proof is perhaps the simplest known proof for Heawood's five-color theorem. One may observe [personal communication from D. Hanson in 1993] the interesting implication of Theorem 36 that if G is a planar graph with a connected bipartite induced subgraph H, then it is always possible to extend a 2-coloring of H to a 5-coloring of G. In addition, Thomassen's proof directly implies a simple linear 5-coloring algorithm, compared to degree 2 polynomial algorithms derived from the traditional proofs of Heawood's five-color theorem.

As pointed out by O.V. Borodin [personal communication in 1993] it should be noted that Voigt's example mentioned above shows that the sphere is the only surface $\mathbf{S}$ for which the maximum list-chromatic number $\chi_\ell(\mathbf{S})$ of graphs embeddable on $\mathbf{S}$ exceeds the maximum chromatic number $\chi(\mathbf{S})$ of graphs on $\mathbf{S}$. That is, for every surface $\mathbf{S}$ other than the sphere, $\chi_\ell(\mathbf{S}) = \chi(\mathbf{S})$. This can be seen by observing that a minimal graph G of list-chromatic number k must have minimum degree $\delta(G) \geq k - 1$. Thus similarly to the derivation of Heawood's upper bound $\chi(\mathbf{S}) \leq H(\varepsilon)$ (see Theorem 2), where ε denotes the Euler characteristic of $\mathbf{S}$, one can derive the stronger bound $\chi_\ell(\mathbf{S}) \leq H(\varepsilon)$. Using Theorem 3, $\chi_\ell(\mathbf{S}) = \chi(\mathbf{S})$ follows, except when $\mathbf{S}$ is the Klein bottle. To show that $\chi_\ell(G) \leq \chi(\text{Klein bottle}) = 6$ if G is embedded on the Klein bottle, one may assume by minimality that the minimum degree of G satisfies $\delta(G) \geq 6$. An easy application of Euler's formula for the Klein bottle implies that G is 6-regular, thus Theorem 35 gives the desired bound on $\chi_\ell(G)$.

In general, the list-chromatic number satisfies $\chi(G) \leq \chi_\ell(G) \leq \text{col}(G)$. In some instances treated above, χ_ℓ behaves more like χ than like col. As a further example of this, χ_ℓ satisfies Theorem 1, as noted by Johnson [1994], whereas col does not satisfy Theorem 1 (see Erdős and Hajnal [1966]). However, Alon [1993] showed with a probabilistic proof that there exists a function g such that if $\chi_\ell(G) \leq s$, then the average degree of G (and of all subgraphs of G) is bounded from above by $g(s)$, and hence col(G) is bounded from above by $g(s) + 1$. Thus a class of graphs has bounded list-chromatic number if and only if it has bounded coloring number. Since col is an easily controlled parameter, this can be used in constructions of classes of graphs G where $\chi(G)$ and $\chi_\ell(G)$ behave very differently. For example, if G_d is d-regular, then $\text{col}(G_d) = d + 1 \to \infty$ as $d \to \infty$, and hence also $\chi_\ell(G_d) \to \infty$. But $\chi(G_d)$ may of course be bounded from above, even by 2.

For the *list-edge-chromatic number* $\chi'_\ell(G)$, where the lists of colors are assigned to the edges of G and each edge must be colored with a color from its list, it is expected that this number is always equal to the edge-chromatic number of G. The following conjecture first appeared in print in a paper by Bollobás and Harris [1985], but it was thought of earlier by V.G. Vizing, R.P. Gupta, and by M.O. Albertson and K.L. Collins, all independently (see Häggkvist and Chetwynd [1992]).

List-coloring Conjecture. *Every graph G satisfies* $\chi'_\ell(G) = \chi'(G)$.

The special case $G = K_{n,n}$ is a classical problem due to J. Dinitz. It is often formulated in terms of an $n \times n$ array of sets all of size n and asks if it is always possible to choose one element from every set so that the chosen elements from each row are distinct as are the chosen elements from each column. F. Galvin gave a very short argument answering the Dinitz problem in the affirmative [personal communications from N. Alon and L. Goddyn in 1994]. The same argument can easily be extended to show that the list-coloring conjecture holds for every bipartite graph G.

Borodin [1990b] proved that $\chi'_\ell = \Delta$ holds for planar graphs of maximum degree $\Delta \geq 14$.

For d-regular d-edge-colorable planar graphs the conjecture $\chi'_\ell = \chi'$ is true, as proved by Ellingham and Goddyn [1993] using Theorem 34. The case $d = 3$ is implied by results of Vigneron [1946] (see also Jaeger [1989] and Kauffman [1990]) and Scheim [1974] combined with Theorem 34. This implies by Theorem 26 that the four-color theorem is equivalent to the statement that every 2-connected 3-regular planar graph has list-edge-chromatic number 3, see Alon [1993], who remarked that this had been observed independently by F. Jaeger and M. Tarsi.

Fleischner and Stiebitz [1992] used Theorem 34 to prove that any 4-regular graph whose edges partition into pairwise disjoint triangles together with a Hamilton cycle is 3-colorable.[2] Theorem 34 in fact gave even more, namely that such a graph is 3-choosable. Thus in some cases when considering an open problem of the form $\chi(G) \leq k$, it may be of advantage to consider the stronger statement $\chi_\ell(G) \leq k$—if it is true! Recent surveys of list colorings were given by Häggkvist and Chetwynd [1992] and by Alon [1993].

1.10. GENERALIZED GRAPH COLORING

Problems, results, and definitions in graph coloring theory have variations, extensions, and/or reformulations ad infinitum, it seems. For example:

Theorem 37 (Mihók [1992]). *Let T be a tree on $k + 1$ vertices ($k \geq 4$) and let G be a k-critical graph not containing T as a subgraph. Then G is the complete k-graph K_k.*

The case of Theorem 37 when T is the k-star (i.e., the complete bipartite graph $K_{1,k}$) is the essential part of Theorem 15 (Brooks' theorem). Thus the statement of Theorem 37 is obtained from Theorem 15 by replacing $K_{1,k}$ by an arbitrary tree on $k + 1$ vertices.

Other variations and extensions in graph coloring theory have been described by Jones [1974], Hedetniemi [1973], and Harary [1985], among others. Following

[2] As observed by Fellows [1990], this is equivalent to a theorem on integer partitions. However, the often stated claim that it was formulated as an open problem by I. Schur seems based on a misunderstanding [personal communication from Fellows in 1994].

Borowiecki and Mihók [1991], we shall briefly describe a language that may be used in formulating problems of graph coloring in a general setting.

Let $P_1, P_2, \ldots, P_k$ be graph properties. A $(P_1, P_2, \ldots, P_k)$-*partition* of a graph G is a partition $(V_1, V_2, \ldots, V_k)$ of the vertex set $V(G)$ such that the subgraph $G[V_i]$ induced by V_i has property P_i for $i = 1, 2, \ldots, k$. If G has a $(P_1, P_2, \cdots, P_k)$-partition, then we say that G has *property* $P_1 \cdot P_2 \cdot \ldots \cdot P_k$. If $P_1 = P_2 = \cdots = P_k = P$, then $P_1 \cdot P_2 \cdot \ldots \cdot P_k = P^k$. Letting O denote the property to be without edges, O^k is the property to be k-colorable. The possible uniqueness of a factorization $P_1 \cdot P_2 \cdot \ldots \cdot P_k$ of a reducible property into irreducible properties seems unexplored.

In particular, it seems fruitful to consider *hereditary properties*, meaning those P that are preserved in all subgraphs of graphs having P. The set of all such properties (considering a property as the set of mutually nonisomorphic graphs having the property) forms a lattice $\boldsymbol{L}$ under inclusion. A hereditary property P is determined uniquely by its associated set $\mathcal{F}(P)$ of forbidden subgraphs defined by

$$\mathcal{F}(P) \;=\; \{G : G \notin P, \text{ but } H \in P \text{ for every proper subgraph } H \text{ of } G\}.$$

For example, $\mathcal{F}(O^k)$ is the class of $(k+1)$-critical graphs.

Define the *completeness* of a hereditary property P as the maximum k for which K_{k+1} has P (only the property consisting of all graphs has infinite completeness). Let $\boldsymbol{L}_k$ denote the elements of $\boldsymbol{L}$ with completeness k. Then $\boldsymbol{L}_k$ is a sublattice of $\boldsymbol{L}$ with least element O_k to have at most $k+1$ vertices, for which $\mathcal{F}(O_k) = \{$the complement of $K_{k+2}\}$, and largest element I_k not to contain K_{k+2}, for which $\mathcal{F}(I_k) = \{K_{k+2}\}$. The property O^{k+1} to be $(k+1)$-colorable is in $\boldsymbol{L}_k$; likewise, the property S_k to have maximum degree Δ at most k (with $\mathcal{F}(S_k) = \{K_{1,k+1}\}$), and for a particular tree T on $k+2$ vertices the property not to contain T as a subgraph. Theorems 11 and 37 explore some relations between these elements of $\boldsymbol{L}$. We will mention some further relations that are connected to usual graph coloring and can be stated naturally as generalized graph coloring results.

Let C_p be the property to be p-degenerate (i.e., to have coloring number at most $p+1$). Then it is not hard to see that for all $p, q \geq 0$,

$$C_{p+q+1} \;\subseteq\; C_p \cdot C_q.$$

A similarly expressed but less trivial result is

Theorem 38 (Lovász [1966]). *Let S_p be the property to have maximum degree at most p. Then for $p, q \geq 0$,*

$$S_{p+q+1} \subseteq S_p \cdot S_q.$$

Theorem 38 implies a Brooks-type result for graphs G without a K_r subgraph $(r \geq 4)$:

$$\chi(G) \leq \frac{r-1}{r}(\Delta(G) + 2)$$

found independently by Borodin and Kostochka [1977], Catlin [1978a], and Lawrence [1978]. Another very interesting result is

Theorem 39 (Bollobás and Manvel [1979]). *With properties S_p, C_p, and I_p defined as above, and with $p, q \geq 1$ satisfying $pq > 1$,*

$$S_{p+q} \cap I_{p+q-1} \subseteq (C_{p-1} \cap S_p) \cdot (C_{q-1} \cap S_q).$$

Note that for $p, q \geq 1$, since the chromatic number never exceeds the coloring number,

$$(C_{p-1} \cap S_p) \cdot (C_{q-1} \cap S_q) \subseteq C_{p-1} \cdot C_{q-1} \subseteq O^p \cdot O^q = O^{p+q}.$$

Theorem 39 may thus be regarded as yet another extension of Brooks' Theorem 11 for the case $\Delta \geq 3$, which we can reformulate as follows.

Theorem 40 (Brooks [1941]). *With properties S_p and I_p defined as above and for $n \geq 3$,*

$$S_n \cap I_{n-1} \subseteq O^n.$$

1.11. FINAL REMARKS

This brief introduction to graph coloring hopefully indicates that graph coloring theory has come a long way since its beginnings almost one and a half century ago. But very many challenging unsolved problems remain. So let us now embark on our journey into the jungle of these problems, remembering the words of the Danish poet and designer Piet Hein:

Problems worthy
of attack
prove their worth
by hitting back

BIBLIOGRAPHY

[1984] Aigner M. *Graphentheorie. Eine Entwicklung aus dem 4-Farben Problem.* B.G. Teubner Verlagsgesellschaft, 1984. English translation, BCS Associates, 1987.

[1974] Aksionov V.A. On continuation of 3-coloring of planar graph (in Russian). *Metody Diskret. Analiz.* **26**, 3–19, 1974.

[1979] Albertson M.O. and J.P. Hutchinson. The three excluded cases of Dirac's map-color theorem. *Ann. New York Acad. Sci.* **319**, 7–17, 1979.

[1980a] Albertson M.O. and J.P. Hutchinson. Hadwiger's conjecture for graphs on the Klein bottle. *Discrete Math.* **29**, 1–11, 1980.

[1993] Alon N. Restricted colorings of graphs. In: K. Walker, editor, *Surveys in Combinatorics: Proc. 14th British Combinatorial Conference*, pages 1–33. Cambridge University Press, 1993.

[1992] Alon N. and J.H. Spencer. *The Probabilistic Method.* Wiley, 1992 (with an appendix on open problems by P. Erdős).

[1992] Alon N. and M. Tarsi. Colorings and orientations of graphs. *Combinatorica* **12**, 125–134, 1992.

[1983] Andersen L.D. and A.J.W. Hilton. Thank Evans! *Proc. London Math. Soc. (3)* **47**, 507–522, 1983.

[1976a] Appel K. and W. Haken. Every planar map is four colorable. *Bull. Amer. Math. Soc.* **82**, 711–712, 1976.

[1977a] Appel K. and W. Haken. Every planar map is four colorable. Part I: Discharging. *Illinois J. Math.* **21**, 429–490, 1977.

[1977b] Appel K. and W. Haken. The solution of the four-color map problem. *Sci. Amer.* **237**, 108–121, Oct. 1977.

[1978] Appel K. and W. Haken. The four color problem. In: L.A. Steen, editor, *Mathematics Today: Twelve Informal Essays*, pages 153–180. Springer-Verlag, 1978. Vintage Books, 1980.

[1986] Appel K. and W. Haken. The four color proof suffices. *Math. Intelligencer* **8**(1), 10–20, 1986.

[1989] Appel K. and W. Haken. *Every Planar Map Is Four Colorable*, volume 98 of *Contemporary Mathematics Series*. Amererican Mathematical Society, 1989.

[1977] Appel K., W. Haken, and J. Koch. Every planar map is four colorable. Part II: Reducibility. *Illinois J. Math.* **21**, 491–567, 1977.

[1983] Barnette D. *Map Coloring, Polyhedra, and the Four-Color Problem.* Dolciani Mathematical Expositions No. 8. Mathematics Association of America, 1983.

[1970] Beineke L.W. Characterizations of derived graphs. *J. Combin. Theory* **9**, 129–135, 1970.

[1960] Berge C. Les problèmes de colorations en théorie des graphes. *Publ. Inst. Statist. Univ. Paris* **9**, 123–160, 1960.

[1961] Berge C. Färbung von Graphen, deren sämtliche bzw. deren ungerade Kreise starr sind. *Wiss. Z. Martin-Luther-Univ. Halle–Wittenberg Math.-Natur. Reihe* **10**, 114, 1961.

[1963a] Berge C. Perfect graphs. In: *Six Papers on Graph Theory*, pages 1–21. Indian Statistical Institute, Calcutta, 1963.

[1963b] Berge C. Sur une conjecture relative au problème des codes optimaux de Shannon (Resumé). In: *Comm. 14ème assemblée générale de l'URSI, Tokyo, 1963*, volume XIII–6, pages 317–318. Union Radio Scientifique Internationale, Bruxelles, 1963.

[1966] Berge C. Une application de la théorie des graphes à un problème de codage. In: E.R. Caianello, editor, *Automata Theory*, pages 25–34. Academic Press, 1966.

[1984] Berge C. and V. Chvátal. Introduction. In: C. Berge and V. Chvátal, editors, *Topics on Perfect Graphs*, volume 21 of *Annals of Discrete Mathematics*, pages vii–xiv. North-Holland, 1984.

[1947] Bernhart A. Six rings in minimal five color maps. *Amer. J. Math.* **69**, 391–412, 1947.

[1988] Bigalke H.-G. *Heinrich Heesch*, volume 3 of *Vita Mathematica*. Birkhäuser Verlag, 1988.

[1976] Biggs N.L., E.K. Lloyd, and R.J. Wilson. *Graph Theory 1736–1936*. Clarendon Press, 1976.

[1912] Birkhoff G.D. A determinant formula for the number of ways of coloring a map. *Ann. of Math.* **14**, 42–46, 1912.

[1946] Birkhoff G.D. and D.C. Lewis. Chromatic polynomials. *Trans. Amer. Math. Soc.* **60**, 355–451, 1946.

[1978a] Bollobás B. *Extremal Graph Theory*. Academic Press, 1978.

[1985] Bollobás B. *Random Graphs*. Academic Press, 1985.

[1988] Bollobás B. The chromatic number of random graphs. *Combinatorica* **8**, 49–55, 1988.

[1980] Bollobás B., P.A. Catlin, and P. Erdős. Hadwiger's conjecture is true for almost every graph. *European J. Combin.* **1**, 195–199, 1980.

[1985] Bollobás B. and A.J. Harris. List-colourings of graphs. *Graphs Combin.* **1**, 115–127, 1985.

[1979] Bollobás B. and B. Manvel. Optimal vertex partitions. *Bull. London Math. Soc.* **11**, 113–116, 1979.

[1976] Bondy J.A. Diconnected orientations and a conjecture of Las Vergnas. *J. London Math. Soc. (2)* **14**, 277–282, 1976.

[1979a] Borodin O.V. On acyclic colorings of planar graphs. *Discrete Math.* **25**, 211–236, 1979.

[1990b] Borodin O.V. Generalization of Kotzig's theorem and assigned edge-colorings of planar graphs (in Russian). *Math. Z.* **48**, 22–28, 1990. Translation in *Math. Notes* **48**, 1186–1190, 1990.

[1993a] Borodin O.V. Four problems on planar graphs raised by Branko Grünbaum. *Contemporary Math.* **147**, 149–156, 1993.

[1994a] Borodin O.V. A new proof of Grünbaum's 3 colour theorem. Preprint, Nottingham University and the Russian Academy of Sciences in Novosibirsk, 1994.

[1977] Borodin O.V. and A.V. Kostochka. On an upper bound of a graph's chromatic number, depending on the graph's degree and density. *J. Combin. Theory Ser. B* **23**, 247–250, 1977.

[1991] Borowiecki M. and P. Mihók. Hereditary properties of graphs. In: V.R. Kulli, editor, *Advances in Graph Theory*. Vishwa International Publishers, 1991.

[1941] Brooks R.L. On colouring the nodes of a network. *Proc. Cambridge Phil. Soc.* **37**, 194–197, 1941.

[1969] Brown W.G. and J.W. Moon. Sur les ensembles de sommets indépendentes dans les graphes chromatiques minimaux. *Canad. J. Math.* **21**, 274–278, 1969.

[1951] de Bruijn N.G. and P. Erdős. A colour problem for infinite graphs and a problem in the theory of relations. *Nederl. Akad. Wetensch. Proc. Ser. A* **54**, 371–373, 1951 (*Indag. Math.* **13**).

[1978a] Catlin P.A. A bound on the chromatic number of a graph. *Discrete Math.* **22**, 81–83, 1978.

[1878] Cayley A. Open problem. *Proc. London Math. Soc.* **9**, 148, 1878.

[1879] Cayley A. On the colouring of maps. *Proc. Roy. Geog. Soc. (New Ser.)* **1**, 259–261, 1879.

[1984b] Chvátal V. A semi-strong perfect graph conjecture. In: C. Berge and V. Chvátal, editors, *Topics on Perfect Graphs*, volume 21 of *Annals of Discrete Mathematics*, pages 279–280. North-Holland, 1984.

[1947] Descartes B. A three-colour problem. *Eureka* **9**, 1947.

[1948b] Descartes B. Solutions to problems in Eureka No. 9. *Eureka* **10**, 1948.

[1954] Descartes B. Solution to advanced problem No. 4526. *Amer. Math. Monthly* **61**, 532, 1954.

[1951] Dirac G.A. Note on the colouring of graphs. *Math. Z.* **54**, 347–353, 1951.

[1952a] Dirac G.A. A property of 4-chromatic graphs and some remarks on critical graphs. *J. London Math. Soc.* **27**, 85–92, 1952.

[1952b] Dirac G.A. Map colour theorems. *Canad. J. Math.* **4**, 480–490, 1952.

[1952c] Dirac G.A. Some theorems on abstract graphs. *Proc. London Math. Soc. (3)* **2**, 69–81, 1952.

[1953] Dirac G.A. The structure of k-chromatic graphs. *Fund. Math.* **40**, 42–55, 1953.

[1957b] Dirac G.A. Short proof of a map-colour theorem. *Canad. J. Math.* **9**, 225–226, 1957.

[1957c] Dirac G.A. A theorem of R.L. Brooks and a conjecture of H. Hadwiger. *Proc. London Math. Soc. (3)* **7**, 161–195, 1957.

[1960] Dirac G.A. Trennende Knotenpunktmengen und Reduzibilität abstrakter Graphen mit Anwendung auf das Vierfarbenproblem. *J. Reine Angew. Math.* **204**, 116–131, 1960.

[1961] Dirac G.A. On rigid circuit graphs. *Abh. Math. Sem. Univ. Hamburg* **25**, 71–76, 1961.

[1963] Dirac G.A. Percy John Heawood. *J. London Math. Soc.* **38**, 263–277, 1963.

[1964a] Dirac G.A. On the structure of 5- and 6-chromatic abstract graphs. *J. Reine Angew. Math.* **214/215**, 43–52, 1964.

[1974] Dirac G.A. The number of edges in critical graphs. *J. Reine Angew. Math.* **268/269**, 150–164, 1974.

[1965a] Edmonds J. Minimum partition of a matroid into independent subsets. *J. Res. Nat. Bur. Stand.* **69B**, 67–72, 1965.

[1993] Ellingham M.N. and L. Goddyn. List edge colorings of some regular planar multigraphs. Manuscript, 1993. Submitted to *Combinatorica*.

[1958] Erdős P. Remarks on a theorem of Ramsey. *Bull. Res. Council Israel Sect. F* **7**, 21–24, 1957–58.

[1959] Erdős P. Graph theory and probability. *Canad. J. Math.* **11**, 34–38, 1959.

[1966] Erdős P. and A. Hajnal. On chromatic number of graphs and set-systems. *Acta Math. Acad. Sci. Hungar.* **17**, 61–99, 1966.

[1960] Erdős P. and A. Rényi. On the evolution of random graphs. *Magyar Tud. Akad. Mat. Kutató Int. Közl.* **5**, 17–61, 1960.

[1979] Erdős P., A.L. Rubin, and H. Taylor. Choosability in graphs. In: *Proc. West Coast Conference on Combinatorics, Graph Theory and Computing, Arcata, 1979, Congr. Num., 26*, pages 125–157, 1979.

[1736] Euler L. Solutio problematis ad geometriam situs pertinensis. *Comm. Acad. Sci. Imp. Petropol.* **8**, 128–140, 1736 (1741).

[1990] Fellows M.R. Transversals of vertex partitions in graphs. *SIAM J. Discrete Math.* **3**, 206–215, 1990.

[1969] Finck H.-J. and H. Sachs. Über eine von H.S. Wilf angegebene Schranke für die chromatische Zahl endlicher Graphen. *Math. Nachr.* **39**, 373–386, 1969.

[1992] Fleischner H. and M. Stiebitz. A solution to a colouring problem of P. Erdős. *Discrete Math.* **101**, 39–48, 1992.

[1934] Franklin P. A six-color problem. *J. Math. Phys.* **13**, 363–369, 1934.

[1971] Fulkerson D.R. Blocking and anti-blocking pairs of polyhedra. *Math. Programming* **1**, 168–194, 1971.

[1972] Fulkerson D.R. Anti-blocking polyhedra. *J. Combin. Theory Ser. B* **12**, 50–71, 1972.

[1958] Gallai T. Maximum–minimum Sätze über Graphen. *Acta Math. Acad. Sci. Hungar.* **9**, 395–434, 1958.

[1959] Gallai T. Über extreme Punkt- und Kantenmengen. *Ann. Univ. Sci. Budapest Eötvös Sect. Math.* **2**, 133–138, 1959.

[1962] Gallai T. Graphen mit triangulierbaren ungeraden Vielecken. *Magyar Tud. Akad. Mat. Kutató Int. Közl.* **7**, 3–36, 1962.

[1963a] Gallai T. Kritische Graphen I. *Publ. Math. Inst. Hungar. Acad. Sci.* **8**, 165–192, 1963.

[1963b] Gallai T. Kritische Graphen II. *Publ. Math. Inst. Hungar. Acad. Sci.* **8**, 373–395, 1963.

[1968] Gallai T. On directed paths and circuits. In: P. Erdős and G. Katona, editors, *Theory of Graphs*, pages 115–118. Academic Press, 1968.

[1991] Galvin F. and P. Komjáth. Graph colorings and the axiom of choice. *Period. Math. Hungar.* **22**, 71–75, 1991.

[1979] Garey M.R. and D.S. Johnson. *Computers and Intractability: A Guide to the Theory of **NP**-Completeness*. W.H. Freeman and Company, 1979.

[1976] Garey M.R., D.S. Johnson, and L.J. Stockmeyer. Some simplified ***NP***-complete graph problems. *Theoret. Comput. Sci.* **1**, 237–267, 1976.

[1951] Gottschalk W.H. Choice functions and Tychonoff's theorem. *Proc. Amer. Math. Soc.* **2**, 172, 1951.

[1990] Graham R.L., B.L. Rothschild, and J.H. Spencer. *Ramsey Theory*, 2nd ed. Wiley, 1990 (first edition, 1980).

[1981] Grötschel M., L. Lovász, and A. Schrijver. The ellipsoid method and its consequences in combinatorial optimization. *Combinatorica* **1**, 169–197, 1981.

[1959] Grötzsch H. Ein Dreifarbensatz für dreikreisfreie Netze auf der Kugel. *Wiss. Z. Martin-Luther-Univ. Halle–Wittenberg Math.-Natur. Reihe* **8**, 109–120, 1959.

[1963] Grünbaum B. Grötzsch's theorem on 3-colorings. *Michigan Math. J.* **10**, 303–310, 1963.

[1973] Grünbaum B. Acyclic colorings of planar graphs. *Israel J. Math.* **14**, 390–408, 1973.

[1963] Gupta R.P. A note on a theorem on k-colouring the vertices of a graph. In: *Six Papers on Graph Theory*, pages 30–32. Indian Statistical Institute, Calcutta, 1963.

[1966] Gupta R.P. The chromatic index and the degree of a graph. *Notices Amer. Math. Soc.* **13**, abstract 66T–429, 1966.

[1943] Hadwiger H. Über eine Klassifikation der Streckenkomplexe. *Vierteljahrsch. Naturforsch. Ges. Zürich* **88**, 133–142, 1943.

[1992] Häggkvist R. and A.G. Chetwynd. Some upper bounds on the total and list chromatic numbers of multigraphs. *J. Graph Theory* **16**, 503–516, 1992.

[1958] Hajnal A. and J. Surányi. Über die Auflösung von Graphen in vollständige Teilgraphen. *Ann. Univ. Sci. Budapest Eötvös Sect. Math.* **1**, 113–121, 1958.

[1970] Hajnal A. and E. Szemerédi. Proof of a conjecture of Erdős. In: P. Erdős, A. Rényi, and V.T. Sós, editors, *Combinatorial Theory and Its Applications, Vol. II*, volume 4 of *Colloquia Mathematica Societatis János Bolyai*, pages 601–623. North-Holland, 1970.

[1961] Hajós G. Über eine Konstruktion nicht n-färbbarer Graphen. *Wiss. Z. Martin-Luther-Univ. Halle–Wittenberg Math.-Natur. Reihe* **10**, 116–117, 1961.

[1967] Halin R. Unterteilungen vollständiger Graphen in Graphen mit unendlicher chromatischer Zahl. *Abh. Math. Sem. Univ. Hamburg* **31**, 156–165, 1967.

[1985] Harary F. Conditional colorability in graphs. In: F. Harary and J.S. Maybee, editors, *Graphs and Applications: Proceedings of the First Colorado Symposium on Graph Theory*, pages 127–136. Wiley, 1985.

[1890] Heawood P.J. Map colour theorem. *Quart. J. Pure Appl. Math.* **24**, 332–338, 1890.

[1898] Heawood P.J. On the four-colour map theorem. *Quart. J. Pure Appl. Math.* **29**, 270–285, 1898.

[1973] Hedetniemi S.T. On hereditary properties of graphs. *J. Combin. Theory Ser. B* **14**, 94–99, 1973.

[1969] Heesch H. *Untersuchungen zum Vierfarben-problem*. Number 810/810a/810b in B.I. Hochschulscripten. Bibliographisches Institut, 1969.

[1891] Heffter L. Über das Problem der Nachbargebiete. *Math. Ann.* **38**, 477–508, 1891.

[1976a] Jaeger F. On nowhere-zero flows in multigraphs. In: C. St. J.A. Nash-Williams and J. Sheehan, editors, *Proc. Fifth British Combinatorial Conference, Aberdeen, 1975*, pages 373–378. Utilitas Mathematics Publications Inc., 1976.

[1988] Jaeger F. Nowhere-zero flow problems. In: L.W. Beineke and R.J. Wilson, editors, *Selected Topics in Graph Theory*, volume 3, pages 71–95. Academic Press, 1988.

[1989] Jaeger F. On the Penrose number of cubic diagrams. *Discrete Math.* **74**, 85–97, 1989.

[1992] Johnson D.S. The NP-completeness column: an ongoing guide. *J. Algorithms* **13**, 502–524, 1992.

[1994] Johnson P.D. The choice number of the plane. *Geombinatorics* **III**, 122–128, 1994.

[1974] Jones R.P. Hereditary properties and P-chromatic numbers. In: T.P. McDonough and V.C. Mavron, editors, *Combinatorics*, pages 83–88. Cambridge University Press, 1974.

[1993] Kainen P.C. Is the four color theorem true? *Geombinatorics* **III**, 41–56, 1993.

[1972] Karp R. Reducibility among combinatorial problems. In: R.E. Miller and J.W. Thatcher, editors, *Complexity of Computer Computations*, pages 85–104. Plenum Press, 1972.

[1990] Kauffman L.H. Map coloring and the vector cross product. *J. Combin. Theory Ser. B* **48**, 145–154, 1990.

[1954] Kelly J.B. and L.M. Kelly. Paths and circuits in critical graphs. *Amer. J. Math.* **76**, 786–792, 1954.

[1879] Kempe A.B. On the geographical problem of four colours. *Amer. J. Math.* **2**, 193–200, 1879.

[1983] Kierstead H.A. and J.H. Schmerl. Some applications of Vizing's theorem to vertex colorings of graphs. *Discrete Math.* **45**, 277–285, 1983.

[1975] Kilpatrick P.A. *Tutte's first colour-cycle conjecture*. Ph.D. thesis, Cape Town, 1975.

[1916] König D. Über Graphen und ihre Anwendung auf Determinantentheorie und Mengenlehre. *Math. Ann.* **77**, 453–465, 1916.

[1936] König D. *Theorie der endlichen und unendlichen Graphen.* Akademische Verlagsgesellschaft M.B.H. Leipzig, 1936. Reprinted by Chelsea 1950 and by B.G. Teubner 1986. English translation published by Birkhäuser 1990.

[1982] Kostochka A.V. The minimum Hadwiger number for graphs with a given mean degree of vertices (in Russian). *Metody Diskret. Analiz.* **38**, 37–58, 1982.

[1972a] Kronk H.V. and J. Mitchem. On Dirac's generalization of Brooks' theorem. *Canad. J. Math.* **24**, 805–807, 1972.

[1978] Lawrence J. Covering the vertex set of a graph with subgraphs of smaller degree. *Discrete Math.* **21**, 61–68, 1978.

[1970] Lick D.R. and A.T. White. k-degenerate graphs. *Canad. J. Math.* **22**, 1082–1096, 1970.

[1966] Lovász L. On decomposition of graphs. *Studia Sci. Math. Hungar.* **1**, 237–238, 1966.

[1968] Lovász L. On chromatic number of finite set-systems. *Acta Math. Acad. Sci. Hungar.* **19**, 59–67, 1968.

[1972a] Lovász L. A characterization of perfect graphs. *J. Combin. Theory Ser. B* **13**, 95–98, 1972.

[1972b] Lovász L. Normal hypergraphs and the perfect graph conjecture. *Discrete Math.* **2**, 253–267, 1972.

[1973a] Lovász L. Coverings and colorings of hypergraphs. In: *Proc. 4th S–E Conference on Combinatorics, Graph Theory and Computing, Boca Raton, Congr. Num., 8*, pages 3–12, 1973.

[1973b] Lovász L. Independent sets in critical chromatic graphs. *Studia Sci. Math. Hungar.* **8**, 165–168, 1973.

[1982] Lovász L. Bounding the independence number of a graph. In: A. Bachem, M. Grötschel, and B. Korte, editors, *Bonn Workshop on Combinatorial Optimization*, volume 16 of *Annals of Discrete Mathematics*, pages 213–223. North-Holland, 1982.

[1983a] Lovász L. Perfect graphs. In: L.W. Beineke and R.J. Wilson, editors, *Selected Topics in Graph Theory*, volume 2, pages 55–87. Academic Press, 1983.

[1994] Lovász L. Stable sets and polynomials. *Discrete Math.* **124**, 137–153, 1994.

[1988] Lubotzky A., R. Phillips, and P. Sarnak. Ramanujan graphs. *Combinatorica* **8**, 261–277, 1988.

[1993] Lund C. and M. Yannakakis. On the hardness of approximating minimization problems. In: *Proc. 25th ACM Symposium on Theory of Computing*, pages 286–293. ACM, New York, 1993.

[1991] Massey W.S. *A Basic Course in Algebraic Topology.* Springer-Verlag, 1991.

[1968] Matula D.W. A min–max theorem for graphs with application to graph coloring. *SIAM Rev.* **10**, 481–482, 1968.

[1976] Meyniel H. On the perfect graph conjecture. *Discrete Math.* **16**, 339–342, 1976.

[1992] Mihók P. An extension of Brooks' theorem. In: J. Nešetřil and M. Fiedler, editors, *Fourth Czechoslovakian Symposium on Combinatorics, Graphs and Complexity*, volume 51 of *Annals of Discrete Mathematics*, pages 235–236. North-Holland, 1992.

[1962] Minty G.J. A theorem on n-coloring the points of a linear graph. *Amer. Math. Monthly* **63**, 623–624, 1962.

[1977] Müller V., V. Rödl, and D. Turzík. On critical 3-chromatic hypergraphs. *Acta Math. Acad. Sci. Hungar.* **29**, 273–281, 1977.

[1955] Mycielski J. Sur le coloriage des graphes. *Colloq. Math.* **3**, 161–162, 1955.

[1979] Nešetřil J. and V. Rödl. A short proof of the existence of highly chromatic hypergraphs without short cycles. *J. Combin. Theory Ser. B* **27**, 225–227, 1979.

[1967] Ore O. *The Four-Color Problem.* Academic Press, 1967.

[1891] Petersen J. Die Theorie der regulären graphs. *Acta Math.* **15**, 193–220, 1891.

[1992] Prömel H.-J. and A. Steger. Almost all Berge graphs are perfect. *Combin. Probab. Comput.* **1**, 53–79, 1992.

[1949] Rado R. Axiomatic treatment of rank in infinite sets. *Canad. J. Math.* **1**, 337–343, 1949.

[1968] Read R.C. An introduction to chromatic polynomials. *J. Combin. Theory* **4**, 52–71, 1968.

[1988] Read R.C. and W.T. Tutte. Chromatic polynomials. In: L.W. Beineke and R.J. Wilson, editors, *Selected Topics in Graph Theory*, volume 3, pages 15–42. Academic Press, 1988.

[1987] Reed B.A. A semi-strong perfect graph theorem. *J. Combin. Theory Ser. B* **43**, 223–240, 1987.

[1954] Ringel G. Bestimmung der Maximalzahl der Nachbargebiete auf nichtorientierbaren Flächen. *Math. Ann.* **127**, 181–214, 1954.

[1959] Ringel G. *Färbungsprobleme auf Flächen und Graphen.* VEB Deutscher Verlag der Wissenschaften, 1959.

[1974] Ringel G. *Map Color Theorem.* Springer-Verlag, 1974.

[1968] Ringel G. and J.W.T. Youngs. Solution of the Heawood map-coloring problem. *Proc. Natl. Acad. Sci. USA* **60**, 438–445, 1968.

[1993a] Robertson N., P.D. Seymour, and R. Thomas. Hadwiger's conjecture for K_6-free graphs. *Combinatorica* **13**, 279–361, 1993.

[1967] Roy B. Nombre chromatique et plus longs chemins d'un graphe. *Rev. Française Automat. Informat. Recherche Opérationelle Sér. Rouge* **1**, 127–132, 1967.

[1972] Saaty T.L. Thirteen colorful variations on Guthrie's four-color conjecture. *Amer. Math. Monthly* **79**, 2–43, 1972.

[1977] Saaty T.L. and P.C. Kainen. *The Four-Color Problem.* McGraw–Hill, 1977. Dover, 1986.

[1989] Sachs H. and M. Stiebitz. Colour-critical graphs with vertices of low valency. In: L.D. Andersen, I.T. Jakobsen, C. Thomassen, B. Toft, and P.D. Vestergaard, editors, *Graph Theory in Memory of G.A. Dirac*, volume 41 of *Annals of Discrete Mathematics*, pages 371–396. North-Holland, 1989.

[1974] Scheim D.E. The number of edge 3-colorings of planar cubic graphs as a permanent. *Discrete Math.* **8**, 377–382, 1974.

[1981a] Seymour P.D. Nowhere-zero 6-flows. *J. Combin. Theory Ser. B* **30**, 130–135, 1981.

[1949] Shannon C.E. A theorem on coloring the lines of a network. *J. Math. Phys.* **28**, 148–151, 1949.

[1956] Shannon C.E. The zero error capacity of a noisy channel. *I.R.E. Trans. on Inform. Theory* **IT–2**, 8–19, 1956.

[1972] Simonovits M. On colour-critical graphs. *Studia Sci. Math. Hungar.* **7**, 67–81, 1972.

[1993a] Steinberg R. The state of the three color problem. In: J. Gimbel, J.W. Kennedy, and L.V. Quintas, editors, *Quo Vadis, Graph Theory?* volume 55 of *Annals of Discrete Mathematics*, pages 211–248. North-Holland, 1993.

[1993b] Steinberg R. An update on the state of the three color problem. In: J.W. Kennedy and L.V. Quintas, editors, *Graph Theory Notes of New York XXV*, pages 9–12. New York Academy of Sciences, 1993.

[1985] Stiebitz M. *Beiträge zur Theorie der färbungskritischen Graphen.* Dissertation zu Erlangung des akademischen Grades Dr.sc.nat., Technische Hochscule Ilmenau, 1985.

[1987b] Stiebitz M. Subgraphs of colour-critical graphs. *Combinatorica* **7**, 303–312, 1987.

[1973] Stockmeyer L.J. Planar 3-colorability is ***NP***-complete. *SIGACT News* **5**:3, 19–25, 1973.

[1878] Sylvester J.J. On an application of the new atomic theory to the graphical representation of the invariants and covariants of binary quantics, with three appendices. *Amer. J. Math.* **1**, 64–125, 1878.

[1968] Szekeres G. and H.S. Wilf. An inequality for the chromatic number of a graph. *J. Combin. Theory* **4**, 1–3, 1968.

[1880] Tait P.G. On the colouring of maps. *Proc. Roy. Soc. Edinburgh Sect. A* **10**, 501–503, 729, 1878–80.

[1993c] Thomassen C. Grötzsch's 3-color theorem and its counterparts for the torus and the projective plane. Manuscript, 1993. To appear in *J. Combin. Theory Ser. B.*

[1993e] Thomassen C. Every planar graph is 5–choosable. Technical report, The Technical University of Denmark, November 1993. *J. Combin. Theory Ser. B* **62**, 180–181, 1994.

[1910] Tietze H. Einige Bemerkungen über das Problem des Kartenfärbens auf einseitigen Flächen. *Jahresber. Deutsch. Math.-Verein.* **19**, 155–159, 1910.

[1970a] Toft B. On the maximal number of edges of critical k-chromatic graphs. *Studia Sci. Math. Hungar.* **5**, 461–470, 1970.

[1972b] Toft B. Two theorems on critical 4-chromatic graphs. *Studia Sci. Math. Hungar.* **7**, 83–89, 1972.

[1974a] Toft B. Color-critical graphs and hypergraphs. *J. Combin. Theory Ser. B* **16**, 145–161, 1974.

[1975a] Toft B. On colour-critical hypergraphs. In: A. Hajnal, R. Rado, and V.T. Sós, editors, *Infinite and Finite Sets*, volume 10 of *Colloquia Mathematica Societatis János Bolyai*, pages 1445–1457. North-Holland, 1975.

[1985] Toft B. Some problems and results related to subgraphs of colour critical graphs. In: R. Bodendiek, H. Schumacher, and G. Walther, editors, *Graphen in Forschung und Unterricht: Festschrift K. Wagner*, pages 178–186. Barbara Franzbecker Verlag, 1985.

[1954] Tutte W.T. A contribution to the theory of chromatic polynomials. *Canad. J. Math.* **6**, 80–91, 1954.

[1970b] Tutte W.T. More about chromatic polynomials and the golden ratio. In: R. Guy, H. Hanani, N.W. Sauer, and J. Schönheim, editors, *Combinatorial Structures and Their Applications*, pages 439–453. Gordon and Breach, 1970.

[1946] Vigneron L. Remarques sur les réseaux cubiques de classe 3 associés au problème des 4 couleurs. *C.R. Acad. Sci. Paris* **223**, 770–772, 1946.

[1962] Vitaver L.M. Determining a minimal vertex-coloring of a graph by means of Boolean powers of the incidence matrix (in Russian). *Dokl. Akad. Nauk SSSR* **147**, 758–759, 1962.

[1964] Vizing V.G. On an estimate of the chromatic class of a p-graph (in Russian). *Metody Diskret. Analiz.* **3**, 25–30, 1964.

[1965a] Vizing V.G. Critical graphs with given chromatic class (in Russian). *Metody Diskret. Analiz.* **5**, 9–17, 1965.

[1965b] Vizing V.G. The chromatic class of a multigraph (in Russian). *Kibernetika (Kiev)* no. 3, 29–39, 1965. English translation in *Cybernetics* **1**, 32–41.

[1976] Vizing V.G. Vertex colorings with given colors (in Russian). *Metody Diskret. Analiz.* **29**, 3–10, 1976.

[1993] Voigt M. List colourings of planar graphs. *Discrete Math.* **120**, 215–219, 1993.

[1937] Wagner K. Über eine Eigenschaft der ebenen Komplexe. *Math. Ann.* **114**, 570–590, 1937.

[1970] Wagner K. *Graphentheorie*. Number 248/248a* in B.I. Hochschultaschenbücher. Bibliographisches Institut, 1970.

[1975] Weinstein J. Excess in critical graphs. *J. Combin. Theory Ser. B* **18**, 24–31, 1975.

[1976] Welsh D.J.A. *Matroid Theory*. Academic Press, 1976.

[1932a] Whitney H. A logical expansion in mathematics. *Bull. Amer. Math. Soc.* **38**, 572–579, 1932.

[1932b] Whitney H. The coloring of graphs. *Ann. of Math.* **33**, 688–718, 1932.

[1976] Wilson J. New light on the origin of the four-colour conjecture. *Hist. Math.* **3**, 329–330, 1976.

[1978] Woodall D.R. and R.J. Wilson. The Appel–Haken proof of the four-color theorem. In: L.W. Beineke and R.J. Wilson, editors, *Selected Topics in Graph Theory*, pages 83–101. Academic Press, 1978.

[1983] Younger D.H. Integer flows. *J. Graph Theory* **7**, 349–357, 1983.

[1949] Zykov A.A. On some problems of linear complexes (in Russian). *Mat. Sbornik N.S.* **24**, 163–188, 1949. English translation in *Amer. Math. Soc. Transl.* **79**, 1952. Reissued in: *Translation Series 1* **7**, *Algebraic Topology*, 418–449 (American Mathematical Society 1962).

2

Planar Graphs

2.1. FOUR-COLOR THEOREM

Does there exist a short proof of the four-color theorem, that every planar graph is 4-colorable, in which all the details can be checked by hand by a competent mathematician in, say, two weeks?

Is it possible by a short argument at least to exhibit a number N such that if there is a 5-chromatic planar graph, then there is a 5-chromatic planar graph of at most N vertices? That is, is there a short argument to demonstrate that the four-color problem is a finite problem?

Does there exist a graph coloring algorithm A (possibly derived from a complicated proof of the four-color theorem) such that A, with a planar graph G as input, produces a 4-coloring of G in at most $c \cdot |V(G)|^d$ steps, where c is a constant and $d < 2$?

Finally: Is there a short argument that proves the existence of a polynomial algorithm to decide if a given planar graph is 4-colorable? ■

The celebrated four-color problem, due to F. Guthrie, asks if every planar graph is 4-colorable. A proposed solution by Kempe [1879] stood for more than a decade until it was refuted by Heawood [1890] in his first paper. For a brief sketch of the history of the problem see Chapter 1.

The proof of the four-color theorem by Appel and Haken [1977a] and Appel, Haken, and Koch [1977] was based on the same basic idea as Kempe's proof, that is, to find a set of unavoidable and reducible configurations (see Theorem 5 in Chapter 1). But where Kempe's configurations were vertices of degree at most 5, Appel and Haken's initial set had 1936 configurations (Appel and Haken [1977a, 1989] announced that proofs with only 1482, 1405, and 1256 configurations are possible). The unavoidability of the set of configurations was obtained by an elaborate "discharging procedure" and their reducibility established by computer calculations.

The four-color proof was the first example of a mathematical proof relying heavily on the use of computers. It is interesting to note that in the early days of the development of electronic computers, M.H.A. Newman at the University of Manchester wrote a letter dated 8 February 1946 to J. von Neumann at Princeton University stating that with the advent of fast electronic computers "*... mathematical problems of an entirely different kind from those so far tackled by machines might be tried, e.g. testing out (say) the 4-color theorem* [sic] ..." (Newman [1946]). (A team headed by Newman built an electronic computer that became operational in mid–1948 in Manchester, England. Presumably it was the first working electronic stored-program computer; see, e.g., Randell [1980].) The use of computers in mathematical proofs should not necessarily be avoided; in fact, one would tend to trust a computer more than a human being testing a very large number of logically clear cases. However, the four-color theorem seems such a simple statement that one would want a more elegant and more easily checkable proof than the one presented by Appel and Haken. Gardner [1980] remarked: "*The proof is an extraordinary achievement.... To most mathematicians, however, the proof of the four-color conjecture* [sic] *is deeply unsatisfactory.*"

Regarding the time involved in checking parts of their proof by hand, Appel and Haken [1978] wrote: "*A person could easily check the part of the discharging procedure that did not involve reducibility computations in a month or two.*"

N. Robertson, D.P. Sanders, P.D. Seymour, and R. Thomas [personal communication from N. Robertson and P.D. Seymour in 1994] have recently obtained a new improved proof of the four-color theorem, essentially using the same approach as Appel, Haken, and Koch. This proof, based on a simpler discharging procedure, involves less than 700 configurations.

Concerning the question of finiteness; for all surfaces (compact 2-dimensional manifolds) other than the sphere, the coloring problem is finite, as shown by Dirac [1956, 1957a] and by Schumacher and Wagner [1985]. For example, for a surface **S** of Euler characteristic $\varepsilon \leq -2$ it is very easy to prove that the maximum chromatic number among all graphs embeddable on **S** equals the maximum chromatic number among such graphs of size at most (-6ε). If no simple proof of the four-color theorem for the plane can be obtained, it would be very desirable to have at least a simple proof of the finiteness. An answer to the following question would provide a step in this direction:

Does there exist a number $N > 0$ so that for all $n \geq N$ every configuration (without separating triangles) consisting of an n-ring containing m interior vertices, where $m > 3n/2 - 6$, is reducible[1] *or contains a smaller reducible subconfiguration?*

In essence, this question was asked by Appel and Haken [1977a], except that they stated it without any lower bound N for ring size. Their conjecture, that the answer

[1] Here "reducible" means, more precisely, "D-reducible"—for a discussion of the concept of D-reducibility introduced by H. Heesch, see Whitney and Tutte [1975]. Informally, if a D-reducible configuration is contained in a plane triangulation, one may simply delete all the vertices interior to the bounding outer ring of the configuration and consider a possible 4-coloring of the remaining graph. If any such 4-coloring exists, then the original graph can also be 4-colored, either by extending the existing 4-coloring directly to the vertices in the interior, or by modifying the coloring by some sequence of color-pair (or "Kempe-chain") interchanges (involving only the vertices of the configuration ring and the vertices exterior to it) to obtain some other coloring which can be thus extended.

to the question in its original stronger form is affirmative, would strengthen the four-color theorem (take $n = 3$, say, or remove a vertex of small degree n from any plane triangulation and consider the remaining n-ring configuration). If the weaker version of Appel and Haken's question as formulated above can be answered in the affirmative, then proving finiteness of the four-color problem might be possible: A suitable extension of a separator theorem proved by Lipton and Tarjan [1979] (see also Alon, Seymour, and Thomas [1994]) might be used to find reducible configurations in all sufficiently large plane triangulations. This possibility was brought to our attention by P.D. Seymour [personal communication in 1994]. A related question of unavoidability in plane triangulations of certain sets of "likely to be reducible" configurations was solved by Appel and Haken [1976b] and independently by Stromquist [1975].

The algorithmic version of the four-color problem was brought to our attention by P. Ungar [personal communication in 1986]. To say that a problem is "simple," as seen from an algorithmic viewpoint, means that it can be solved in a polynomial number of steps, and preferably such that the polynomial has a low degree d. Since every planar graph has vertices of degree at most 5 it is easy to give a linear (i.e., $d = 1$) 6-coloring algorithm for planar graphs: Arrange the vertices in a smallest last order $x_1, x_2, \ldots, x_i, \ldots, x_n$ (i.e., x_i is of smallest degree in $G - x_n - x_{n-1} - \cdots - x_{i+1}$) and color the vertices sequentially in the order $x_1, x_2, \ldots, x_n$.

Heawood's proof of the five-color theorem can similarly be translated into a 5-coloring algorithm with at most $c \cdot |V(G)|^2$ steps for any planar graph G. In fact, linear 5-coloring algorithms for planar graphs were presented by Matula, Shiloach, and Tarjan [1980, 1981] and independently by Chiba, Nishizeki, and Saito [1981]. A particularly simple linear 5-coloring algorithm is implicit in a recent new proof of the five-color theorem given by C. Thomassen (see Theorem 36 in Chapter 1).

It has been widely stated (e.g., Matula, Shiloach and Tarjan [1980], Frederickson [1984], Boyar and Karloff [1987], Nishizeki and Chiba [1988], or Chrobak and Yung [1989]) that the proof by Appel, Haken, and Koch can be translated into a polynomial algorithm with $d = 2$ for 4-coloring an arbitrary planar graph. This misconception, as it seems, may possibly be due to some of the early descriptions of the proof—for example, an informal discussion given by Haken [1977] that omitted certain critical details concerning the logical structure of the proof. We will now address these critical points in a little more detail.

The proof by Appel, Haken, and Koch contains a particular treatment of non-induced configurations as a special case. Suppose that G contains a configuration C belonging to the unavoidable set U. Because of the particular concept of "configuration containment" used in the proof, it may be the case that distinct vertices and/or edges of C have been identified in G, or that there exist edges of G joining vertices that are nonadjacent in C. When either case occurs, the usual reduction methods do not necessarily apply. However, the limited ring size and diameter of the configurations in U ensures that a short separating cycle (of length at most 6, which is crucial) can be detected in G. Results of Bernhart [1947] must then be applied to reducing such a cycle. An immediate algorithmic implementation of A. Bernhart's reduction unfortunately produces an algorithm A with possibly an exponential worst-case behavior. Essentially, the reduction may require consideration of 4-colorings of as many as 92 smaller reduced graphs in order to obtain eventually a 4-coloring of G. Hence, as a rough estimate of the worst-case behavior of A, the algorithm may perform as many as $92^{|V(G)|}$ steps before finishing. Appel and Haken [1989] addressed this problem in an appendix, where they showed that one can in fact obtain a polynomial time algorithm from a modified version of their proof. Due to complications in dealing with noninduced configurations, they succeeded only in constructing an algorithm with $d = 4$.

The newly obtained proof of N. Robertson, D.P. Sanders, P.D. Seymour, and R. Thomas, using a simpler discharging procedure, gives an unavoidable set with configurations that can be reduced in a straightforward manner, avoiding the complications just described. In fact, from this proof a quadratic (i.e., $d = 2$) 4-coloring algorithm for planar graphs can be derived [personal communication from P.D. Seymour in 1994].

In contrast, the existence of a polynomial 3-coloring algorithm A' for 3-colorable planar graphs is unlikely. Such an A' could be used to decide if a given planar graph is 3-colorable or not in a polynomial number of steps. However, this decision problem is ***NP***-complete even for graphs of maximum degree at most 4, as proved by Stockmeyer [1973] and Garey, Johnson, and Stockmeyer [1976] (see Theorem 19 in Chapter 1). See also the book on ***NP***-completeness by Garey and Johnson [1979].

There is another related algorithmic question that has been asked. Is there a polynomial algorithm that takes a planar graph G as input and gives as output the number of distinct 4-colorings of G? However, a positive answer to this question seems extremely unlikely. Vertigan and Welsh [1992] reported negative results on a wide range of combinatorial enumeration problems for planar graphs. In particular, the problem of counting the 4-colorings of a planar graph is ***#P***-complete, even when restricting to plane triangulations with all degrees even. Enumeration problems that are ***#P***-complete are widely considered even more unlikely to be polynomially solvable than their counterparts among the decision problems, the ***NP***-complete problems (for a brief introduction to enumeration problems and the class ***#P***, see Garey and Johnson [1979]).

The last question is due to S. McGuinness [personal communication in 1991], asking for a short argument proving that the problem of deciding if a given planar graph is 4-colorable can be solved in polynomial time. A short proof of the four-color theorem would give such an algorithm—indeed a very trivial one, always answering YES for all input graphs! Perhaps one possible approach to McGuinness' question would be to consider the possibility of a polynomial algorithm for deciding 3-colorability of a 4-regular planar graph. It follows from the theorem of Tait (Theorem 26 in Chapter 1) that the four-color theorem is equivalent to the statement that the (planar and 4-regular) line graph of any 3-regular planar graph without a bridge is 3-colorable. A polynomial algorithm for 3-coloring the line graph of any 3-regular planar graph without a bridge is in fact equivalent to a polynomial algorithm for 4-coloring any planar graph. Thus the polynomial 4-coloring algorithm of Appel and Haken [1989] gives a polynomial algorithm for 3-coloring 2-connected 4-regular planar line graphs. However, not every 2-connected 4-regular planar graph can be 3-colored, as shown by Koester [1985a, 1985b, 1990]. Deciding 3-colorability of a planar graph of maximum degree four is ***NP***-complete (see Theorem 19 in Chapter 1), but this does not seem to imply that deciding 3-colorability of a 4-regular planar graph is ***NP***-complete.

The finiteness question and the question of McGuinness appear to be independent, and both seem considerably weaker than the problem of finding a short proof of the four-color theorem.

2.2. CARTESIAN SEQUENCES

A Cartesian sequence $C_0C_1 \dots C_n$ is a word in four letters, say A,B,C and R, defined recursively by:

(a) two neighboring letters are different, and
(b) the sub-sequence consisting of every even numbered letter $C_0C_2C_4 \dots$ is Cartesian.

For example, ABCARABAC is a Cartesian sequence.

For integers n and $i_0, i_1, \dots, i_m$ such that $0 \leq i_0 < i_1 < i_2 < \cdots < i_m \leq n$, give a direct proof that there exists a Cartesian sequence $C_0C_1 \cdots C_n$ such that also $C_{i_0}C_{i_1} \cdots C_{i_m}$ is Cartesian. ■

This problem does not have the appearance of a graph coloring problem. However, the sisters Blanche and Rose Descartes proved in 1968, using the theorem of Whitney [1931] on Hamilton cycles in 4-connected plane triangulations, that the existence of Cartesian sequences as described above is in fact equivalent to the four-color theorem (see Descartes and Descartes [1968]).

This equivalent formulation of the four-color problem is little known. In a letter to R. Steinberg, written in October 1978 in Paris, B. Descartes expressed her sentiments: "*Malheureusement, nous n'avons pas réussi à démontrer la vérité de l'hypothèse célèbre des quatre couleurs. Malgré cela, je crois que nos résultats sont très intéressants. Cependant, je n'ai vu qu'une seule référence à nos idées dans la littérature, et cette référence les a mal expliquées. Je crois qu'on trouve ici un exemple du préjugé contre les femmes. Personne ne croit que les femmes comprennent bien le mathématiques.*"

We are grateful to C.A.B. Smith [personal communication in 1990] for bringing this problem and the letter from B. Descartes to our attention. We also thank R. Steinberg for giving permission to quote from a private letter.

Another intriguing reformulation of the four-color theorem due to Kauffman [1990], and also based on Whitney's theorem, states that for any two bracketings L and R of a vector cross product expression $v_1 \times v_2 \times \cdots \times v_n$, there exists a solution to $L = R \neq 0$ with each $v_t \in \{i, j, k\}$, the standard unit orthogonal basis of $\boldsymbol{R}^3$.

■ 2.3. INTERSECTION GRAPHS OF PLANAR SEGMENTS

Let L be a finite set of finite line segments in the plane. The intersection graph of L is the graph whose vertices correspond to the elements of L, and xy is an edge of the graph if and only if the two line segments in L corresponding to x and y cross (i.e., they intersect and are nonparallel).

If G is a planar graph, is G isomorphic to the intersection graph of such an L consisting of line segments of at most four different slopes? ■

This question was apparently first asked by West [1991b], who remarked that it is a particularly ambitious question, since it would imply the four-color theorem, but that no counterexample is known. Without loss of generality one may assume that each line segment is either vertical, horizontal, or has slope $+1$ or α for some fixed α (this follows by an affine transformation). It is not clear to us if one may further assume that $\alpha = -1$ without loss of generality.

It is not even known if every planar graph is the intersection graph of some collection of line segments L. This problem was raised by Scheinerman [1984] in his Ph.D. thesis, where he proved that every outerplanar graph (i.e., a graph with a plane embedding so that one of the face boundaries contains all the vertices of the graph) is the intersection graph of some L. He also proved that there exists a representation of any given planar graph as an intersection graph of pairs of line segments; that is, every vertex may correspond not only to one, but also to two line segments.

Hartman, Newman, and Ziv [1991] proved that every bipartite planar graph is the intersection graph of some L consisting only of vertical and horizontal line segments. This was proved by Duchet, Hamidoune, Las Vergnas, and Meyniel [1983] in the more restricted case when one side of the bipartition consists only of vertices of degree two.

The following question was asked by Scheinerman [personal communication in 1993]: If G is a 3-colorable planar graph, is G isomorphic to the intersection graph of an L consisting of line segments of at most three different slopes?

■ 2.4. RINGEL'S EARTH-MOON PROBLEM

Consider maps on two spheres (earth and moon), such that each country consists of one connected part on each sphere. How few colors f_2 do we need to color all such maps? (As usual, neighboring countries get different colors, and the two parts of each country get the same color.) ■

This problem is due to Ringel [1959], who remarked that $8 \leq f_2 \leq 12$. As reported by Gardner [1980] and by Ringel [1985], T. Sulanke proved in 1974 that in fact $9 \leq f_2$.

The dual formulation of the problem asks for the maximum chromatic number of graphs of thickness 2. The thickness of a graph G is defined to be the smallest number of planar graphs whose union is G. Let f_t denote the maximum chromatic number of graphs of thickness t. Then $f_1 = 4$ is the four-color theorem. Sulanke's graph showing $9 \leq f_2$ consists of a 5-cycle completely joined to a K_6. This graph is 9-chromatic and of thickness 2.

Since a graph G of thickness t has at most $(3t \cdot |V(G)| - 6t)$ edges and thus a vertex of degree at most $6t - 1$, it follows (like the six-color theorem for planar graphs) that $f_t \leq 6t$. The thickness $t(K_p)$ of the complete graph K_p is $\lfloor (p+7)/6 \rfloor$, *except* that $t(K_9) = t(K_{10}) = 3$. This was shown in a series of papers by, among others, Beineke and Harary [1965], Mayer [1972], Alekseev and Gonchakov [1976], and Vasak [1976] (see the survey by White and Beineke [1978]). This gives

$$6t - 2 \leq f_t \leq 6t \quad \text{for } t \geq 3.$$

Ringel [1985] remarked that surprisingly this is a little tighter than the known bounds for f_2.

■ 2.5. ORE AND PLUMMER'S CYCLIC CHROMATIC NUMBER

The cyclic chromatic number $\chi_c(G)$ of a 2-connected plane graph G is the minimum number of colors in an assignment of colors to the vertices of G such that for every face-bounding cycle F of G the vertices of F have different colors. Obviously $\Delta^*(G) \leq \chi_c(G)$, where $\Delta^*(G)$ is the size of a largest face.

Is $\chi_c(G) \leq \frac{3}{2} \cdot \Delta^*(G)$?

If G is 3-connected, is $\chi_c(G) \leq \Delta^*(G) + 2$? ■

The cyclic chromatic number was introduced by Ore and Plummer [1969]. They were motivated by the dual problem of coloring maps such that countries with a common border line or a common border point get different colors. Ore and Plummer [1969] proved that $\chi_c(G) \leq 2 \cdot \Delta^*(G)$. Borodin [1992a] improved the bound to $\chi_c(G) \leq 2 \cdot \Delta^*(G) - 3$ for $\Delta^*(G) \geq 8$. The conjecture $\chi_c(G) \leq \frac{3}{2} \cdot \Delta^*(G)$ was implicitly stated by Borodin [1984].

Restricting attention to 3-connected planar graphs G, Plummer and Toft [1987] proved that $\chi_c(G) \leq \Delta^*(G) + 9$ and, among other similar results, that $\chi_c(G) \leq \Delta^*(G) + 4$ if $\Delta^*(G) \geq 42$. They noted that the prism graph and subdivisions thereof would show the bounds suggested above to be best possible. O.V. Borodin [personal communication in 1987] proved among other results that the inequality $\chi_c(G) \leq \Delta^*(G) + 3$ holds whenever $\Delta^*(G) \geq 24$ (Borodin [1990c]).

For $\Delta^*(G) = 3$ the statement $\chi_c(G) \leq 4$ is the four-color theorem. For $\Delta^*(G) = 4$, a theorem of Borodin [1984, 1985, 1989c], proving a conjecture of Ringel [1965] on 1-embeddable graphs, implies that $\chi_c(G) \leq 6$ (see also Problem 2.14).

For $\Delta^*(G) = 5$ and G 3-regular, the problem has also been attacked by Borodin [personal communication in 1993], who found (but did not publish) a complicated sketch of an argument that in this case $\chi_c(G) \leq 7$. If correct, this implies that the square of a 3-regular planar graph G with $\Delta^*(G) = 5$ is 7-colorable, that is, G can be 7-colored such that any two vertices of distance at most 2 are colored differently. This would give a partial answer to a question asked by Wegner [1977], if every 3-regular planar graph has such a 7-coloring (see Problem 2.18).

■ 2.6. VERTEX PARTITIONINGS W.R.T. COLORING NUMBER

Let G be a planar graph. Is it possible to partition the vertex set into two classes V_1 and V_2 such that

(a) V_1 is independent, and
(b) every subgraph of the graph induced by V_2 has minimum degree at most 3? ■

This possible generalization of the five-color theorem is due to Borodin [1976a]. A graph G, every subgraph of which has minimum degree at most d, is said to be d-degenerate and to have coloring number at most $d + 1$, that is,

$$\text{col}(G) = 1 + \max_{H \subseteq G} \min_{x \in V(H)} (d(x, H)).$$

It is easy to see (by a straightforward induction argument, or by coloring G sequentially, or from a critical subgraph of G) that $\chi(G) \leq \text{col}(G)$.

An $(i_1, i_2, \ldots, i_m)$-partition of G is a partitioning of the vertex set of G into m parts $V_1, V_2, \ldots, V_m$ such that the induced subgraph $G[V_j]$ has coloring number $\text{col}(G[V_j]) \leq i_j$ for every $j = 1, 2, \ldots, m$. Thus the problem asks for (1,4)-partitions of all planar graphs. Since, by an application of Euler's formula, every planar graph has coloring number at most 6, it is easy to show by an inductive argument that G has an $(i_1, i_2, \ldots, i_m)$-partition if $\sum_{1 \leq j \leq m} i_j \geq 6$. In particular, every planar graph has a (3,3)-partition. In fact, Chartrand, Geller, and Hedetniemi [1971] observed that it has a partitioning of the vertex set into two classes V_1 and V_2 such that the subgraphs induced by V_1 and V_2 are outerplanar graphs. (A graph is called outerplanar if it can be embedded in the plane with all its vertices on the outer face boundary. The outerplanar graphs form a proper subset of the set of planar graphs of coloring number at most 3.) It is much more difficult to obtain a partitioning of the edge set of a planar graph into two sets E_1 and E_2 such that $G_i = (V(G), E_i)$ for each $i = 1, 2$ is outerplanar, but this has recently been achieved by Heath [1991].

The existence of a partitioning of the edge set of a planar graph into three sets such that the corresponding subgraphs G_i $(i = 1, 2, 3)$ are acyclic follows from the result of J. Edmonds mentioned in Section 1.4. In fact, it follows that there exists a partitioning of the edge set into six sets such that each corresponding graph G_i $(i = 1, 2, \ldots, 6)$ is acyclic and has no path of length 3 (i.e., it is a "star forest"). In this statement, "six" can be reduced to "five" by a result of O.V. Borodin (Theorem 8 in Chapter 1). This was observed by S.L. Hakimi, J. Mitchem, and E.F. Schmeichel [personal communication from N. Alon in 1994], thus answering a question posed by Algor and Alon [1989], who proved that "five" is best possible.

Borodin [1976a] conjectured, similarly to the above, that every planar graph has a (2,3)-partition (i.e., a partition of the vertex set into two classes V_1 and V_2 such that the subgraph G_1 induced by V_1 is a forest, and every subgraph of the graph G_2 induced by V_2 has minimum degree at most 2). This conjecture was recently proved by Thomassen [1993d]. Thomassen [personal communication in 1993] expected that a solution to the (1,4)-partition problem might also be obtainable by similar methods.

More generally, Borodin [1979a] conjectured that a planar graph G has a 5-coloring such that the union of k arbitrary color classes, $k \leq 4$, induces a subgraph of coloring number at most k. In the same paper Borodin proved the existence of a 5-coloring in which any two color classes together induce an acyclic subgraph (i.e., a forest, i.e., a subgraph of coloring number at most 2). This result was conjectured by Grünbaum [1973], who proved that every planar graph has such an acyclic 9-coloring. The number 9 was improved to 8 by Mitchem [1974], to 7 by Albertson and Berman [1976, 1977], to 6 by Kostochka [1976], and finally, to Grünbaum's conjectured value 5 by Borodin [1976b, 1979a, 1993a] (Theorem 8 in Chapter 1). Borodin's proof was a major achievement—being somewhat similar to the known proofs of the four-color theorem it involves around 450 reducible configurations, and it took several years to complete [personal communication in 1993]. The corollary that every planar graph has a (1,2,2)-partition (i.e., a partition into an independent set and two forests) was first obtained by Stein [1970, 1971].

Attempts at generalizing the four-color problem along these lines have been unsuccessful. Planar graphs do not in general have (1,3)-partitions, as can for example be demonstrated from the graph of the icosahedron, a 5-regular triangulation. Furthermore, Chartrand and Kronk [1969] gave an example of a planar graph G without a (2,2)-partition. The dual G^* of G is the famous non-Hamiltonian 3-regular planar graph discovered by Tutte [1946]. Indeed, Stein [1970, 1971] made the striking observation that a plane triangulation G has a (2,2)-partition if and only if the dual G^* has a Hamilton cycle. Finally, Wegner [1973] produced an example of a planar graph G such that every pair of different colors in any 4-coloring of G induces at least one cycle, and hence G has no (1,1,2)-partition. A conjecture by Grünbaum [1973] that a triangle-free planar graph has a 4-coloring in which any two color classes together induce an acyclic subgraph, was disproved by Kostochka and Melnikov [1976].

A weaker question than asking for a (2,2)-partition of a planar graph is the following, asked by Albertson and Berman [1979]: Does every planar graph G contain an induced forest on at least $|V(G)|/2$ vertices? The theorem of Borodin [1976b, 1979a] implies an induced forest on at least $2|V(G)|/5$ vertices. A classical weaker question was posed independently by Vizing [1968] and by P. Erdős (see Bondy and Murty [1976]): If G is planar, is it then true that G contains an independent set of $|V(G)|/4$ vertices? The affirmative answer follows from the four-color theorem, but no direct proof seems to be known. Albertson [1976] proved, independently of the four-color theorem, the similar statement with $|V(G)|/4$ replaced by $2|V(G)|/9$.

A 1-embeddable graph, which can be drawn in the plane such that every edge is crossed by at most one other edge, is 6-colorable, as proved by Borodin [1984, 1985, 1989c]. Borodin [personal communication in 1993] made the suggestion that perhaps it is possible to strengthen this result by showing that every 1-embeddable graph has a (2,2,2)-partition. Archdeacon [1983] conjectured that every 1-embeddable graph has a partition of its vertex set into two sets V_1 and V_2 such that V_1 induces a forest and V_2 induces a planar graph.

■ 2.7. VERTEX PARTITIONINGS W.R.T. MAXIMUM DEGREE

Let G be a planar graph, and let $\Delta_m(G)$ denote the smallest possible value of the sum $\Delta(G[V_1]) + \Delta(G[V_2]) + \cdots + \Delta(G[V_m])$, where $(V_1, V_2, \ldots, V_m)$ is a partitioning of the vertex set of G and $\Delta(G[V_i])$ denotes the maximum degree of the induced subgraph $G[V_i]$ for $i = 1, 2, \ldots, m$.

What is the best possible upper bound $f_2(\Delta)$ on $\Delta_2(G)$ as a function of the maximum degree Δ of G? What is the best possible upper bound $f_3(\Delta)$ on $\Delta_3(G)$? ■

These two questions generalize a question asked by O.V. Borodin [personal communication in 1993] on determining the value of $\lim_{\Delta \to \infty} f_3(\Delta)$, shown to be less than or equal to 6 by a result of Cowen, Cowen, and Woodall [1986]. However, as pointed out by P. Mihók [personal communication in 1993], examples (see below) found by Broere and Mynhardt [1985b] show that the number 6 is, in fact, the right answer to Borodin's original question.

For partitioning into classes V_1 and V_2 there is no upper bound on $\Delta(G[V_1]) + \Delta(G[V_2])$, and consequently, $f_2(\Delta) \to \infty$ as $\Delta \to \infty$. This was proved by Mihók [1983b], who gave examples showing $f_2(\Delta) \geq \sqrt{\Delta}$ for infinitely many values of Δ. The corresponding questions for partitioning into four or more classes have been solved with the four-color theorem, which implies that each class may be chosen as an independent set of vertices, hence $f_m(\Delta) = 0$ for all $m \geq 4$.

A theorem of L. Lovász (Theorem 38 in Chapter 1) implies that $f_m(\Delta) \leq \Delta - m + 1$ without assuming planarity. The facts stated above on partitionings of planar graphs into three or more parts demonstrate that for planar graphs this bound can be greatly improved.

A closely related partitioning problem was studied by Cowen, Cowen, and Woodall [1986] and by Wessel [1988] (see also the comprehensive survey by Woodall [1990]). They considered partitions of $V(G)$ into m parts such that each part induces a subgraph with maximum degree bounded by a constant $k \geq 0$. Hence G is m-colorable if and only if G has such a partition with $k = 0$. Cowen, Cowen, and Woodall proved that every planar graph G allows such a partition with $(m, k) = (3, 2)$ and gave a simple argument, independent of the four-color theorem, to show that G always has a partition with $(m, k) = (4, 1)$. The former result was improved by Poh [1990], who proved a conjecture of Mihók [1983b] and independently of Broere and Mynhardt [1985b], that the vertices of G can be partitioned into three parts, each inducing a forest consisting of a union of disjoint paths. (Goddard [1991] also claimed to have proved this result, but M. Borowiecki pointed out a flaw in the proof [personal communication from P. Mihók in 1993].) On the other hand, as shown by Broere and Mynhardt [1985b], for any fixed $L \geq 2$ there is a planar graph G_L such that $V(G_L)$ cannot be partitioned into three parts, each inducing a subgraph of G_L without an induced path of length L. For L sufficiently large G_L also provides an example of a planar graph with $\Delta_3(G_L) = 6$.

W. Wessel [personal communication in 1988] disproved a conjecture by Cowen, Cowen, and Woodall [1986], suggesting that every planar graph has a partition of the vertex set into three parts, each inducing a union of disjoint paths such that the union of any two of these parts induces an outerplanar subgraph (i.e., with an embedding in the plane such that all its vertices lie on the boundary of a single face). The counterexample by Wessel consists of the graph K_4 with each edge (x, y) replaced by six internally disjoint (x, y)-paths of length two.

Goddard [1991] remarked that it is not known if the vertex set of a planar graph can be partitioned into two sets, one inducing a union of disjoint paths, the other inducing an outerplanar graph. This would strengthen the result by Poh [1990] mentioned above, since Mihók [1983b] has shown that the vertex set of any outerplanar graph can be partitioned into two sets, each inducing a union of disjoint paths. However, Thomassen [1993d] proved that the answer to Goddard's question is negative.

■ 2.8. THE THREE-COLOR PROBLEM

Does there exist an extension of Heawood's three-color theorem, that the vertices of a plane triangulation can be 3-colored if all degrees are even, to a sufficient condition for 3-colorability of a wider class of planar graphs? ■

The theorem of Heawood [1898] (Theorem 9 in Chapter 1) gives a necessary and sufficient condition for 3-colorability of plane triangulations. A "good characterization" in the sense of Edmonds [1965b] for 3-colorability of planar graphs in general seems beyond hope, because it is an ***NP***-complete problem to decide if a planar graph, even of maximum degree at most 4, is 3-colorable. This was shown by Garey, Johnson, and Stockmeyer [1976] (see Theorem 19 in Chapter 1, and for proofs see also the book on ***NP***-completeness by Garey and Johnson [1979]). The words "*even* of maximum degree at most 4" above may be misleading. Edwards [1986] proved, as conjectured by Welsh and Petford [1985], that for a fixed $\alpha > 0$ and graphs G in which all vertices have degree at least $\alpha \cdot |V(G)|$, the question of 3-colorability is solvable in polynomial time.

Kempe [1879] seemed to suggest the following: For each vertex x of a plane graph G, consider the vertices $y \neq x$ so that G contains a face with both x and y on its boundary. If the number of such vertices y is even for all $x \in V(G)$, is G then 3-colorable? That the answer is negative can be seen by examples such as the 4-chromatic plane graph shown in Figure 1.

A necessary and sufficient (but not "good") condition for 3-colorability of planar graphs was obtained by Król [1972, 1973]. The easily proved result of Król says

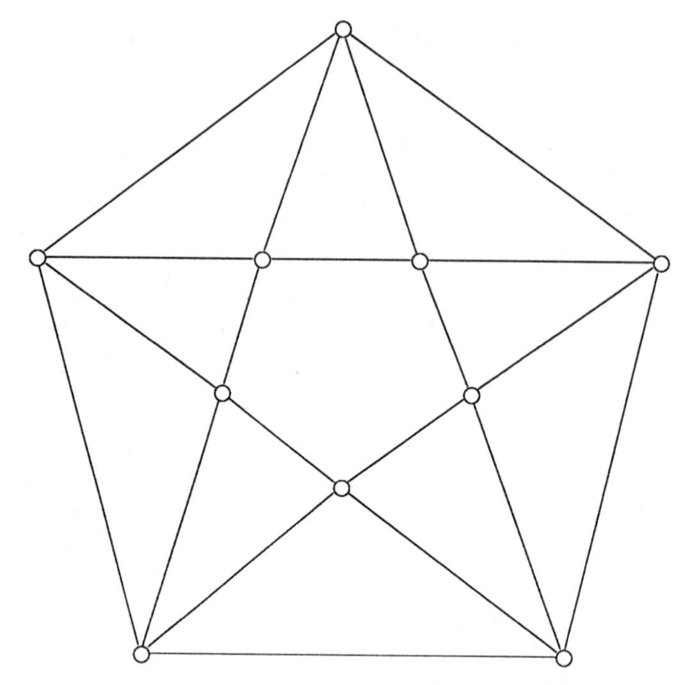

Figure 2.1. A 4-chromatic planar graph. For each vertex x, the union of the face boundaries containing x contains six other vertices.

that a planar graph G is 3-colorable if and only if G is a subgraph of some plane triangulation with all degrees even.

It would be very interesting to extend Heawood's result to include the triangle-free planar graphs. These are 3-colorable by the theorem of Grötzsch [1959] (Theorem 10 in Chapter 1). By the theorems of Grötzsch and Król every triangle-free planar graph can be embedded in a plane triangulation with all degrees even. Is there a simple way to see this directly, thus giving a new simple proof of Grötzsch's theorem?

In contrast, characterizing 2-colorable graphs is easy. A graph G is 2-colorable if and only if all its cycles are even, as proved by D. König (Theorem 14 in Chapter 1). In particular, a plane graph G is 2-colorable if and only if each face is bounded by an even number of edges (counting an edge with the same face on both sides twice as a bounding edge for the face).

Steinberg [1993a, 1993b] gave a thorough survey of the three-color problem.

■ 2.9. STEINBERG'S THREE-COLOR PROBLEM

If G is a planar graph without 4-cycles and without 5-cycles, is G then 3-colorable? ■

According to Aksionov and Melnikov [1978], this question was posed by R. Steinberg in 1975. The question can, of course, also be formulated as follows: If G is a 4-critical planar graph, is it then true that G contains a 4-cycle or a 5-cycle?

A partial result was obtained by Abbott and Zhou [1991a], who showed that a 4-critical plane graph must contain a cycle of length 4 or 5 or a face of size k, where $6 \leq k \leq 11$. They also gave an infinite family of 4-critical planar graphs with no cycles of length 4 and six cycles of length 5, and another infinite family without cycles of length 5 and only four cycles of length 4.

O.V. Borodin [personal communications in 1992 and 1993] has proved the interesting result that every plane graph of minimum degree at least three contains two triangles with a common edge, or a face of size k, where $4 \leq k \leq 9$, or a face of size 10 bounded by vertices all of degree three. A truncated dodecahedron, obtained from the graph of the dodecahedron by replacing each vertex by a triangle, shows that the latter case may occur. The result of Borodin implies (e.g., by Gallai's Theorem 17 in Chapter 1, or by a short direct argument) that a 4-critical plane graph must contain a cycle of length 4 or a face of size k, where $5 \leq k \leq 9$. In particular, every planar graph without k-cycles, where $4 \leq k \leq 9$, is 3-colorable. This result was obtained independently by Sanders and Zhao [1994].

Borodin [1994b] proved his result by a simple discharging argument. Let G be planar of minimum degree at least 3 and without two triangles with a common edge, and fix an embedding of G in the plane. If V and F denote the set of vertices and faces, respectively, it follows by Euler's formula that

$$\sum_{x \in V \cup F} (r(x) - 4) = -8 = \sum_{x \in V \cup F} \mu(x),$$

where $r(x)$ is the degree of x if $x \in V$ and the size of x if $x \in F$, and where $\mu(x) = r(x) - 4$ is the initial "charge" on x. Define a new charge μ^* by modifying μ according to the following rules:

- **(a)** Transfer a charge of $-1/3$ from every facial triangle to each of the three vertices on its boundary.
- **(b)** Transfer $2/3$ from every face of size > 3 to each boundary vertex of degree 3 contained in a triangular face.
- **(c)** Transfer $1/3$ from every face of size > 3 to each boundary vertex of degree 3 not contained in a triangular face.
- **(d)** Transfer $1/3$ from every face of degree > 3 to each boundary vertex of degree 4 contained in two triangular faces.
- **(e)** Transfer $1/3$ from every face of degree > 3 to each boundary vertex of degree 4 contained in a nonneighboring triangular face.

It is clear that μ^* satisfies

$$\sum_{x \in V \cup F} \mu^*(x) = \sum_{x \in V \cup F} \mu(x) = -8 < 0,$$

and it is not hard to check that every $x \in V \cup F$ with $\mu^*(x) < 0$ must be a nontriangular face of G, and moreover x must either be of size at most 9, or of size 10 having only boundary vertices of degree 3 each contained in a neighboring triangle. This finishes the proof.

Grötzsch [1959] proved that if G is a planar graph without triangles, then G is 3-colorable (see Theorem 10 in Chapter 1).

2.10. GRÜNBAUM AND HAVEL'S THREE-COLOR PROBLEM

Let G be a planar graph and let $d(G)$ denote the minimal distance between two triangles in G, given by the number of edges in a shortest path joining two triangles in G. If $d(G)$ is finite, but sufficiently large, say 4 or 5, is G then 3-colorable? ■

If $d(G) = \infty$, then there is at most one triangle in each component of G, and by the theorem of Grötzsch [1959] (Theorem 10 in Chapter 1) and an extension by B. Grünbaum and by Aksionov [1974], G is indeed 3-colorable. Grünbaum [1963] asked if $d(G) \geq 1$ is sufficient, but this was disproved by Havel [1969]. It is known that $d(G) \leq 3$ will not do (Aksionov and Melnikov [1978]), and for $d(G) = 2$ there is a 3-connected example of Ma [1984] showing this.

The smallest known examples of 4-chromatic planar graphs with $d(G) \geq 1$, $d(G) \geq 2$ and $d(G) \geq 3$ have 16, 32, and 44 vertices, respectively. Aksionov and Melnikov [1978] conjectured that smaller examples than these graphs do not exist.

At a conference in Boca Raton in 1986, M.O. Albertson remarked [personal communication] that it might be better to look for other conditions on triangles that force G to be 3-colorable. Perhaps $d(G)$ is the wrong parameter. Borodin [1993a] has expressed that "*...one acquires a conviction that the positive solution of Grünbaum–Havel's problem "hardly" exists.*" Borodin [1994c] explained the difficulties encountered in attempts to solve the problem.

2.11. GRÖTZSCH AND SACHS' THREE-COLOR PROBLEM

Let G be a 4-regular plane graph or multigraph and suppose that G has a cycle decomposition S (i.e., each edge of G is in exactly one cycle of the decomposition) with every pair of adjacent edges on a face always in different cycles of S. Such a graph G arises as a superposition of simple closed curves in the plane with tangencies disallowed.

If the cycles of S can be partitioned into four classes, such that two cycles in the same class are disjoint, is $\chi(G) \leq 3$? ■

Without the condition of S being partitioned into four classes, the question is due to H. Grötzsch and was asked by Sachs [1974, 1978] at several conferences. A

4-chromatic example consisting of five overlapping geometric circles in the plane was discovered by Koester [1984, 1985a, 1985b, 1990], who also found a 4-critical example consisting of seven overlapping circles. On the other hand, Jaeger [1978] proved that $\chi(G) \leq 3$ if the cycles of S can be partitioned into three classes, such that two cycles in the same class are disjoint.

Jaeger [1976b] asked if, without any condition of S being partitionable, G is edge-colorable in 4 colors, that is, if $\chi'(G) = 4$? This question has been studied by Jaeger [1978] and Jaeger and Shank [1981], who among other things proved that if S can be partitioned into four classes as described above, then $\chi'(G) = 4$, and that the corresponding statement with five classes implies the four-color theorem.

■ 2.12. BARNETTE'S CONJECTURE

Let G be a simple 3-colorable planar graph. Is it possible to partition the vertex set of G into two subsets so that each induces an acyclic subgraph of G? ■

Grünbaum [1970b] mentioned that the following conjecture is due to D. Barnette: If G is a 3-connected bipartite 3-regular planar graph, then G has a Hamilton cycle. Barnette's conjecture is true if and only if the answer to the question above is affirmative. This is a consequence of two theorems:

Theorem (Król [1972, 1973]). *Let G be a simple planar graph. Then G is 3-colorable if and only if there exists a simple plane triangulation H containing G as a subgraph, so that all the vertices of H have even degrees.*

Theorem (Stein [1970, 1971]). *Let H be a plane triangulation. Then there exists a partitioning of the vertex set of H into two subsets each of which induces a forest if and only if the dual H^* has a Hamilton cycle.*

We note that for a 3-regular connected plane graph G

$$G \text{ is 3-connected} \Leftrightarrow G \text{ is 3-edge-connected} \Leftrightarrow G^* \text{ is simple,}$$

where G^* is the plane dual of G. Thus the condition of 3-connectivity in Barnette's conjecture corresponds dually to the restriction to simple graphs.

It is not difficult to find examples which show that the condition of 3-connectivity in Barnette's conjecture cannot be omitted. Moreover, Tutte [1946] found an example of a 3-connected planar graph without a Hamilton cycle; hence the conjecture is false for nonbipartite graphs. Finally, a conjecture of Tutte that the condition of planarity can be removed, was disproved by J.D. Horton (see Bondy and Murty [1976]). Bondy and Murty [1976] included Barnette's conjecture in their list of open problems in graph theory. They remarked that P. Goodey has established its truth for plane graphs all of whose faces have size four or six.

P. Mihók [personal communication in 1993] first pointed out to us that Barnette's conjecture is equivalent to a question about coloring planar graphs. He formulated the following problem: Does there exist a planar graph G satisfying $\chi(G) = \varrho(G) = 3$? Here $\varrho(G)$, the point arboricity of G, denotes the minimum number of colors used in a coloring of the vertices of G so that each color class induces an acyclic subgraph. The parameter ϱ, which is clearly a lower bound for χ, was introduced by Chartrand, Kronk, and Wall [1968], who observed that if G is a simple planar graph, then $\varrho(G) \leq 3$. Hence the answer to the question asked by P. Mihók is negative if and only if Barnette's conjecture is true.

■ 2.13. LIST-COLORING PLANAR GRAPHS

Is every 3-colorable planar graph 4-choosable, i.e. for any given assignment of a list of size 4 to every vertex, is it possible to color the graph properly so that each vertex receives a color from its own list?

Does there exist a polynomial algorithm that takes as its input any 2-connected 3-regular planar graph G and a list of three different colors for each edge, and gives as output an edge-coloring of G with each edge having received a color from its list? ■

The second question was asked in the excellent survey by Alon [1993], who explained that every 3-regular 3-edge-colorable planar graph has list-edge-chromatic number $\chi'_\ell = 3$ (see also Section 1.9). The combination of this statement with the four-color theorem implies, by the theorem of P.G. Tait (Theorem 26 in Chapter 1), that it is always possible to color the edges of G as described. However, a polynomial algorithm for finding such a coloring seems unknown.

It was proved, more generally, by Ellingham and Goddyn [1993] that every d-regular d-edge-colorable planar multigraph has list-edge-chromatic number $\chi'_\ell = d$, thus giving an affirmative answer to a special case of the list-coloring conjecture (see Section 1.9). It is not known if an algorithmic proof can be obtained.

Almost all natural questions for list coloring of vertices of planar graphs have been solved, except the first question above, due to T.R. Jensen and M. Stiebitz. Erdős, Rubin, and Taylor [1979] characterized the class of all 2-choosable graphs. Thomassen [1993e] proved with a very elegant argument that every planar graph is 5-choosable, with a proof that can be translated into a simple linear algorithm for finding a list coloring. Voigt [1993] showed that not every planar graph is 4-choosable. Kratochvíl and Tuza [1993] observed that every triangle-free planar graph is 4-choosable, and Voigt [1994] showed that such graphs are not always 3-choosable. Gutner [1994] obtained very strong negative results on the possibility of obtaining polynomial algorithms for deciding 3- and 4-choosability of planar graphs, namely by proving that the problem of deciding if a given planar graph is 4-choosable is ***NP***-hard, and that the problem of deciding if a given triangle-free planar graph is 3-choosable is also ***NP***-hard (see Garey and Johnson [1979] for an introduction to the theory of ***NP***-hard problems). Thomassen [1994] gave an algorithmic proof showing that a planar graph is 3-choosable if each of its cycles has length at least 5, and remarked that this result strengthens the theorem of Grötzsch [1959] (Theorem 10 in Chapter 1) without making use of Euler's formula in the proof.

Alon and Tarsi [1992] proved that every planar bipartite graph is 3-choosable. Their proof was based on algebraic methods (see also Section 1.9), and did not immediately suggest an efficient algorithm for coloring the vertices of a given bipartite planar graph with given lists of size 3. Such an algorithm can be obtained, as noted by Alon [1993], by an application of a theorem of M. Richardson (see Problem 13.9). In contrast, Gutner [1994] proved that when the lists assigned to the vertices of a bipartite planar graph are allowed to have three different lengths 2, 3, and 100, it is an ***NP***-hard problem to decide if a coloring exists.

■ 2.14. KRONK AND MITCHEM'S ENTIRE CHROMATIC NUMBER

The entire chromatic number $\chi_{\mathrm{En}}(G)$ of a plane multigraph G is the minimum number of colors in a coloring of the vertices, edges, and faces of G, where incident or adjacent elements are colored differently (two faces touching only in a vertex may be colored the same, and likewise a face and an edge touching in only one vertex). Is $\chi_{\mathrm{En}}(G) \leq \Delta(G) + 4$, where $\Delta(G)$ denotes the maximum degree of G? ■

An affirmative answer was conjectured by Kronk and Mitchem [1972b, 1973]. Izbicki [1968] reported that M. Neuberger proved the conjecture for 3-regular graphs, assuming the truth of the four-color theorem. Kronk and Mitchem proved this without using the four-color theorem. Hence every 3-regular plane graph has an entire coloring using seven colors.

For Δ sufficiently large, Borodin [1987, 1988, 1992c, 1993b] has shown in a series of papers that the answer is affirmative. Summing up the results, Borodin proved $\chi_{\mathrm{En}}(G) \leq \Delta(G) + 4$ if $\Delta(G) \geq 7$, and moreover $\chi_{\mathrm{En}}(G) \leq \Delta(G) + 2$ when $\Delta(G) \geq 12$. The latter bound is best possible, as shown by the example of the complete bipartite graph $K_{1,n}$—but it is not known if the bound for $\Delta \leq 11$ can be improved. For 2-connected G, Borodin [personal communication in 1993] has obtained $\chi_{\mathrm{En}}(G) \leq \Delta(G) + 1$ if $\Delta(G) \geq 20$. From the results stated above it follows that Kronk and Mitchem's conjecture has been solved except for $\Delta(G) = 4, 5$, or 6.

The first consideration of a problem of this type is due to Ringel [1965], who considered simultaneous colorings of vertices and faces (*not* edges). Ringel's six-color conjecture, that every planar graph can be thus 6-colored, was proved by Borodin [1984, 1985, 1989c]. This follows from the theorem that every graph which is 1-embeddable in the plane (i.e., embeddable such that every edge is crossed by at most one other edge) is vertex 6-colorable in the usual sense. Ringel [1965] proved the case when G is 3-regular. Archdeacon [1983] proved the case when G is triangle-free, and he also proved a five-color theorem for bipartite G.

Melnikov [1975] conjectured that the edges and faces of G can be colored with $\Delta(G) + 3$ colors so that no pair of incident or adjacent elements receive the same color. Borodin [1994d] proved that if $\Delta(G) \geq 10$, then $\Delta(G) + 1$ colors suffice for such a coloring. Melnikov's conjecture remains open for $\Delta(G) \leq 7$.

■ 2.15. NINE-COLOR CONJECTURE

Let G be a 3-regular plane graph. Is it possible to color the faces of G with nine colors in such a fashion that for each edge e of G all (four or fewer) faces incident with the ends of e have pairwise different colors? ■

The conjecture that the answer is affirmative is due to Bouchet, Fouquet, Jolivet, and Riviere [1987], who proved that twelve colors are sufficient. They also conjectured that only seven colors are needed if G is a 3-regular bipartite planar graph (i.e., if G can be face-colored using three colors, see Theorem 9 in Chapter 1).

Borodin [1990a] improved the upper bound on the number of colors to eleven. He considered the dual problem of what he termed "diagonal coloring" of the vertices in a plane triangulation. Borodin [personal communication in 1993] has expressed that a bound of ten colors seems attainable using similar techniques. Bouchet, Fouquet, Jolivet, and Riviere [1987] and Borodin [1992b] also considered the generalization of this problem to other surfaces.

■ 2.16. UNIQUELY COLORABLE GRAPHS

A k-chromatic graph G with exactly one k-coloring, that is, any two k-colorings induce the same partition of the vertex set, is called uniquely k-colorable. Is it possible to give a structural characterization of the uniquely 3-colorable and 4-colorable planar graphs? Can these classes be recognized by polynomial algorithms? ■

Uniquely colorable graphs were introduced and studied by Cartwright and Harary [1968] and Harary, Hedetniemi, and Robinson [1969], and the planar kind by Chartrand and Geller [1969], who proved that a uniquely 3-colorable planar graph has at least two triangles, and that a uniquely 4-colorable planar graph is a triangulation.

A class $\mathcal{C}$ of uniquely 4-colorable planar graphs can be obtained from the complete graph K_4 by repeatedly inserting new vertices of degree 3 in triangular faces. Does $\mathcal{C}$ contain all the uniquely 4-colorable planar graphs? Tutte [1976] mentioned a stronger conjecture by R. Cantoni according to which the graphs in $\mathcal{C}$ are precisely the duals of the 3-regular planar graphs with exactly three Hamilton cycles. By the result on uniquely 4-colorable planar graphs of Chartrand and Geller [1969] mentioned above, and by the classical arguments of P.G. Tait (see Theorem 26 in Chapter 1), the dual of a uniquely 4-colorable planar graph is a uniquely 3-edge-colorable 3-regular planar graph. It is not hard to see that every uniquely 3-edge-colorable 3-regular graph has exactly three Hamilton cycles, since the union of any two color classes must form such a cycle. Hence Cantoni's conjecture would imply that a 3-regular planar graph G is uniquely 3-edge-colorable if and only if G has exactly three Hamilton cycles. Is there some way to prove this weaker statement directly? Thomason [1982] gave examples to show that the condition of planarity cannot be removed.

Uniquely k-colorable graphs in general were studied by among others Nešetřil [1972], Osterweil [1974], and Bollobás [1978b]. The existence of uniquely k-colorable graphs without short cycles was obtained by Nešetřil [1973], E. Artzy and Erdős [1974], and Bollobás and Sauer [1976]. A more general existence theorem was obtained by Müller [1975, 1979] (see Problem 17.13 for a formulation of this interesting result).

2.17. DENSITY OF 4-CRITICAL PLANAR GRAPHS

What are the values of

$$L = \limsup_{G} \frac{|E(G)|}{|V(G)|} \qquad \text{and} \qquad l = \liminf_{G} \frac{|E(G)|}{|V(G)|}$$

where the range of G is the set of all 4-critical planar graphs? ■

G.A. Dirac and Gallai [1964] asked the question about L and conjectured that the answer is $L = 2$. Moreover, they made a stronger conjecture, saying that if G is a 4-critical planar graph, then $|E(G)| \leq 2 \cdot |V(G)| - 2$, thus strengthening a conjecture of G.A. Dirac that G must contain at least two vertices of degree 3. The minimum degree of G is at least 3, since G is 4-critical. Hence if G satisfies the conjectured inequality, it follows that G must contain at least four vertices of degree 3.

Counterexamples to these conjectures were produced by Koester [1985a, 1985b, 1990], who found examples of 4-regular 4-critical planar graphs. Such graphs G satisfy $|E(G)| = 2 \cdot |V(G)|$, and the smallest known example has 80 edges and 40 vertices. Abbott and Zhou [1991b] proved that examples exist for all $|V(G)| \geq 40$ for which $|V(G)|$ is congruent to 1 modulo 3, and they asked if any examples exist where $|V(G)|$ is not congruent to 1 modulo 3. Grünbaum [1988] used Koester's graph in a construction to prove that $L \geq 79/39$. He also proved that $L = \sup_G |E(G)|/|V(G)|$ (taken over all 4-critical planar graphs G), and $|E(G)|/|V(G)| < L$ for any 4-critical planar graph G. The lower bound has been improved to $L \geq 39/19$ by Koester [1991] and independently Abbott and Zhou [1991b]. Koester [1991] also proved the upper bound $L \leq 5/2$. Combining the results of Grünbaum and Koester gives the inequality $2|E(G)| < 5|V(G)|$ for all 4-critical planar graphs G. It follows that G contains a vertex of degree 3 or 4, a result shown independently by Abbott, Katchalski, and Zhou [to appear].

Families of 4-critical planar graphs with a low density of edges can be found by repeated applications of the construction of G. Hajós (see Theorem 18 in Chapter 1). For any fixed number $n \geq 4$, except $n = 5$, the construction gives a 4-critical planar graph G with $|V(G)| = n$ and

$$|E(G)| = \begin{cases} 5n/3 & \text{for } n \equiv 0 \pmod 3 \\ (5n-2)/3 & \text{for } n \equiv 1 \pmod 3 \\ (5n+2)/3 & \text{for } n \equiv 2 \pmod 3, n \neq 5. \end{cases}$$

Thus we have $l \leq 5/3$; does equality hold? Are the values for $|E(G)|$ given above the best possible? For nonplanar graphs they are not; it is possible to obtain 4-critical graphs G with $|V(G)| = n \equiv 2 \pmod 3$, $n \geq 8$ and $|E(G)| = (5n-1)/3$, using the same construction but starting from a (unique and nonplanar) 4-critical graph with 8 vertices and 13 edges (see Toft [1974b]).

The lower bound $l \geq 3/2$ is immediate from the condition of 4-criticality. The theorem of T. Gallai that a k-critical noncomplete graph G on n vertices satisfies

$$|E(G)| \geq (k-1)n + \frac{k-3}{k^2-3}n$$

(Theorem 17 in Chapter 1) provides the slightly better bound $l \geq 20/13$. Can this bound be further improved? Note that the bound has been derived without assuming planarity of G.

■ 2.18. SQUARE OF PLANAR GRAPHS

> Can the vertices of a planar graph of maximum degree 3 be colored with seven colors such that if two vertices have distance at most 2, then they are colored differently? ■

Equivalently, this asks for the chromatic number of the square of a planar graph of maximum degree at most 3.

The problem is due to Wegner [1977], who proved that eight colors are sufficient and that seven colors would be the best possible. The problem was brought to our attention by O.V. Borodin [personal communication in 1992]. It has also been mentioned by Gionfriddo [1986].

Borodin [personal communication in 1993] considered the problem for 3-regular graphs without faces of size greater than 5. He found, but did not publish, a complicated sketch of an argument proving that the squares of such graphs can always be 7-colored. If correct, it would also solve, for this restricted class of graphs, a problem of Ore and Plummer [1969] on the cyclic chromatic number of plane graphs (i.e., the smallest possible number of colors used in a coloring so that two vertices adjacent to a common face must receive different colors; see also Problem 2.5).

We do not know a 3-connected 3-regular planar example that requires seven colors; hence one might ask if six colors suffice in this case. Wegner [1977] observed that a 3-connected 3-regular planar graph of diameter two has at most six vertices. More generally, when asking the same question for graphs of maximum degree Δ,

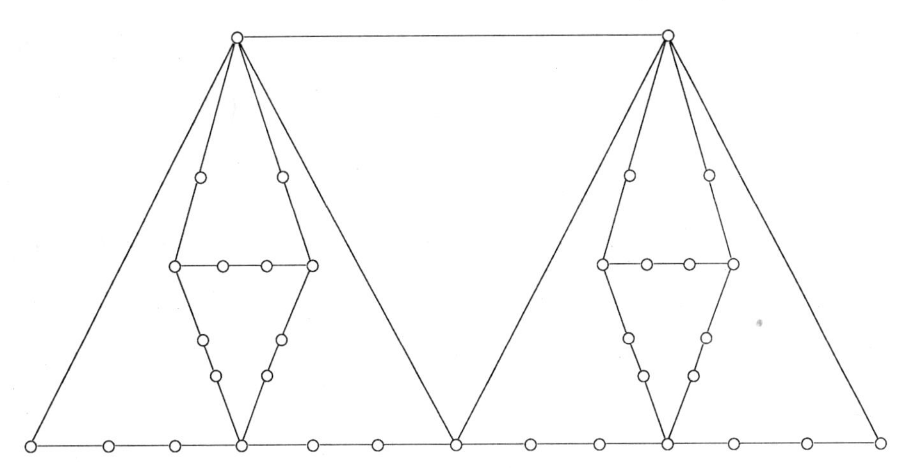

Figure 2.2. A planar graph needing five colors when assigning different colors to every pair of vertices at distance 3.

where $\Delta > 3$, Wegner [1977] conjectured that the maximum number χ_Δ of colors needed is

$$\chi_\Delta = \begin{cases} \Delta + 5 & \text{for } 4 \leq \Delta \leq 7 \\ \left\lfloor \dfrac{3\Delta}{2} \right\rfloor + 1 & \text{for } \Delta \geq 8, \end{cases}$$

and he proved that these numbers cannot be replaced by smaller values.

For larger distances $d > 2$ the problem was studied by Wegner [1977] in the planar case, and it has been studied also for nonplanar graphs by Kramer and Kramer [1969a, 1969b] and Kramer [1972]. Kramer and Kramer [1986] more recently considered the case $d = 3$ for bipartite 3-regular planar graphs and proved that for this problem eight colors are necessary and sufficient (assuming the four-color theorem).

A variation of the problem has also been studied where it is only required that vertices of distance equal to d be colored differently. E. Sampathkumar has presented a 5-color theorem and a 4-color conjecture for planar graphs and every odd distance d, as reported in *Graph Theory Newslett.* **6** (No. 3, 1977). A 3-regular purported counterexample to the conjecture presented in *Graph Theory Newslett.* **6** (No. 5, 1977) can actually be 3-colored. However, we found that the graph of Figure 2 serves as a counterexample, since it needs five colors for $d = 3$. For even distance d, there is no bound on the number of colors, as can be shown from subdivisions of complete bipartite graphs $K_{1,n}$.

BIBLIOGRAPHY

[to appear] Abbott H.L., M. Katchalski, and B. Zhou. On a conjecture of G.A. Dirac concerning 4-critical planar graphs. To appear in *Discrete Math.*

[1991a] Abbott H.L. and B. Zhou. On small faces in 4-critical planar graphs. *Ars Combin.* **32**, 203–207, 1991.

[1991b] Abbott H.L. and B. Zhou. The edge density of 4-critical planar graphs. *Combinatorica* **11**, 185–189, 1991.

[1974] Aksionov V.A. On continuation of 3-coloring of planar graph (in Russian). *Metody Diskret. Analiz.* **26**, 3–19, 1974.

[1978] Aksionov V.A. and L.S. Melnikov. Essay on the theme: the three-color problem. In: A. Hajnal and V.T. Sós, editors, *Combinatorics*, volume 18 of *Colloquia Mathematica Societatis János Bolyai*, pages 23–34. North-Holland, 1978.

[1976] Albertson M.O. A lower bound for the independence number of a planar graph. *J. Combin. Theory Ser. B* **20**, 84–93, 1976.

[1976] Albertson M.O. and D.M. Berman. The acyclic chromatic number. In: *Proc. 7th S–E Conference on Combinatorics, Graph Theory and Computing, Baton Rouge, 1976, Congr. Num., 17*, pages 51–60, 1976.

[1977] Albertson M.O. and D.M. Berman. Every planar graph has an acyclic 7-coloring. *Israel J. Math.* **28**, 169–174, 1977.

[1979] Albertson M.O. and D.M. Berman. A conjecture on planar graphs. In: J.A. Bondy and U.S.R. Murty, editors, *Graph Theory and Related Topics*, page 357. Academic Press, 1979.

[1976] Alekseev V.B. and V.S. Gonchakov. Thickness of an arbitrary complete graph (in Russian). *Mat. Sb.* **101(143)**, 212–230, 1976.

[1989] Algor I. and N. Alon. The star arboricity of graphs. *Discrete Math.* **75**, 11–22, 1989.

[1993] Alon N. Restricted colorings of graphs. In: K. Walker, editor, *Surveys in Combinatorics: Proc. 14th British Combinatorial Conference*, pages 1–33. Cambridge University Press, 1993.

[1994] Alon N., P.D. Seymour, and R. Thomas. Planar separators. *SIAM J. Discrete Math.* **7**, 184–193, 1994.

[1992] Alon N. and M. Tarsi. Colorings and orientations of graphs. *Combinatorica* **12**, 125–134, 1992.

[1976b] Appel K. and W. Haken. The existence of unavoidable sets of geographically good configurations. *Illinois J. Math.* **20**, 218–297, 1976.

[1977a] Appel K. and W. Haken. Every planar map is four colorable. Part I: Discharging. *Illinois J. Math.* **21**, 429–490, 1977.

[1978] Appel K. and W. Haken. The four color problem. In: L.A. Steen, editor, *Mathematics Today: Twelve Informal Essays*, pages 153–180. Springer-Verlag, 1978. Vintage Books, 1980.

[1989] Appel K. and W. Haken. *Every Planar Map Is Four Colorable*, volume 98 of *Contemporary Mathematics Series*. American Mathematical Society, 1989.

[1977] Appel K., W. Haken, and J. Koch. Every planar map is four colorable. Part II: Reducibility. *Illinois J. Math.* **21**, 491–567, 1977.

[1983] Archdeacon D. Coupled colorings of planar maps. In: *Proc. 14th S–E Conference on Combinatorics, Graph Theory and Computing, Boca Raton, 1983, Congr. Num., 39*, pages 89–93, 1983.

[1965] Beineke L.W. and F. Harary. The thickness of the complete graph. *Canad. J. Math.* **17**, 850–859, 1965.

[1947] Bernhart A. Six rings in minimal five color maps. *Amer. J. Math.* **69**, 391–412, 1947.

[1978b] Bollobás B. Uniquely colorable graphs. *J. Combin. Theory Ser. B* **25**, 54–61, 1978.

[1976] Bollobás B. and N.W. Sauer. Uniquely colourable graphs with large girth. *Canad. J. Math.* **28**, 1340–1344, 1976.

[1976] Bondy J.A. and U.S.R. Murty. *Graph Theory with Applications*. Macmillan, 1976.

[1976a] Borodin O.V. On decomposition of graphs into degenerate subgraphs (in Russian). *Metody Diskret. Analiz.* **28**, 3–11, 1976.

[1976b] Borodin O.V. A proof of B. Grünbaum's conjecture on acyclic 5-colorability of planar graphs (in Russian). *Dokl. Akad. Nauk SSSR* **231**, 18–20, 1976.

[1979a] Borodin O.V. On acyclic colorings of planar graphs. *Discrete Math.* **25**, 211–236, 1979.

[1984] Borodin O.V. Solution of Ringel's problem on vertex-face coloring of plane graphs and coloring of 1–planar graphs (in Russian). *Metody Diskret. Analiz.* **41**, 12–26, 1984.

[1985] Borodin O.V. Simultaneous colorings of plane graphs and coloring of 1–embedded graphs (in Russian). In: *Graphen und Netzwerke: Theorie und Anwendungen. 30. Intern. Wiss. Koll. Tech. Hochsch. Ilmenau, 1985*, pages 19–20.

[1987] Borodin O.V. Consistent colorings of graphs on the plane (in Russian). *Metody Diskret. Analiz.* **45**, 21–27, 1987.

[1988] Borodin O.V. Consistent coloring of vertices, edges and faces of plane graphs (in Russian). *Metody Diskret. Analiz.* **47**, 27–37, 1988.

[1989c] Borodin O.V. A new proof of the 6 color theorem. Manuscript, 1989. To appear in *J. Graph Theory.*

[1990a] Borodin O.V. Diagonal 11-coloring of plane triangulations. *J. Graph Theory* **14**, 701–704, 1990.

[1990c] Borodin O.V. Structure, contraction and cyclic coloring of 3-polytopes. Manuscript, 1990. Submitted to *J. Graph Theory.*

[1992a] Borodin O.V. Cyclic coloring of plane graphs. *Discrete Math.* **100**, 281–289, 1992.

[1992b] Borodin O.V. Diagonal coloring of the vertices of triangulations. *Discrete Math.* **102**, 95–96, 1992.

[1992c] Borodin O.V. Structural theorem on plane graphs with application to the entire coloring. Manuscript, 1992. Submitted to *J. Graph Theory.*

[1993a] Borodin O.V. Four problems on plane graphs raised by Branko Grünbaum. *Contemporary Math.* **147**, 149–156, 1993.

[1993b] Borodin O.V. The structure of edge neighborhoods in planar graphs with application in coloring (in Russian). *Math. Z.* **53**, 35–47, 1993.

[1994b] Borodin O.V. Structural properties of plane graphs without adjacent triangles and an application to the 3-colouring. Preprint, Nottingham University and the Russian Academy of Sciences in Novosibirsk, 1994. Submitted to *J. Graph Theory.*

[1994c] Borodin O.V. Irreducible graphs in the Grünbaum-Havel 3-colour problem. Preprint, Nottingham University and the Russian Academy of Sciences in Novosibirsk, 1994.

[1994d] Borodin O.V. Simultaneous coloring of edges and faces of plane graphs. *Discrete Math.* **128**, 21–33, 1994.

[1987] Bouchet A., J.-L. Fouquet, J.-L. Jolivet, and M. Riviere. On a special face colouring of cubic graphs. *Ars Combin.* **24**, 67–76, 1987.

[1987] Boyar J.F. and H.J. Karloff. Coloring planar graphs in parallel. *J. Algorithms* **8**, 470–479, 1987.

[1985b] Broere I. and C. Mynhardt. Generalized colorings of outerplanar and planar graphs. In: Y. Alavi, G. Chartrand, L. Lesniak, D.R. Lick, and C.E. Wall, editors, *Graph Theory with Applications to Algorithms and Computer Science*, pages 151–161. Wiley, 1985.

[1968] Cartwright D. and F. Harary. On the coloring of signed graphs. *Elem. Math.* **23**, 85–89, 1968.

[1969] Chartrand G. and D.P. Geller. On uniquely colorable planar graphs. *J. Combin. Theory* **6**, 271–278, 1969.

[1971] Chartrand G., D.P. Geller, and S.T. Hedetniemi. Graphs with forbidden subgraphs. *J. Combin. Theory* **10**, 12–41, 1971.

[1969] Chartrand G. and H.V. Kronk. The point-arboricity of planar graphs. *J. London Math. Soc.* **44**, 612–616, 1969.

[1968] Chartrand G., H.V. Kronk, and C.E. Wall. The point-arboricity of a graph. *Israel J. Math.* **6**, 169–175, 1968.

[1981] Chiba N., T. Nishizeki, and N. Saito. A linear 5-coloring algorithm of planar graphs. *J. Algorithms* **2**, 317–327, 1981.

[1989] Chrobak M. and M. Yung. Fast algorithms for edge-coloring planar graphs. *J. Algorithms* **10**, 35–51, 1989.

[1986] Cowen L.J., R.H. Cowen, and D.R. Woodall. Defective colourings of graphs in surfaces: Partitions into subgraphs of bounded valency. *J. Graph Theory* **10**, 187–195, 1986.

[1968] Descartes B. and R. Descartes. La coloration des cartes. *Eureka* **31**, 29–31, 1968.

[1956] Dirac G.A. Map colour theorems related to the Heawood colour formula. *J. London Math. Soc.* **31**, 460–471, 1956.

[1957a] Dirac G.A. Map colour theorems related to the Heawood colour formula II. *J. London Math. Soc.* **32**, 436–455, 1957.

[1983] Duchet P., Y. Hamidoune, M. Las Vergnas, and H. Meyniel. Representing a planar graph by vertical lines joining different levels. *Discrete Math.* **46**, 319–321, 1983.

[1965b] Edmonds J. Paths, trees and flowers. *Canad. J. Math.* **17**, 449–467, 1965.

[1986] Edwards K. The complexity of colouring problems on dense graphs. *Theoret. Comput. Sci.* **43**, 337–343, 1986.

[1993] Ellingham M.N. and L. Goddyn. List edge colorings of some regular planar multigraphs. Manuscript, 1993. Submitted to *Combinatorica.*

[1974] Erdős P. Some new applications of probability methods to combinatorial analysis and graph theory. In: *Proc. 5th S–E Conference on Combinatorics, Graph Theory and Computing, Boca Raton, Congr. Num., 10*, pages 39–54, 1974.

[1979] Erdős P., A.L. Rubin, and H. Taylor. Choosability in graphs. In: *Proc. West Coast Conference on Combinatorics, Graph Theory and Computing, Arcata, 1979, Congr. Num., 26*, pages 125–157, 1979.

[1984] Frederickson G.N. On linear-time algorithms for five-coloring planar graphs. *Inform. Process. Lett.* **19**, 219–224, 1984.

[1964] Gallai T. Critical graphs. In: *Theory of Graphs and Its Applications: Proc. Symposium, Smolenice, June 1963*, pages 43–45. Czechoslovak Academy of Science, 1964.

[1980] Gardner M. Mathematical games. *Sci. Amer.* **242**(2), 14–21, 1980.

[1979] Garey M.R. and D.S. Johnson. *Computers and Intractability: A Guide to the Theory of* ***NP****-Completeness*. W.H. Freeman and Company, 1979.

[1976] Garey M.R., D.S. Johnson, and L.J. Stockmeyer. Some simplified ***NP***-complete graph problems. *Theoret. Comput. Sci.* **1**, 237–267, 1976.

[1986] Gionfriddo M. A short survey on some generalized colourings of graphs. *Ars Combin.* **30**, 275–284, 1986.

[1991] Goddard W. Acyclic colorings of planar graphs. *Discrete Math.* **91**, 91–94, 1991.

[1959] Grötzsch H. Ein Dreifarbensatz für dreikreisfreie Netze auf der Kugel. *Wiss. Z. Martin-Luther-Univ. Halle–Wittenberg Math.-Natur. Reihe* **8**, 109–120, 1959.

[1963] Grünbaum B. Grötzsch's theorem on 3-colorings. *Michigan Math. J.* **10**, 303–310, 1963.

[1970b] Grünbaum B. Polytopes, graphs and complexes. *Bull. Amer. Math. Soc.* **76**, 1131–1201, 1970.

[1973] Grünbaum B. Acyclic colorings of planar graphs. *Israel J. Math.* **14**, 390–408, 1973.

[1988] Grünbaum B. The edge-density of 4-critical planar graphs. *Combinatorica* **8**, 137–139, 1988.

[1994] Gutner S. The complexity of planar graph choosability. Manuscript, 1994.

[1977] Haken W. An attempt to understand the four color problem. *J. Graph Theory* **1**, 193–206, 1977.

[1969] Harary F., S.T. Hedetniemi, and R.W. Robinson. Uniquely colorable graphs. *J. Combin. Theory* **6**, 264–270, 1969.

[1991] Hartman I.B.-A., I. Newman, and R. Ziv. On grid intersection graphs. *Discrete Math.* **87**, 42–52, 1991.

[1969] Havel I. On a conjecture of B. Grünbaum. *J. Combin. Theory* **7**, 184–186, 1969.

[1991] Heath L.S. Edge coloring planar graphs with two outerplanar subgraphs. In: *Proc. of the Second Annual ACM-SIAM Symposium on Discrete Algorithms*, pages 195–202, 1991.

[1890] Heawood P.J. Map colour theorem. *Quart. J. Pure Appl. Math.* **24**, 332–338, 1890.

[1898] Heawood P.J. On the four-colour map theorem. *Quart. J. Pure Appl. Math.* **29**, 270–285, 1898.

[1968] Izbicki H. Verallgemeinerte Farbenzahlen. In: H. Sachs, H.-J. Voss, and H. Walther, editors, *Beiträge zur Graphentheorie*, pages 81–84. B.G. Teubner Verlagsgesellschaft, 1968.

[1976b] Jaeger F. Problem. In: C. St. J.A. Nash-Williams and J. Sheehan, editors, *Proc. Fifth British Combinatorial Conference, Aberdeen, 1975*, pages 682–683. Utilitas Mathematics Publications, Inc., 1976.

[1978] Jaeger F. Sur les graphes couvert par leurs bicycles et la conjecture des quatre couleurs. In: J.-C. Bermond, J.-C. Fournier, M. Las Vergnas, and D. Sotteau, editors, *Problèmes Combinatoires et Théorie des Graphes*, pages 243–247. Édition Centre National Recherche Scientifique, Paris, 1978.

[1981] Jaeger F. and H. Shank. On the edge-coloring problem for a class of 4–regular maps. *J. Graph Theory* **5**, 269–275, 1981.

[1990] Kauffman L.H. Map coloring and the vector cross product. *J. Combin. Theory Ser. B* **48**, 145–154, 1990.

[1879] Kempe A.B. On the geographical problem of four colours. *Amer. J. Math.* **2**, 193–200, 1879.

[1984] Koester G. Bemerkung zu einem Problem von H. Grötzsch. *Wiss. Z. Univ. Halle* **33**, 129, 1984.

[1985a] Koester G. Coloring problems on a class of 4–regular planar graphs. In: H. Sachs, editor, *Graphs, Hypergraphs and Applications: Proc. Conference on Graph Theory, Eyba, 1984*, pages 102–105. B.G. Teubner Verlagsgesellschaft, 1985.

[1985b] Koester G. Note to a problem of T. Gallai and G.A. Dirac. *Combinatorica* **5**, 227–228, 1985.

[1990] Koester G. 4-critical 4–valent planar graphs constructed with crowns. *Math. Scand.* **67**, 15–22, 1990.

[1991] Koester G. On 4-critical planar graphs with high edge density. *Discrete Math.* **98**, 147–151, 1991.

[1976] Kostochka A.V. Acyclic 6-coloring of planar graphs (in Russian). *Metody Diskret. Analiz.* **28**, 40–56, 1976.

[1976] Kostochka A.V. and L.S. Melnikov. Note to the paper of Grünbaum on acyclic colorings. *Discrete Math.* **14**, 403–406, 1976.

[1972] Kramer F. Sur le nombre chromatique $K(p,G)$ des graphes. *Rev. Française Automat. Inform. Reche. Opér. Sér. Rouge* **6**, 67–70, 1972.

[1969a] Kramer F. and H. Kramer. Un problème de coloration des sommets d'un graphe. *C.R. Acad. Sci. Paris Sér. A–B* **268**, A46–A48, 1969.

[1969b] Kramer F. and H. Kramer. Ein Färbungsproblem der Knotenpunkte eines Graphen bezüglich der Distanz *p*. *Rev. Roumaine Math. Pures Appl.* **14**, 1031–1038, 1969.

[1986] Kramer F. and H. Kramer. On the generalized chromatic number. In: A. Barlotti, M. Biliotti, A. Cossu, G. Korchmaros, and G. Tallini, editors, *Combinatorics '84*, pages 275–284. North-Holland, 1986.

[1993] Kratochvíl J. and Z. Tuza. Algorithmic complexity of list colorings. Manuscript, 1993. To appear in *Discrete Appl. Math.*

[1972] Król M. On a sufficient and necessary condition of 3-colorability of a planar graph, I. *Prace Nauk. Inst. Mat. Fiz. Teoret. PWr.* **6**, 37–40, 1972.

[1973] Król M. On a sufficient and necessary condition of 3-colorability of a planar graph, II. *Prace Nauk. Inst. Mat. Fiz. Teoret PWr.* **9**, 49–54, 1973.

[1972b] Kronk H.V. and J. Mitchem. The entire chromatic number of a normal graph is at most seven. *Bull. Amer. Math. Soc.* **78**, 799–800, 1972.

[1973] Kronk H.V. and J. Mitchem. A seven-color theorem on the sphere. *Discrete Math.* **5**, 253–260, 1973.

[1979] Lipton R.J. and R.E. Tarjan. A separator theorem for planar graphs. *SIAM J. Appl. Math.* **36**, 177–189, 1979.

[1984] Ma Han-Kun A counterexample to the conjecture of Aksionov and Melnikov on non-3-colorable planar graphs. *J. Combin. Theory Ser. B* **36**, 218–219, 1984.

[1980] Matula D.W., Y. Shiloach, and R.E. Tarjan. Two linear-time algorithms for five-coloring a planar graph. Technical Report STAN–CS–80–830, Computer Science Dept., Stanford University, 1980.

[1981] Matula D.W., Y. Shiloach, and R.E. Tarjan. Analysis of two linear-time algorithms for five-coloring a planar graph. In: *Proc. 12th S–E Conference on Combinatorics, Graph Theory and Computing, Baton Rouge, 1981, Congr. Num., 33*, page 407, 1981 (abstract).

[1972] Mayer J. Décomposition de K_{16} en trois graphes planaires. *J. Combin. Theory Ser. B* **13**, 71, 1972.

[1975] Melnikov L.S. Problem 9 In: M. Fiedler, editor, *Recent Advances in Graph Theory. Proc. Symposium, Prague, June 1974*, page 543. Academia Praha, 1975.

[1983b] Mihók P. On vertex partition numbers of graphs. In: M. Fiedler, editor, *Graphs and Other Combinatorial Topics*, pages 183–188. B.G. Teubner Verlagsgesellschaft, 1983.

[1974] Mitchem J. Every planar graph has an acyclic 8-coloring. *Duke Math. J.* **41**, 177–181, 1974.

[1975] Müller V. On colorable critical and uniquely colorable critical graphs. In: M. Fiedler, editor, *Recent Advances in Graph Theory, Proc. Symposium, Prague, June 1974*, pages 385–386. Academia Praha, 1975.

[1979] Müller V. On colorings of graphs without short cycles. *Discrete Math.* **26**, 165–176, 1979.

[1972] Nešetřil J. On critical uniquely colorable graphs. *Arch. Math. (Basel)* **23**, 210–213, 1972.

[1973] Nešetřil J. On uniquely colorable graphs without short cycles. *Časopis Pěst. Mat.* **98**, 122–125, 1973.

[1946] Newman M.H.A. Letter to J. von Neumann. The von Neumann Archive, Library of Congress, February 8, 1946.

[1988] Nishizeki T. and N. Chiba. *Planar Graphs: Theory and Algorithms*, volume 32 of *Annals of Discrete Mathematics*. North Holland, 1988.

[1969] Ore O. and M.D. Plummer. Cyclic coloration of plane graphs. In: W.T. Tutte, editor, *Recent Progress in Combinatorics*, pages 287–293. Academic Press, 1969.

[1974] Osterweil L.J. Some classes of uniquely 3-colorable graphs. *Discrete Math.* **8**, 59–69, 1974.

[1987] Plummer M.D. and B. Toft. Cyclic coloration of 3–polytopes. *J. Graph Theory* **11**, 507–515, 1987.

[1990] Poh K.S. On the linear vertex-arboricity of a planar graph. *J. Graph Theory* **14**, 73–75, 1990.

[1980] Randell B. The COLOSSUS. In: N. Metropolis, J. Howlett, and G.-C. Rota, editors, *A History of Computing in the Twentieth Century*, pages 47–92. Academic Press, 1980.

[1959] Ringel G. *Färbungsprobleme auf Flächen und Graphen.* VEB Deutscher Verlag der Wissenschaften, 1959.

[1965] Ringel G. Ein Sechsfarbenproblem auf der Kugel. *Abh. Math. Sem. Univ. Hamburg* **29**, 107–117, 1965.

[1985] Ringel G. 250 Jahre Graphentheorie. In: R. Bodendiek, H. Schumacher, and G. Walther, editors, *Graphen in Forschung und Unterricht: Festschrift K. Wagner*, pages 136–152. Barbara Franzbecker Verlag, 1985.

[1974] Sachs H. Problem. *Math. Balkanica* **4**, 536, 1974.

[1978] Sachs H. A three-colour conjecture of Grötzsch. In: J.-C. Bermond, J.-C. Fournier, M. Las Vergnas, and D. Sotteau, editors, *Problèmes Combinatoires et Théorie des Graphes*, page 441. Édition Centre National Recherche Scientifique, 1978.

[1994] Sanders D.P. and Y. Zhao. A note on the three color problem. Manuscript, 1994. Submitted to *Graphs Combin.*

[1984] Scheinerman E.R. *Intersection classes and multiple intersection parameters of graphs*. Ph.D. thesis, Princeton University, 1984.

[1985] Schumacher H. and K. Wagner. Zur Finitisierung von Färbungsproblemen. *Mitt. Math. Ges. Hamburg* **11**, 335–340, 1985.

[1970] Stein S.K. B-sets and coloring problems. *Bull. Amer. Math. Soc.* **76**, 805–806, 1970.

[1971] Stein S.K. B-sets and planar maps. *Pacific J. Math.* **37**(1), 217–224, 1971.

[1993a] Steinberg R. The state of the three color problem. In: J. Gimbel, J.W. Kennedy, and L.V. Quintas, editors, *Quo Vadis, Graph Theory?* volume 55 of *Annals of Discrete Mathematics*, pages 211–248. North-Holland, 1993.

[1993b] Steinberg R. An update on the state of the three color problem. In: J.W. Kennedy and L.V. Quintas, editors, *Graph Theory Notes of New York XXV*, pages 9–12. New York Academy of Sciences, 1993.

[1973] Stockmeyer L.J. Planar 3-colorability is ***NP***-complete. *SIGACT News* **5**:3, 19–25, 1973.

[1975] Stromquist W.R. *Some aspects of the four color problem*. Ph.D. thesis, Harvard University, 1975.

[1982] Thomason A.G. Cubic graphs with three Hamiltonian cycles are not always uniquely edge colorable. *J. Graph Theory* **6**, 219–221, 1982.

[1993d] Thomassen C. Decomposing planar graphs. Technical report, The Technical University of Denmark, 1993.

[1993e] Thomassen C. Every planar graph is 5–choosable. Technical report, The Technical University of Denmark, 1993. *J. Combin. Theory Ser. B* **62**, 180–181, 1994.

[1994] Thomassen C. 3-list-coloring planar graphs of girth 5. Technical report, The Technical University of Denmark, February 1994.

[1974b] Toft B. On critical subgraphs of colour-critical graphs. *Discrete Math.* **7**, 377–392, 1974.

[1946] Tutte W.T. On Hamiltonian circuits. *J. London Math. Soc.* **21**, 98–101, 1946.

[1976] Tutte W.T. Hamiltonian circuits. In: *Colloquio Internazionale sulle Teorie Combinatorie, Roma, Tomo I*, pages 193–199. Accademia Nazionale dei Lincei, 1976.

[1976] Vasak J.M. The thickness of the complete graph. *Notices Amer. Math. Soc.* **23**, A–479, 1976.

[1992] Vertigan D.L. and D.J.A. Welsh. The computational complexity of the Tutte plane: the bipartite case. *Combin. Probab. Comput.* **1**, 181–187, 1992.

[1968] Vizing V.G. Some unsolved problems in graph theory (in Russian). *Uspekhi Mat. Nauk* **23**, 117–134, 1968. English translation in *Russian Math. Surveys* **23**, 125–141.

[1993] Voigt M. List colourings of planar graphs. *Discrete Math.* **120**, 215–219, 1993.

[1994] Voigt M. A not 3-choosable planar graph without 3-cycles. Manuscript, 1994. To appear in *Discrete Math.*

[1973] Wegner G. Note on a paper of B. Grünbaum on acyclic colorings. *Israel J. Math.* **14**, 409–412, 1973.

[1977] Wegner G. Graphs with given diameter and a coloring problem. Technical report, University of Dortmund, 1977.

[1985] Welsh D.J.A. and D.M. Petford. A randomized attack on an ***NP***-complete problem. Oxford, 1985.

[1988] Wessel W. Kritische Graphen. In: *Proc. 33rd Int. Wiss. Koll. Tech. Hochsch. Ilmenau, October 1988.*

[1991b] West D.B. Open problems. *SIAM J. Discrete Math. Newslett.* **2**(1), 10–12, 1991.

[1978] White A.T. and L.W. Beineke. Topological graph theory. In: L.W. Beineke and R.J. Wilson, editors, *Selected Topics in Graph Theory*. Academic Press, 1978.

[1931] Whitney H. A theorem on graphs. *Ann. of Math.* **32**, 378–390, 1931.

[1975] Whitney H. and W.T. Tutte. Kempe chains and the four colour problem. In: D.R. Fulkerson, editor, *Studies in Graph Theory, Part II*, pages 378–413. The Mathematical Association of America, 1975.

[1990] Woodall D.R. Improper colourings of graphs. In: R. Nelson and R.J. Wilson, editors, *Graph Colourings*, volume 218 of *Pitman Research Notes in Mathematics Series*, pages 45–63. Longman, 1990.

3

Graphs on Higher Surfaces

■ 3.1. HEAWOOD'S EMPIRE PROBLEM

Let M be a map on a surface $\mathbf{S}$ of Euler characteristic ε, and let each country of the map have at most m disjoint connected parts. Heawood [1890] proved (compare Theorem 2 in Chapter 1) that the countries of M can be colored by at most

$$H(\varepsilon, m) = \left\lfloor \frac{1}{2} \cdot \left(6m + 1 + \sqrt{(6m+1)^2 - 24\varepsilon}\right) \right\rfloor$$

colors such that neighboring countries get different colors and all parts of the same country get the same color, except possibly in the case $\varepsilon = 2$ and $m = 1$ (the four-color problem). For which surfaces $\mathbf{S}$ are $H(\varepsilon, m)$ colors necessary for some maps on $\mathbf{S}$? ■

The higher surfaces, that is, the compact 2-dimensional manifolds, can be classified as follows. The sphere with g handles attached $\mathbf{S}_g$ has Euler characteristic $\varepsilon = 2 - 2g$. The projective plane with g handles $\mathbf{P}_g$ has Euler characteristic $\varepsilon = 1 - 2g$, and the Klein bottle with g handles $\mathbf{K}_g$ has Euler characteristic $\varepsilon = -2g$. The surfaces $\mathbf{S}_g$ are orientable, whereas $\mathbf{P}_g$ and $\mathbf{K}_g$ are nonorientable.

The topological prerequisite for Heawood's formula is that a graph G (with at least 3 vertices and without loops and multiple edges) embedded on $\mathbf{S}$ satisfies

$$|E(G)| \leq 3 \cdot |V(G)| - 3 \cdot \varepsilon(\mathbf{S})$$

where $E(G)$ and $V(G)$ are the sets of edges and vertices of G, respectively. The derivation of Heawood's formula from this can be found in Ringel [1959] and in Hutchinson [1993], among others.

For the sphere ($\varepsilon = 2$) with $m = 2$ and for the torus ($\varepsilon = 0$) with $m = 1$, Heawood [1890] showed by examples that $H(2, 2) = 12$ and $H(0, 1) = 7$ colors are necessary for some maps.

By embedding a complete graph of size $H(2-2g,1)$ on $\mathbf{S}_g$ for $1 \leq g \leq 7$, Heffter [1891] answered the question affirmatively for these orientable surfaces and $m = 1$. Extracts from the original papers of Heawood and Heffter are contained in the excellent book by Biggs, Lloyd, and Wilson [1976]. Still for the case $m = 1$, Dirac [1952b, 1957b] proved for every map on a surface $\mathbf{S}$ of Euler characteristic $\varepsilon = 0$ or $\varepsilon \leq -2$ that if it needs $H(\varepsilon, 1)$ colors, then it must contain $H(\varepsilon, 1)$ mutually neighboring countries (Theorem 4 in Chapter 1). The cases $\varepsilon = 1$ and $\varepsilon = -1$ did not yield to Dirac's arithmetic, but these were covered by Albertson and Hutchinson [1979]. The idea of such a result and a proof for the torus were both originally due to P. Ungar. An embedding of a map with $H(\varepsilon, 1)$ mutually neighboring countries is possible for any given surface $\mathbf{S}$ of Euler characteristic $\varepsilon \leq 1$, except the Klein bottle, by the remarkable theorems of Ringel [1954] and Ringel and Youngs [1968] (see Theorem 3 in Chapter 1) for the nonorientable and orientable cases, respectively (the excellent monographs by Ringel [1959, 1974] contain expositions of these and related results).

A large number of affirmative answers have been obtained for $m \geq 2$ as well. These are summarized in Table 3.1.

As for the case of the torus and $m \geq 2$, a construction by H. Taylor was announced by Taylor [1981b], by Jackson and Ringel [1983, 1985], and by White [1984], but we have not been able to find any literature that describes the result in detail. A research problem posed by Taylor [1979] may suggest a feasible idea for obtaining such a construction. At any rate, a complete solution for the torus was contained in the more general result of Jackson and Ringel [1985].

Table 3.1 Known cases where $H(\varepsilon, m)$ colors are necessary for m-pires on $\mathbf{S}$.

m	$\mathbf{S}$	Due to
2	Sphere	Heawood [1890]
3,4	Sphere	H. Taylor, see Gardner [1980]
≥ 2	Sphere	Jackson and Ringel [1984b]
2	Torus	Ringel [1959]
≥ 2	Torus	H. Taylor (see above)
≥ 2 and even	Orientable and $H(\varepsilon, m) \equiv 1 \pmod{12}$	Jackson and Ringel [1985]
≥ 2 and odd	Orientable and $H(\varepsilon, m) \equiv 4, 7 \pmod{12}$	Jackson and Ringel [1985]
≥ 2	Projective plane	Jackson and Ringel [1983]
≥ 3	Klein bottle	Jackson and Ringel [1985]
≥ 2	Nonorientable, $\neq$ Klein bottle, and $H(\varepsilon, m) \equiv 1, 4, 7 \pmod{12}$	Jackson and Ringel [1985]
2	Klein bottle	Borodin [1989a]

As can be seen, there are unsolved problems for $m = 2$, both in the orientable and nonorientable cases, the simplest being $\mathbf{S}_3$ and $\mathbf{P}_2$, respectively. For $m = 3$ the simplest unsolved cases are $\mathbf{S}_3$ and $\mathbf{K}_2$. Only a single case is known where the answer is not affirmative. For $m = 1$ Franklin [1934] proved that 6 colors are sufficient to color every map on the Klein bottle, whereas $H(0, 1) = 7$. Jackson and Ringel [1985] suspected a negative answer for the Klein bottle also for $m = 2$, but Borodin [1989a] showed this not to be so by giving an embedding of K_{13} as a 2-pire on the Klein bottle, thus showing that $H(0, 2) = 13$ colors are needed.

3.2. GRÜNBAUM'S 3-EDGE-COLOR CONJECTURE

Is it true for every triangulation (without loops and without multiple edges) of an orientable surface that it is possible to label the edges by three colors in such a fashion that the edges of each triangle have three different colors? ■

An affirmative answer was conjectured by Grünbaum [1969], who remarked that the conjecture cannot be generalized to nonorientable surfaces, but that if four colors are available, a labeling is possible for all surfaces. This follows, in dual formulation, from the theorem of V.G. Vizing (Theorem 29 in Chapter 1) that every simple graph G has an edge coloring using $\Delta(G) + 1$ colors in such a fashion that each vertex is incident with edges of different colors.

The Petersen graph can be embedded on the projective plane such that the projective dual is a triangulation of the projective plane without a labeling in three colors. Hence the natural extension of Grünbaum's conjecture to nonorientable surfaces would fail already for the projective plane.

The Petersen graph also has an embedding on the torus such that the toroidal dual is a triangulation without a labeling in three colors. However, this triangulation has multiple edges. Grünbaum's conjecture implies that an embedding of a snark on an orientable surface has a dual with multiple edges. (A "snark" is a nontrivial 4-edge-chromatic 3-regular graph, see Isaacs [1975] and Gardner [1976]. Here "nontrivial" means without triangles and without edge cuts of fewer than four edges, except three edges incident to the same vertex.) Embedding properties of some of the known snarks have been studied by Tinsley and Watkins [1985], and by Szekeres [1975], who proved that the snarks of Isaacs [1975] are not counterexamples to Grünbaum's conjecture.

For the plane, the equivalent dual of the conjecture says that in a 3-connected 3-regular plane graph G the edges can be 3-colored in such a fashion that each vertex is incident with three edges of three different colors. P.G. Tait (Theorem 26 in Chapter 1) proved that this statement is equivalent to the existence of a usual 4-coloring of the regions into which G divides the plane (see the book by Biggs, Lloyd, and Wilson [1976]). Since the four-color problem can be reduced to 3-regular plane maps, Grünbaum's conjecture for the plane is thus equivalent to the four-color theorem.

3.3. ALBERTSON'S FOUR-COLOR PROBLEM

Let **S** be a surface of Euler characteristic ε. Does there exist a constant, $M(\varepsilon)$, such that for every graph G embeddable on **S**, all but $M(\varepsilon)$ of the vertices of G can be 4-colored? In particular, if G is toroidal, can all but three of the vertices of G be 4-colored? ■

This question was first asked by Albertson [1981], but related questions have been considered by, among others, Berman and Paul [1989].

Since a graph G on **S** has at most $3 \cdot |V(G)| - 3 \cdot \varepsilon$ edges, and hence minimum degree at most 6 so long as $|V(G)| > (-6\varepsilon)$, it follows from a smallest last order of the vertices and a sequential coloring that all but (-6ε) of the vertices of G can be 7-colored. By a slightly more involved argument using induction and the theorem of R.L. Brooks (Theorem 11 in Chapter 1), it can be proved that all but $(-9\varepsilon + 1)$ of the vertices of G can be 6-colored. (Outline of argument: If there is a vertex of degree at most 5, then OK by induction. If all degrees are greater than or equal to 6, then delete all vertices of degree at least 7 and one vertex from each K_7 in the remaining graph.) Thus with 4 replaced by 6 the answer to Albertson's problem is affirmative. With 4 replaced by 5 the answer is also affirmative. Thomassen [1993b] has shown that every graph on **S** can be 5-colored except for at most $2^{13-2\varepsilon}$ vertices. Berman and Paul [1992] proved, using the four-color theorem, that every graph embedded on $\mathbf{S}_g$, the sphere with g handles, can be (improperly) 5-colored with at most $74g - 36$ edges joining two vertices of the same color. With 4 replaced by 3 the answer is negative, since graphs on **S** can contain arbitrarily many disjoint copies of K_4.

An affirmative answer to Albertson's question immediately implies the four-color theorem, since arbitrarily many copies of a hypothetical 5-chromatic planar graph would be embeddable on **S**. Berman and Paul [1992] suggested a weaker version of Albertson's question, not directly implying the four-color theorem: For the sphere with g handles $\mathbf{S}_g$, is there a number N such that every graph on $\mathbf{S}_g$ has an improper 4-coloring without a monochromatic tree (i.e., a tree with all its vertices colored the same) of more than N vertices? With "tree" replaced by "path" the answer is affirmative as shown by Berman and Paul [1989].

Albertson and Hutchinson [1978] proved that $g \cdot \sqrt{2n}$ vertices can always be removed from a graph G with n vertices embedded on $\mathbf{S}_g$ to leave a planar graph. Thus $\sqrt{2n}$ vertices can be removed from a toroidal graph on n vertices to leave a 4-colorable graph (by the four-color theorem). The result by Albertson and Hutchinson [1978] has been improved by Hutchinson and Miller [1987], who showed that $g \cdot \sqrt{2n}$ can be replaced by $26 \cdot \sqrt{gn}$.

It is tempting to suggest that all locally planar graphs on any surface are 4-colorable (with some suitable definition of locally planar). However, Fisk [1978] constructed 5-chromatic graphs that are embeddable on all surfaces except the sphere and furthermore seem to satisfy every reasonable local planarity definition (see Albertson and Stromquist [1982]). Albertson and Stromquist [1982] considered a particularly natural notion of "local planarity," and they proved that a graph is 5-colorable if it can be embedded on the torus in such a way that all its noncontractible cycles are long (of length at least 8), a best possible result by Fisk's examples (a cycle is "contractible" if its deletion disconnects the surface in such a way that one of the resulting components is homeomorphic to a plane disc). Thomassen [1993a] characterized all 6-chromatic graphs embeddable on the torus and proved that they have noncontractible triangles, thus answering a question asked by Albertson and Hutchinson [1980b].

Hutchinson [1984] defined "locally planar" to mean that all edges are "short" (relative to the dimensions of the surface) and proved that such locally planar graphs on any surface are 5-colorable. This result was generalized by Thomassen [1993b], who proved that there exists a number $n(\mathbf{S})$ such that every graph embedded on $\mathbf{S}$ without noncontractible cycles of length at most $n(\mathbf{S})$ can be 5-colored. Related results were derived by quite elementary techniques by Fisk and Mohar [1994].

■ 3.4. IMPROPER COLORINGS

Let $\mathbf{S}$ be a surface of Euler characteristic ε. What is the smallest number, $K(\varepsilon)$, such that for all graphs G embeddable on $\mathbf{S}$ there is an improper 3-coloring in which every vertex is adjacent to at most $K(\varepsilon)$ vertices of the same color as itself? ■

The conjecture that such a constant $K(\varepsilon)$ exists for every $\varepsilon \leq 2$ was made by Cowen, Cowen, and Woodall [1986] and proved by Archdeacon [1987], who gave the bound

$$K(\varepsilon) \leq \max\{15, \lceil -3\varepsilon/2 \rceil - 1\}.$$

O.V. Borodin [personal communication in 1993] suggested that it might be possible to obtain an improvement of this upper bound. Cowen, Cowen, and Woodall [1986] proved for the plane, with $\varepsilon = 2$, that the right answer is $K(2) = 2$. They also considered the corresponding question for improper k-colorings in general and proved that every graph embedded on a surface $\mathbf{S}$ with $\varepsilon \leq 2$ has an improper 4-coloring in which every vertex is adjacent to at most $\max\{14, \lceil -4\varepsilon/3 \rceil - 1\}$ vertices of the same color. For improper 2-colorings no similar bound exists for any surface. This follows from a construction by Mihók [1983b].

The result of Cowen, Cowen, and Woodall [1986] mentioned above, together with the four-color theorem (see Section 1.2), implies that every plane graph has an improper k-coloring in which every vertex is adjacent to at most d vertices of the same color as itself if and only if ($k \geq 3$ and $d \geq 2$) or ($k \geq 4$ and $d \geq 0$). Woodall [1990] asked as an unsolved problem to find analogous conditions for toroidal graphs or graphs embeddable on the projective plane.

The problem of vertex partitioning a graph on a surface into at most k parts with each part inducing a d-degenerate subgraph has been solved by Lick and White [1972] and Borodin [1976a].

■ 3.5. NUMBER OF 6-CRITICAL GRAPHS ON A SURFACE

Let **S** be an orientable or nonorientable surface. Is the number of nonisomorphic 6-critical graphs embeddable on **S** finite? ■

The question was raised by Thomassen [1993a], who proved that there are exactly four 6-critical graphs on the torus (see Problem 3.6). This answered a question of Albertson and Stromquist [1982]. Albertson and Hutchinson [1980b] also studied 6-chromatic toroidal graphs.

If G is k-critical and embeddable on a surface **S** of Euler characteristic ε, then (since the minimum degree of G is at least $k-1$)

$$(k-1) \cdot |V(G)| \leq 2 \cdot |E(G)| \leq 6 \cdot |V(G)| - 6 \cdot \varepsilon.$$

Thus for $k \geq 8$,

$$|V(G)| \leq -6\varepsilon/(k-7).$$

Since 8-chromatic graphs can only be embedded on surfaces with $\varepsilon \leq -2$ by the formula of Heawood [1890] (Theorem 2 in Chapter 1), the inequality is only of interest for such surfaces. Hence for all $k \geq 8$ the number of k-critical graphs on a fixed surface with $\varepsilon \leq -2$ is finite. Coloring problems for higher surfaces are thus often of a finite nature. It was Dirac [1956, 1957a] who first noted and used this finiteness.

Mohar [1993] and Thomassen [1993a] independently observed that the number of 7-critical graphs on **S** is finite, arguing as follows: T. Gallai proved that the vertices of degree $k-1$ in a k-critical graph induce a subgraph whose blocks are either odd cycles or complete graphs (Theorem 17 in Chapter 1). Thus considering an embedding of a 7-critical graph different from the complete graph K_7, it cannot have a vertex v of degree 6 adjacent to six other vertices of degree 6 so that all the faces incident to v are bounded by triangles. However, Mohar and Thomassen noted that every graph of minimum degree at least 6 that can be embedded on **S** and has sufficiently many vertices must contain such a vertex v. Mohar [1993] gave the upper bound -60ε for the number of vertices of a 7-critical graph on a surface of Euler characteristic $\varepsilon \leq -1$.

In fact, the "moreover" part of Gallai's theorem (Theorem 17 in Chapter 1) easily implies that a 7-critical graph different from K_7 on a surface of Euler characteristic $\varepsilon \leq -1$ has at most -69ε vertices.

The result of Mohar [1993] and Thomassen [1993a] immediately implies that for any surface there is a polynomial algorithm to decide for a given graph embeddable on the surface if it is 6-colorable. A more direct argument for this, giving a linear algorithm, was presented by Edwards [1992].

Constructions by Fisk [1978] (see also Albertson and Stromquist [1982]) show that the number of 5-critical graphs on any surface except the sphere is infinite. Fisk's examples consist of triangulations of the surface with all vertices of even degree, except two adjacent vertices of odd degree, and he showed that no such graph is 4-colorable.

■ 3.6. TOROIDAL POLYHEDRA

A toroidal polyhedron consists of plane convex polygons joined along their edges, without bending or twisting the polygons out of shape, such that their union is topologically equivalent to the surface of a torus. Can the faces (i.e., the polygons) of a toroidal polyhedron be colored with four colors so that two adjacent faces are always colored differently? ■

This question was asked by Barnette [1985], who proved with a simple argument that six colors suffice, and conjectured that four would also suffice. C. Thomassen [personal communication in 1994] has proved a five color theorem by showing that no 6-critical graph embeddable on the torus can be realized as a subgraph of the dual graph of a toroidal polyhedron, where the vertices of the dual graph correspond to the faces of the polyhedron, and the edges to the pairs of adjacent faces. This was obtained as an application of a theorem of Thomassen [1993a] that there exist precisely four different 6-critical toroidal graphs: the complete graph on 6 vertices K_6, the graph on 8 vertices obtained by joining a 5-cycle completely to a K_3, a graph on 9 vertices obtained by applying Hajós' construction (see Section 1.4) to two copies of K_4 followed by joining the resulting graph completely to a K_2, and finally a 6-regular graph on 11 vertices described by Albertson and Hutchinson [1980b].

Without the geometrical restriction on the embedding, there exist maps on the torus that require seven colors, as shown by Heawood [1890]. Thus, with the restriction one can do better.

Croft, Falconer, and Guy [1991] included Barnette's problem as Problem B24 in their list, and they asked in general for the number of colors needed to color the faces of non-convex polyhedra of other topological surfaces. For which surfaces do the polyhedra require fewer colors than arbitrary maps?

Of course, toroidal polyhedra are not convex. For arbitrary convex polyhedra four colors always suffice, by the four-color theorem.

C. Thomassen [personal communication in 1994] has pointed out that it is possible to arrange five plane polygons, not all convex, joined along their edges such that their union forms part of the surface of a torus, with the five corresponding faces pairwise adjacent. Thus for the variation of the problem when the faces are not required to be convex, a five-color theorem would be best possible. It is however not obvious to us if six colors always suffice in this case.

3.7. POLYNOMIAL COLORING OF EMBEDDED GRAPHS

Let **S** be a fixed orientable or nonorientable surface and let $k > 0$ be a fixed number. Does there exist a polynomial algorithm for deciding if a given graph embedded on **S** can be k-colored? ■

The question was asked by Thomassen [1993a], who proved that the answer is affirmative for $k \geq 6$. The problem is open for $k = 4$ except on the sphere (see Section 1.2), and for $k = 5$ except on the sphere, the projective plane (Albertson and Hutchinson [1979]), and the torus (Thomassen [1993a]). For $k = 3$ the decision problem is ***NP***-complete even for the sphere (see Theorem 19 in Chapter 1). For triangle-free graphs the answer is affirmative for all $k \geq 4$ [personal communication from J. Gimbel in 1994]. The problem of the existence of a polynomial algorithm for deciding 3-colorability of triangle-free graphs on a given surface has been studied by J. Gimbel and C. Thomassen. It seems open for all surfaces other than the sphere. For the sphere the existence follows from Grötzsch's theorem (Theorem 10 in Chapter 1). Thomassen [1993a] asked if it is true that only a finite number of 6-critical graphs can be embedded on **S** (Problem 3.5). An affirmative answer to this question would imply an affirmative answer to the question above for $k = 5$.

It seems slightly more difficult to find polynomial algorithms for k-coloring a graph embedded on **S**. Appel and Haken [1989] first obtained a polynomial algorithm for 4-coloring graphs on the sphere, and a quadratic algorithm for this problem was obtained by N. Robertson, D.P. Sanders, P.D. Seymour and R. Thomas [personal communication from N. Robertson and P.D. Seymour in 1994]. The proof by Thomassen [1993a] leads to a polynomial algorithm for finding a 5-coloring of a graph on the torus if such a coloring exists.

3.8. SPARSE EMBEDDED GRAPHS

What is the largest possible chromatic number of a graph that can be embedded on a surface **S** and in which every cycle has length at least g? In particular, if G is embedded on the Klein bottle and every cycle in G has length at least 5, is G then 3-colorable? Similarly, what is the maximum chromatic number of a graph G that can be embedded on a surface **S** and for which the number of triangles in G is bounded by a constant t? ■

For graphs embedded on the torus, the question corresponding to the question above for graphs embedded on the Klein bottle is due to Kronk and White [1972], who proved that the answer for the torus is affirmative with 3 replaced by 4, and also with 5 replaced by 6. The answer is negative with 5 replaced by 4, as can be seen from the triangle-free 4-chromatic graph by Mycielski [1955]. The similar statements also hold for graphs embeddable on the projective plane and the Klein bottle, as proved by Woodburn [1989], who asked the 3-colorability question for both these surfaces.

Thomassen [1993c] proved that the answer to Kronk and White's question for graphs on the torus is affirmative. He also proved that the answer is affirmative for graphs on the projective plane. For both surfaces Thomassen [1993c] even allowed cycles of length 3 or 4 in the graph, but each such cycle must be noncontractible (i.e., it does not bound a region homeomorphic to a plane disc). A similar strengthening for the Klein bottle cannot be obtained, since the complete graph K_4 has an embedding without contractible cycles. A theorem of Kronk [1972] implies that every graph embeddable without contractible triangles on the torus or the Klein bottle can be 4-colored.

Kronk [1972] studied the chromatic numbers of triangle-free graphs on surfaces in general. Some of his results were improved by Cook [1975], in particular sufficient conditions for 3-colorability. It has recently been investigated, if every graph that can be embedded on the double torus (i.e., the sphere with two handles) and whose cycles are all of length at least 6 can be 3-colored. A solution to this problem has recently been announced by J. Gimbel [personal communication in 1994], who remarked that perhaps the answer is affirmative even with "6" replaced by "5." General results in this direction were obtained by Fisk and Mohar [1994].

Sherman [1993] investigated the maximum chromatic number $\chi(\mathbf{S}, t)$ of graphs with at most t triangles on a surface $\mathbf{S}$. She proved $\chi(\text{torus}, 15) = 5$ and $\chi(\text{double torus}, 7) = 5$. It is easy, from a theorem of G.A. Dirac (Theorem 16 in Chapter 1) and Euler's formula, to see that $\chi(\text{torus}, 1) \leq 4$ (and the 4-chromatic triangle-free graph of Mycielski [1955] shows that in fact equality holds). The largest value of t for which $\chi(\text{torus}, t) = 4$ seems unknown—it is known to be at least 2 and at most 9. Thomassen [1993a] proved that there are exactly four different 6-critical graphs on the torus. An examination of these graphs reveals that a 6-chromatic toridal graph has at least 20 triangles. Thus a best possible bound is $\chi(\text{torus}, 19) \leq 5$ (the complete graph K_5 shows that equality holds, and K_6 shows $\chi(\text{torus}, 20) \geq 6$).

Graphs embeddable with all faces even-sided have been studied by Hartsfield and Ringel [1989], Youngs [1991], and Hutchinson [1994]. Best possible upper bounds for χ are known for an infinite number of cases, but not for all.

The motivation for the results and open problems mentioned above is the theorem of H. Grötzsch, and its extension by B. Grünbaum and V.A. Aksionov (see Theorem 10 in Chapter 1 and the discussion following it), that every planar graph with at most three triangles is 3-colorable.

■ 3.9. RINGEL'S 1-CHROMATIC NUMBER

The 1-chromatic number, $\chi_1(\mathbf{S})$, of a surface $\mathbf{S}$ is the maximum chromatic number $\chi(G)$ for all graphs G 1-embeddable on $\mathbf{S}$ (i.e., embeddable on $\mathbf{S}$ such that no edge is crossed over by more than one other edge). Known bounds are:

$$\chi_1(\mathbf{S}_g) \leq \left\lfloor \frac{1}{2} \cdot \left(9 + \sqrt{64g + 17}\right) \right\rfloor \quad \text{for} \quad g \geq 1,$$

and for even $\chi_1(\mathbf{S}_g)$,

$$\chi_1(\mathbf{S}_g) \le \left\lfloor \frac{1}{3} \cdot \left(13 + \sqrt{144g + 25}\right) \right\rfloor \quad \text{for} \quad g \ge 1.$$

Are these bounds together best possible? ■

Ringel [1981] proved these Heawood-type bounds and posed the problem also for nonorientable surfaces. He gave the following Heawood-type upper bound for a nonorientable surface $\mathbf{N}_q$ of Euler characteristic $\varepsilon = 2 - q$ $(q \ge 1)$.

$$\chi_1(\mathbf{N}_q) \le \left\lfloor \frac{1}{2} \cdot \left(9 + \sqrt{32q + 17}\right) \right\rfloor,$$

where $\mathbf{N}_q = \mathbf{P}_{(q-1)/2}$ for q odd and $\mathbf{N}_q = \mathbf{K}_{(q-2)/2}$ for q even.

For the sphere $\mathbf{S}_0$ Ringel [1965] proved that $6 \le \chi_1(\mathbf{S}_0) \le 7$ and Borodin [1984, 1985, 1989c] proved that $\chi_1(\mathbf{S}_0) = 6$. Exact answers are known for a few more cases in addition to the sphere: $\chi_1(\text{torus}) = \chi_1(\text{Klein bottle}) = 9$ and $\chi_1(\mathbf{S}_{83}) = 41$ are due to Ringel [1981], and $\chi_1(\text{projective plane}) = 7$ is due to Schumacher [1984]. Korzhik [1991] proved the following results: $\chi_1(\mathbf{S}_{26}) = 25, 37 \le \chi_1(\mathbf{S}_{75}) \le 39, 39 \le \chi_1(\mathbf{S}_{79}) \le 40$, $\chi_1(\mathbf{N}_{247}) = 49$, and $39 \le \chi_1(\mathbf{N}_{158}) \le 40$.

Korzhik [1994] obtained the following lower bounds (at the First Estonian Conference on Graphs and Applications, May 1991, dedicated to the 70th birthday of F. Harary, Korzhik announced slightly weaker bounds, with the constants 34 replaced by 58 and 54, respectively):

$$\chi_1(\mathbf{S}_g) \ge \left\lfloor \frac{1}{2} \cdot \left(9 + \sqrt{64g + 17}\right) \right\rfloor - 34$$

for orientable surfaces, and

$$\chi_1(\mathbf{N}_q) \ge \left\lfloor \frac{1}{2} \cdot \left(9 + \sqrt{32q + 17}\right) \right\rfloor - 34$$

for nonorientable surfaces $\mathbf{N}_q$. Thus Ringel's bounds are asymptotically almost best possible. Korzhik also announced that the Heawood-type upper bound for nonorientable surfaces $\mathbf{N}_q$ is exact for $q = 6, 28, 66, 72, 128, 200$ (see Korzhik [1993, 1994]).

The similarly defined 2-chromatic number, $\chi_2(\mathbf{S})$, was studied by Schumacher [1990], who obtained the bound

$$\chi_2(\mathbf{S}) \le \left\lfloor \frac{1}{2} \cdot \left(11 + \sqrt{121 - 40\varepsilon}\right) \right\rfloor$$

when $\mathbf{S}$ has Euler-characteristic ε. He also gave the exact answers $\chi_2(\text{projective plane}) = \chi_2(\text{torus}) = \chi_2(\text{Klein bottle}) = 10$ and $\chi_2(\mathbf{S}_5) = 16$.

■ 3.10. BORODIN'S CONJECTURE ON DIAGONAL COLORING

Let **S** be a surface of Euler characteristic ε and G be a graph embedded as a triangulation on **S** (i.e., all the faces of the embedding are triangles). A diagonal k-coloring of G is a coloring of the vertices of G with k colors, so that for each edge e of G all (four or fewer) vertices belonging to the boundary of one of the two faces incident to e have different colors. Let $\chi_{\text{diag}}(\mathbf{S})$ denote the diagonal chromatic number of **S**, defined as the minimum k for which every triangulation of **S** has a diagonal k-coloring. Is it true that

$$\chi_{\text{diag}}(\mathbf{S}) = \left\lfloor \frac{1}{2} \cdot \left(13 + \sqrt{169 - 48\varepsilon}\right) \right\rfloor$$

except if **S** is the sphere? ■

Borodin [1992b] proved that the right-hand side of this expression gives an upper bound for $\chi_{\text{diag}}(\mathbf{S})$, thus improving an earlier bound obtained by Bouchet, Fouquet, Jolivet, and Riviere [1987], and he conjectured that the improved bound is best possible. For the torus $\mathbf{S}_1$ with $\varepsilon = 0$ the answer is affirmative, since Bouchet, Fouquet, Jolivet, and Riviere [1987] proved that $\chi_{\text{diag}}(\mathbf{S}_1) = 13$.

For the sphere, Bouchet, Fouquet, Jolivet, and Riviere [1987] proved (in a dual formulation) that $\chi_{\text{diag}}(\mathbf{S}_0) \leq 12$ and conjectured that $\chi_{\text{diag}}(\mathbf{S}_0) = 9$ (Problem 2.15). Borodin [1990a] improved the upper bound to $\chi_{\text{diag}}(\mathbf{S}_0) \leq 11$.

■ 3.11. ACYCLIC COLORINGS

A k-coloring of a graph G is called acyclic if every subgraph of G induced by the union of two color classes is a forest (i.e., without cycles). Determine for every surface **S** the minimum number $a(\mathbf{S})$ such that every graph embeddable on **S** has an acyclic $a(\mathbf{S})$-coloring. Is $a(\mathbf{S})$ always equal to $\chi(\mathbf{S})$, the maximum chromatic number of graphs embeddable on **S**, except for the sphere and the Klein bottle? ■

Acyclic colorings were defined and studied by Grünbaum [1973], who proved that every planar graph has an acyclic 9-coloring. This was improved to 8 by Mitchem [1974], to 7 by Albertson and Berman [1976, 1977], to 6 by Kostochka [1976], and finally to Grünbaum's conjectured value 5 by Borodin [1976b, 1979a] (Theorem 8 in Chapter 1). Kostochka and Melnikov [1976] proved that the value 5 cannot be improved even for bipartite graphs with coloring number at most 3 (i.e., those in which every nonempty subgraph has minimum degree at most 2).

Albertson and Berman [1976], and independently O.V. Borodin, observed that no surface was known, except the sphere, for which $a(\mathbf{S})$ is strictly larger than the usual chromatic number $\chi(\mathbf{S})$. Borodin conjectured that $a(\mathbf{S}) = \chi(\mathbf{S})$, except when $\mathbf{S}$ is the sphere. However, there is in fact at least one further exception, namely the Klein bottle, as noted by Mohar, Robertson, and Sanders [1994]. They observed that the complete 8-graph K_8 when removing the edges of three vertex-disjoint paths of lengths 1, 2, and 2, respectively, shows that $a(\text{Klein bottle}) \geq 7$ (whereas $\chi(\text{Klein bottle}) = 6$, see Theorem 3 in Chapter 1 and the accompanying discussion). Albertson and Berman [1976] proved that $a(\text{torus}) \leq 8$, and Mohar, Robertson, and Sanders [1994] that $a(\text{projective plane}) \leq 7$ (but $\chi(\text{torus}) = 7$ and $\chi(\text{projective plane}) = 6$, see Theorem 3 in Chapter 1). More generally, Albertson and Berman [1978] proved that $a(\mathbf{S}_g) \leq 4g + 4$ for the sphere with g handles. This was improved to $a(\mathbf{S}_g) \leq c \cdot g^{2/3}$ for a positive constant c, with a similar extension to nonorientable surfaces, by Mohar, Robertson, and Sanders [1994].

■ 3.12. COCHROMATIC NUMBERS

The cochromatic number, $z(G)$, of a graph G is the minimum number of subsets into which $V(G)$ can be partitioned so that each subset induces either an independent set of vertices or a complete graph. Determine $z(\mathbf{S})$, the maximum $z(G)$ for graphs G embedded on a surface $\mathbf{S}$. ■

The problem is due to Straight [1979, 1980], who determined the answers for the sphere [$z(\mathbf{S}_0) = 4$], the projective plane [$z(\mathbf{P}_0) = 5$], the Klein bottle [$z(\mathbf{K}_0) = 6$], the projective plane with one handle [$z(\mathbf{P}_1) = 6$], and the Klein bottle with one handle [$z(\mathbf{K}_1) = 7$]. Straight [1979] conjectured for the torus $\mathbf{S}_1$ that $z(\mathbf{S}_1) = 5$. This was proved by Thomassen [1993b], who moreover proved that a coloring with five colors can always be chosen such that at most one color class consists of a complete graph. Straight [1979] conjectured more generally that $z(\mathbf{S})$ equals the maximum n for which the disjoint union of $K_1, K_2, K_3, \ldots, K_n$ is embeddable on $\mathbf{S}$, and he observed that this number is certainly a lower bound for $z(\mathbf{S})$. This conjecture was disproved by Gimbel [1986].

Recently, $z(\mathbf{S})$ has been studied by J. Gimbel and C. Thomassen [personal communication in 1994]. See also Problem 17.3 for more information on cochromatic numbers.

■ 3.13. GRAPHS ON PSEUDO-SURFACES

The pseudo-surface $\mathbf{S}(0; n, 2)$ is obtained from the sphere $\mathbf{S}_0$ by identifying pairs of points x_i and y_i for $i = 1, 2, \ldots, n$, where $x_1, x_2, \ldots, x_n$, $y_1, y_2, \ldots, y_n$ are all distinct. With pairs replaced by triples, one obtains the pseudo-surface $\mathbf{S}(0; n, 3)$, and so on. With $\mathbf{S}_0$ replaced by the sphere with g handles $\mathbf{S}_g$ one obtains the pseudo-surface $\mathbf{S}(g; n, 2)$, and so on. What are the maximum chromatic numbers for graphs embeddable on such pseudo-surfaces? ■

This problem for $\mathbf{S}(0; n, 2)$ was introduced and studied by Dewdney [1972] and mentioned by White [1973]. When a graph is embedded on a pseudo-surface it is normally assumed, as Dewdney [1972] did, that the singular points on the surface cannot be interior points on curves representing edges. Hence if the embedding of a graph G contains one of the singular points, then it is a vertex of G. With this understanding, the maximum chromatic number χ_n^* for graphs embeddable on $\mathbf{S}(0; n, 2)$ was determined by Borodin and Melnikov [1974] (see also White [1984]). This can be done using standard arguments and splitting results proved by Borodin and Melnikov [1974] and independently by Hartsfield, Jackson, and Ringel [1985].

Dewdney [1972] observed that the bound $\chi_n^* \leq 4 + n$ always holds, and that it holds with equality for $n = 1, 2$. White [1978] reported that M. O'Bryan and J. Williamson in 1973 proved equality also for $n = 3, 4$. Supposing that equality holds in $\chi_5^* \leq 9$, a 9-critical graph G on $\mathbf{S}(0; 5, 2)$ satisfies

$$8 \cdot |V(G)| \leq 2 \cdot |E(G)| \leq 6 \cdot (|V(G)| + 5) - 12$$

because the minimum degree is at least 8 and G can be made planar by splitting at most 5 vertices. Hence $|V(G)| \leq 9$ and $G = K_9$. But it is known that K_9 needs 6 splittings of vertices to become planar. Thus $\chi_5^* \leq 8$ and, in fact, K_8 embeds on $\mathbf{S}(0; 5, 2)$, hence $\chi_5^* = 8$. More generally, a k-critical graph on $\mathbf{S}(0; n, 2)$ satisfies

$$(k - 1) \cdot |V(G)| \leq 6 \cdot (|V(G)| + n) - 12.$$

Hence for $k \geq 7$,

$$(k - 7) \cdot k \leq 6n - 12,$$

implying that

$$k \leq \left\lfloor \frac{1}{2} \cdot \left(7 + \sqrt{1 + 24n}\right) \right\rfloor. \qquad (*)$$

The right-hand side of (*) gives χ_n^* for $6 \leq n \leq 12$, as can be deduced from the splitting results of Borodin and Melnikov [1974] and Hartsfield, Jackson, and Ringel [1985]. [A.K. Dewdney, in *Math. Reviews* **54** (#5031) 1977, reported that T. Sulanke obtained this result independently.] An interesting case is $\chi_{12}^* = 12$, where (*) gives $\chi_{12}^* \leq 12$, and an example of Heawood [1890] of a planar map with 12 mutually neighboring countries each with two parts gives $\chi_{12}^* \geq 12$. In a k-critical graph with $k \geq 13$ each vertex has degree at least 12, and hence the splitting of some vertices into pairs of new vertices gives a graph of average degree at least 6 and thus a nonplanar graph. Therefore, $\chi_n^* = 12$ for all $n \geq 12$.

A variation is to look for the maximum chromatic number of graphs embeddable on at least one surface obtained from the sphere by successively identifying n pairs of points, not necessarily distinct (i.e., points to be identified may themselves have been obtained by earlier identifications). The results of Hartsfield, Jackson, and Ringel

[1985] show this number for $n \geq 6$ to be the right-hand side of (*). For $n \leq 5$ the number is equal to χ_n^*.

Another variation is the case where edges of embedded graphs may pass through singular points. Let for this case the maximum chromatic number for graphs embeddable on $\mathbf{S}(0; n, 2)$ be denoted by χ_n^{**}. Clearly, $\chi_n^{**} \geq \chi_n^*$ holds for all n, and equality holds for $n \leq 12$. For $n \geq 13$, (*) remains an upper bound; an improved bound can, however, be obtained as follows. Let G be embedded as described; then

$$(k-1) \cdot |V(G)| \leq 6 \cdot (|V(G)| + s_1) - 12 + 2s_2,$$

where s_1 is the number of vertices of G situated in singular points and s_2 the number of edges passing through singular points. Note that a planar graph can be obtained from G by deleting s_2 edges and splitting each of s_1 vertices into a pair of new vertices. Thus

$$\begin{aligned}(k-1) \cdot |V(G)| &\leq 6 \cdot |V(G)| + 4s_1 - 12 + 2 \cdot (s_1 + s_2) \\ &\leq 10 \cdot |V(G)| - 12 + 2n,\end{aligned}$$

and therefore for $k \geq 11$,

$$(k-11) \cdot k \leq -12 + 2n,$$

implying

$$k \leq \left\lfloor \frac{1}{2} \cdot \left(11 + \sqrt{73 + 8n}\right) \right\rfloor. \qquad (**)$$

Borodin and Melnikov [1974] have shown that (**) gives the correct value of χ_n^{**} for all $n \geq 12$. From the derivation above the same conclusion follows for a surface obtained from the sphere by successive identification of n not necessarily distinct pairs of points. Borodin and Melnikov furthermore considered the similar problem, with n distinct pairs, for higher orientable or nonorientable surfaces and proved that for any fixed $\mathbf{S}$, the corresponding Heawood-type bound is the best possible, provided that n is sufficiently large, depending on $\mathbf{S}$ (see also Borodin [1989a]). Splitting results of Hartsfield [1986] become relevant for coloring problems on pseudo-surfaces obtained by identifying pairs of points on the torus.

For pairs of points on the sphere it appears that all the natural variations of the problem have been solved. With triples of points identified instead of pairs the problem seems only sparsely studied. As noted by Kainen [1972], it follows from the four-color theorem that the maximum chromatic number of graphs embeddable on a pseudo-surface $\mathbf{S}(0; 1, t)$ for $t \geq 2$ is equal to 5.

For a given $\mathbf{S}(g; n, t)$ and a given k, one may ask if there exists a polynomial algorithm to decide k-colorability of graphs embeddable on $\mathbf{S}(g; n, t)$. In particular, the cases $g = 0$ and $k = 4, 5$ seem interesting. A positive answer for $g < g_0$, with $t = 2$ and $k = 5$ would seem to provide (by arguments of Thomassen [1993b]) a positive answer to a question by Thomassen [1993a]: Does there exist a polynomial algorithm to decide 5-colorability of graphs embeddable on $\mathbf{S}_{g_0}$?

Different types of pseudo-surfaces are the spindle surfaces. A spindle surface can be obtained from a graph G by replacing each edge xy of G by a topological sphere with x and y as two distinct points. These surfaces were studied by Bodendiek, Schumacher, and Wagner [1985], by Schumacher and Wagner [1985], who described coloring properties of spindle surfaces, and by Wagner and Bodendiek [1990].

BIBLIOGRAPHY

[1981] Albertson M.O. Open problem 2. In: G. Chartrand, Y. Alavi, D.L. Goldsmith, L. Lesniak-Foster, and D.R. Lick, editors, *The Theory and Applications of Graphs*, page 609. Wiley, 1981.

[1976] Albertson M.O. and D.M. Berman. The acyclic chromatic number. In: *Proc. 7th S–E Conference on Combinatorics, Graph Theory and Computing, Baton Rouge, 1976, Congr. Num., 17*, pages 51–60, 1976.

[1977] Albertson M.O. and D.M. Berman. Every planar graph has an acyclic 7-coloring. *Israel J. Math.* **28**, 169–174, 1977.

[1978] Albertson M.O. and D.M. Berman. An acyclic analogue to Heawood's theorem. *Glasgow Math. J.* **19**, 163–166, 1978.

[1978] Albertson M.O. and J.P. Hutchinson. On the independence ratio of a graph. *J. Graph Theory* **2**, 1–8, 1978.

[1979] Albertson M.O. and J.P. Hutchinson. The three excluded cases of Dirac's map-color theorem. *Ann. New York Acad. Sci.* **319**, 7–17, 1979.

[1980b] Albertson M.O. and J.P. Hutchinson. On six-chromatic toroidal graphs. *Proc. London Math. Soc. (3)* **41**, 533–556, 1980.

[1982] Albertson M.O. and W.R. Stromquist. Locally planar toroidal graphs are 5-colorable. *Proc. Amer. Math. Soc.* **84**, 449–456, 1982.

[1989] Appel K. and W. Haken. *Every Planar Map Is Four Colorable*, volume 98 of *Contemporary Mathematics Series*. American Mathematical Society, 1989.

[1987] Archdeacon D. A note on defective colorings of graphs in surfaces. *J. Graph Theory* **11**, 517–519, 1987.

[1985] Barnette D. Coloring polyhedral manifolds. In: J.E. Goodman, E. Lutwak, J. Malkevitch, and R. Pollack, editors, *Discrete Geometry and Convexity*, pages 192–195. New York Academy of Science, 1985.

[1989] Berman K.A. and J.L. Paul. A 4-color theorem for surfaces of genus g. *Proc. Amer. Math. Soc.* **105**, 513–522, 1989.

[1992] Berman K.A. and J.L. Paul. The bounded chromatic number for graphs of genus g. *J. Combin. Theory Ser. B* **56**, 183–196, 1992.

[1976] Biggs N.L., E.K. Lloyd, and R.J. Wilson. *Graph Theory 1736–1936*. Clarendon Press, 1976.

[1985] Bodendiek R., H. Schumacher, and K. Wagner. Über Graphen auf Flächen und Spindelflächen. In: R. Bodendiek, H. Schumacher, and G. Walther, editors, *Graphen in Forschung und Unterricht: Festschrift K. Wagner*, pages 18–47. Barbara Franzbecker Verlag, 1985.

[1976a] Borodin O.V. On decomposition of graphs into degenerate subgraphs (in Russian). *Metody Diskret. Analiz.* **28**, 3–11, 1976.

[1976b] Borodin O.V. A proof of B. Grünbaum's conjecture on acyclic 5-colorability of planar graphs (in Russian). *Dokl. Akad. Nauk SSSR* **231**, 18–20, 1976.

[1979a] Borodin O.V. On acyclic colorings of planar graphs. *Discrete Math.* **25**, 211–236, 1979.

[1984] Borodin O.V. Solution of Ringel's problem on vertex-face coloring of plane graphs and coloring of 1–planar graphs (in Russian). *Metody Diskret. Analiz.* **41**, 12–26, 1984.

[1985] Borodin O.V. Simultaneous colorings of plane graphs and coloring of 1–embedded graphs (in Russian). In: *Graphen und Netzwerke: Theorie und Anwendungen. 30. Intern. Wiss. Koll. Tech. Hochsch. Ilmenau, 1985*, pages 19–20.

[1989a] Borodin O.V. Representing K_{13} as a 2–pire map on the Klein bottle. *J. Reine Angew. Math.* **393**, 132–133, 1989.

[1989c] Borodin O.V. A new proof of the 6 color theorem. Manuscript, 1989. To appear in *J. Graph Theory*.

[1990a] Borodin O.V. Diagonal 11-coloring of plane triangulations. *J. Graph Theory* **14**, 701–704, 1990.

[1992b] Borodin O.V. Diagonal coloring of the vertices of triangulations. *Discrete Math.* **102**, 95–96, 1992.

[1974] Borodin O.V. and L.S. Melnikov. The chromatic number of a pseudo-surface (in Russian). *Metody Diskret. Analiz.* **24**, 8–20, 1974.

[1987] Bouchet A., J.-L. Fouquet, J.-L. Jolivet, and M. Riviere. On a special face colouring of cubic graphs. *Ars Combin.* **24**, 67–76, 1987.

[1975] Cook R.J. Chromatic number and girth. *Period. Math. Hungar.* **6**, 103–107, 1975.

[1986] Cowen L.J., R.H. Cowen, and D.R. Woodall. Defective colourings of graphs in surfaces: Partitions into subgraphs of bounded valency. *J. Graph Theory* **10**, 187–195, 1986.

[1991] Croft H.T., K.J. Falconer, and R.K. Guy. *Unsolved Problems in Geometry*. Springer-Verlag, 1991.

[1972] Dewdney A.K. The chromatic number of a class of pseudo-2–manifolds. *Manuscripta Math.* **6**, 311–320, 1972.

[1952b] Dirac G.A. Map colour theorems. *Canad. J. Math.* **4**, 480–490, 1952.

[1956] Dirac G.A. Map colour theorems related to the Heawood colour formula. *J. London Math. Soc.* **31**, 460–471, 1956.

[1957a] Dirac G.A. Map colour theorems related to the Heawood colour formula II. *J. London Math. Soc.* **32**, 436–455, 1957.

[1957b] Dirac G.A. Short proof of a map-colour theorem. *Canad. J. Math.* **9**, 225–226, 1957.

[1992] Edwards K. The complexity of some graph colouring problems. *Discrete Appl. Math.* **36**, 131–140, 1992.

[1978] Fisk S. The non-existence of colorings. *J. Combin. Theory Ser. B* **24**, 247–248, 1978.

[1994] Fisk S. and B. Mohar. Coloring graphs without short non-bounding cycles. *J. Combin. Theory Ser. B* **60**, 268–276, 1994.

[1934] Franklin P. A six-color problem. *J. Math. Phys.* **13**, 363–369, 1934.

[1976] Gardner M. Mathematical games. *Sci. Amer.* **234** and **235**, 1976. April issue, pp. 126–130; September issue, pp. 210–211.

[1980] Gardner M. Mathematical games. *Sci. Amer.* **242**(2), 14–21, 1980.

[1986] Gimbel J. Three extremal problems in cochromatic theory. *Rostock. Math. Kolloq.* **30**, 73–78, 1986.

[1969] Grünbaum B. Conjecture 6. In: W.T. Tutte, editor, *Recent Progress in Combinatorics*, page 343. Academic Press, 1969.

[1973] Grünbaum B. Acyclic colorings of planar graphs. *Israel J. Math.* **14**, 390–408, 1973.

[1986] Hartsfield N. The toroidal splitting number of the complete graph K_n. *Discrete Math.* **62**, 35–47, 1986.

[1985] Hartsfield N., B. Jackson, and G. Ringel. The splitting number of the complete graph. *Graphs Combin.* **1**, 311–329, 1985.

[1989] Hartsfield N. and G. Ringel. Minimal quadrangulations of orientable surfaces. *J. Combin Theory Ser. B* **46**, 84–95, 1989.

[1890] Heawood P.J. Map colour theorem. *Quart. J. Pure Appl. Math.* **24**, 332–338, 1890.

[1891] Heffter L. Über das Problem der Nachbargebiete. *Math. Ann.* **38**, 477–508, 1891.

[1984] Hutchinson J.P. A five-color theorem for graphs on surfaces. *Proc. Amer. Math. Soc.* **90**, 497–504, 1984.

[1993] Hutchinson J.P. Coloring ordinary maps, maps of empires, and maps of the moon. *Math. Mag.* **66**, 211–226, 1993.

[1994] Hutchinson J.P. Three-coloring graphs embedded on surfaces with all faces even-sided. Manuscript, 1994. Submitted.

[1987] Hutchinson J.P. and G.L. Miller. On deleting vertices to make a graph of positive genus planar. In: *Discrete Algorithms and Complexity Theory*, pages 81–98. Academic Press, 1987.

[1975] Isaacs R. Infinite families of non-trivial trivalent graphs which are not Tait colorable. *Amer. Math. Monthly* **82**, 221–239, 1975.

[1983] Jackson B. and G. Ringel. Maps of *m*–pires on the projective plane. *Discrete Math.* **46**, 15–20, 1983.

[1984b] Jackson B. and G. Ringel. Solution of Heawood's empire problem in the plane. *J. Reine Angew. Math.* **347**, 146–153, 1984.

[1985] Jackson B. and G. Ringel. Heawood's empire problem. *J. Combin. Theory Ser. B* **38**, 168–178, 1985.

[1972] Kainen P.C. On the chromatic number of certain 2–complexes. In: *Proc. 3rd S–E Conference on Combinatorics, Graph Theory and Computing, Baton Rouge, 1972, Congr. Num., 6*, pages 291–295, 1972.

[1991] Korzhik V.P. *Current graphs and their application to constructing specific embeddings of graphs* (in Russian). Ph.D. thesis, Moscow State University, 1991.

[1993] Korzhik V.P. Some new results on the 1-chromatic number of surfaces. In: M. Kilp and U. Nummert, editors, *Proc. First Estonian Conference on Graphs and Applications*, pages 162–166. Tartu University, 1993.

[1994] Korzhik V.P. A lower bound for the one-chromatic number of a surface. *J. Combin. Theory Ser. B* **61**, 40–56, 1994.

[1976] Kostochka A.V. Acyclic 6-coloring of planar graphs (in Russian). *Metody Diskret. Analiz.* **28**, 40–56, 1976.

[1976] Kostochka A.V. and L.S. Melnikov. Note to the paper of Grünbaum on acyclic colorings. *Discrete Math.* **14**, 403–406, 1976.

[1972] Kronk H.V. The chromatic number of triangle-free graphs. In: Y. Alavi, D.R. Lick, and A.T. White, editors, *Graph Theory and Applications*, volume 303 of *Lecture Notes in Mathematics*, pages 179–181. Springer-Verlag, 1972.

[1972] Kronk H.V. and A.T. White. A 4-color theorem for toroidal graphs. *Proc. Amer. Math. Soc.* **34**, 83–86, 1972.

[1972] Lick D.R. and A.T. White. The point partition numbers of closed 2-manifolds. *J. London Math. Soc. (2)* **4**, 577–583, 1972.

[1983b] Mihók P. On vertex partition numbers of graphs. In: M. Fiedler, editor, *Graphs and Other Combinatorial Topics*, pages 183–188. B.G. Teubner Verlagsgesellschaft, 1983.

[1974] Mitchem J. Every planar graph has an acyclic 8-coloring. *Duke Math. J.* **41**, 177–181, 1974.

[1993] Mohar B. 7-critical graphs of bounded genus. *Discrete Math.* **112**, 279–281, 1993.

[1994] Mohar B., N. Robertson, and D.P. Sanders. On acyclic colorings of graphs on surfaces. Manuscript, April 1994.

[1955] Mycielski J. Sur le coloriage des graphes. *Colloq. Math.* **3**, 161–162, 1955.

[1954] Ringel G. Bestimmung der Maximalzahl der Nachbargebiete auf nichtorientierbaren Flächen. *Math. Ann.* **127**, 181–214, 1954.

[1959] Ringel G. *Färbungsprobleme auf Flächen und Graphen*. VEB Deutscher Verlag der Wissenschaften, 1959.

[1965] Ringel G. Ein Sechsfarbenproblem auf der Kugel. *Abh. Math. Sem. Univ. Hamburg* **29**, 107–117, 1965.

[1974] Ringel G. *Map Color Theorem*. Springer-Verlag, 1974.

[1981] Ringel G. A nine color theorem for the torus and the Klein bottle. In: G. Chartrand, Y. Alavi, D.L. Goldsmith, L. Lesniak-Foster, and D.R. Lick, editors, *The Theory and Applications of Graphs*, pages 507–515. Wiley, 1981.

[1968] Ringel G. and J.W.T. Youngs. Solution of the Heawood map-coloring problem. *Proc. Natl. Acad. Sci. USA* **60**, 438–445, 1968.

[1984] Schumacher H. Ein 7–Farbensatz 1–einbettbarer Graphen auf der projektiven Ebene. *Abh. Math. Sem. Univ. Hamburg* **54**, 5–14, 1984.

[1990] Schumacher H. On 2–embeddable graphs. In: R. Bodendiek and R. Henn, editors, *Topics in Combinatorics and Graph Theory*, pages 651–661. Physica Heidelberg, 1990.

[1985] Schumacher H. and K. Wagner. Zur Finitisierung von Färbungsproblemen. *Mitt. Math. Ges. Hamburg* **11**, 335–340, 1985.

[1993] Sherman M. *Colorability of graphs having bounded number of triangles and bounded genus*. Master's thesis, North Dakota State University, 1993.

[1979] Straight H.J. Cochromatic number and genus of a graph. *J. Graph Theory* **3**, 43–51, 1979.

[1980] Straight H.J. Note on the cochromatic number of several surfaces. *J. Graph Theory* **4**, 115–117, 1980.

[1975] Szekeres G. Non-colorable trivalent graphs. In: A.P. Street and W.D. Wallis, editors, *Combinatorial Mathematics III*, volume 452 of *Lecture Notes in Mathematics*, pages 227–233. Springer-Verlag, 1975.

[1979] Taylor H. Problem 5. In: *Proc. West Coast Conference on Combinatorics, Graph Theory and Computing, Arcata, 1979, Congr. Num., 26*, page 320, 1979.

[1981b] Taylor H. Synchronization patterns and related problems in combinatorial analysis and graph theory. Technical Report 509, University of Southern California Electronic Science Laboratory, 1981.

[1993a] Thomassen C. Five-coloring graphs on the torus. Technical report, The Technical University of Denmark, January 1993. *J. Combin. Theory Ser. B* **62**, 11–33, 1994.

[1993b] Thomassen C. Five-coloring maps on surfaces. *J. Combin. Theory Ser. B*, **59**, 89–105, 1993.

[1993c] Thomassen C. Grötzsch's 3-color theorem and its counterparts for the torus and the projective plane. Manuscript 1993. To appear in *J. Combin. Theory Ser. B*.

[1985] Tinsley F.C. and J.J. Watkins. A study of snark embeddings. In: F. Harary and J.S. Maybee, editors, *Graphs and Applications: Proceedings of the First Colorado Symposium on Graph Theory*, pages 317–332. Wiley, 1985.

[1990] Wagner K. and R. Bodendiek. *Graphentheorie II*. BI Wissenschaftsverlag, 1990.

[1973] White A.T. *Graphs, Groups and Surfaces*. North-Holland, 1973.

[1978] White A.T. The proof of the Heawood conjecture. In: L.W. Beineke and R.J. Wilson, editors, *Selected Topics in Graph Theory*, pages 51–82. Academic Press, 1978.

[1984] White A.T. *Graphs, Groups and Surfaces, Completely revised and enlarged edition*, volume 8 of *North-Holland Mathematics Studies*. North-Holland, 1984.

[1990] Woodall D.R. Improper colourings of graphs. In: R. Nelson and R.J. Wilson, editors, *Graph Colourings*, volume 218 of *Pitman Research Notes in Mathematics Series*, pages 45–63. Longman, 1990.

[1989] Woodburn R.L. A 4-color theorem for the Klein bottle. *Discrete Math.* **76**, 271–276, 1989.

[1991] Youngs D.A. 4-chromatic projective graphs. Manuscript, 1991. To appear in *J. Graph Theory*.

4

Degrees

■ 4.1. THE COLORING NUMBER

The coloring number col(G) of a graph G on n vertices $x_1, x_2, \ldots, x_n$ is defined by

$$\text{col}(G) = \min_p \max_i \{d(x_{p(i)}, G_{p(i)}) + 1\},$$

where the minimum is taken over all permutations p of $\{1, 2, \ldots, n\}$ and $G_{p(i)}$ is the subgraph of G induced by $x_{p(1)}, x_{p(2)}, \ldots, x_{p(i)}$. The coloring number satisfies

$$\chi(G) \leq \text{col}(G) \leq \Delta(G) + 1,$$

where $\Delta(G)$ is the maximum degree of G. Several questions can be posed, among them the following four:

1. For which graphs G is $\chi(G) = \text{col}(G)$?
2. More restrictive: Which graphs satisfy $\chi(G') = \text{col}(G')$ for all induced subgraphs G' of G?
3. Which graphs are k-critical with respect to col; that is, for which graphs G is $\text{col}(G') < \text{col}(G) = k$ for all proper subgraphs G' of G?
4. More restrictive: Which graphs G satisfy $\text{col}(G') < \text{col}(G) = \chi(G)$ for all proper subgraphs G' of G? ■

The coloring number was first defined (and in particular studied for infinite graphs) by Erdős and Hajnal [1966]. For finite graphs of size n the coloring number seems at first sight to be a "difficult" parameter because the minimum is taken over $n!$ possible permutations. However, as explained in Theorem 12 and the discussion of it in Chapter 1, there is a very simple polynomial algorithm to compute col(G) for any graph G. This was first discovered by Matula [1968] and independently by Finck and Sachs [1969].

Moreover, col(G) is equal to the upper bound for the chromatic number given by Szekeres and Wilf [1968]:

$$\mathrm{SW}(G) = \max_{H \subseteq G} \min_{x \in V(H)} (d(x,H) + 1).$$

This fact was proved by Halin [1967] and independently by Matula [1968], by Finck and Sachs [1969], and by Lick and White [1970] (Theorem 12 in Chapter 1). Lick and White [1970] defined a graph G to be k-degenerate if and only if every subgraph H of G has minimum degree at most k. Thus by the results stated above, G is k-degenerate if and only if $\mathrm{col}(G) \leq k + 1$. Simões-Pereira [1976] gave a survey of k-degenerate graphs.

The background for question 1 is the theorem of Brooks [1941] (Theorem 11 in Chapter 1), characterizing those G for which $\chi(G) = \Delta(G) + 1$. It is easy to see that $\mathrm{col}(G) = \Delta(G) + 1$ if and only if G has a $\Delta(G)$-regular connected component. Thus a characterization of those G for which $\chi(G) = \mathrm{col}(G)$ would provide a substantial extension of Brooks' theorem. However, there seems to be so many graphs with $\chi(G) = \mathrm{col}(G)$ that no reasonable characterization is possible. Also, it follows from a result of Garey, Johnson, and Stockmeyer [1976] that it is an ***NP***-complete problem to decide if a given planar graph G with $\mathrm{col}(G) = 4$ satisfies $\chi(G) < 4$. This decision problem also becomes ***NP***-complete when "planar graph" is replaced by "line graph"; this follows from a result of Holyer [1981]. Thus one cannot expect that it should be possible to give a simple characterization of graphs satisfying $\chi(G) = \mathrm{col}(G)$.

One may perhaps still expect that if further restrictions are added on other parameters of G it will be possible to characterize the graphs for which $\chi(G) = \mathrm{col}(G)$. A.V. Kostochka investigated the problem when restricting to graphs all of whose cycles have length at least g, where g is fixed, and he proved in 1980[1] that such examples exist for all g [personal communication from Kostochka in 1993].

Question 2 is due to O.V. Borodin, who posed it in 1978. It is similar in spirit to the definition of the perfect graphs of C. Berge (see Section 1.6), but as Borodin remarked [personal communication in 1985] it seems a simpler concept: Berge's perfect graphs are defined in terms of two "difficult" graph parameters ω and χ, whereas Borodin's perfect graphs are defined in terms of one "difficult" (χ) and one "easy" (col) parameter.

A graph G as in question 2 was called "ω-perfect" by Gasparyan and Markosyan [1990] (see also Reed [1990]), who noted that clearly, no such graph contains an even cycle as an induced subgraph, and they proved that if G is not ω-perfect, then G contains an even cycle or an even cycle with exactly one diagonal as an induced subgraph. They pointed out that the situation for ω-perfect graphs is in some sense more complicated than for Berge's perfect graphs; the complement of an ω-perfect graph is not always ω-perfect. However, they proved that G and G's complement

[1]Kostochka published this in a paper distributed by the Russian Academy of Sciences in Novosibirsk, but it is not generally accessible.

$\overline{G}$ are both ω-perfect if and only if neither G nor $\overline{G}$ contains an even cycle as an induced subgraph. In comparison, the strong perfect graph conjecture of C. Berge (Section 1.6) suggests that G is perfect (in Berge's sense) if and only if neither G nor $\overline{G}$ contains an odd cycle of length at least 5 as an induced subgraph.

Christen and Selkow [1979] characterized other classes of "perfect" graphs defined similarly in terms of (a) χ and Γ, (b) ω and Γ, (c) χ and ψ, and (d) ω and ψ, where ω is the size of a largest complete subgraph, Γ is the Grundy number (see Problem 10.4), and ψ is the achromatic number (see Problem 10.5). These numbers satisfy $\omega \leq \chi \leq \Gamma \leq \psi$; however, Γ and ψ have no direct relation to the coloring number (e.g., a tree has coloring number 2 and can have arbitrarily large Γ, and two disjoint 5-cycles completely joined by edges has coloring number 8 and $\Gamma = \psi = 6$).

For question 3, again there seems to be too many graphs that a simple characterization is possible. But of course these graphs are polynomially recognizable. Borodin has informed us [personal communication in 1985] that the best characterization seems to go along the following lines: G is $(k+1)$-critical w.r.t. col if and only if (i) G is connected, (ii) $\delta(G) = k$, (iii) the vertices of degree at least $k+1$ in G are independent, and (iv) $\text{col}(G - x) \leq k$ for all x of degree k in G. A similar characterization was given by Mihók [1981b]. Mihók studied $(n, k-1)$-critical graphs. Such a graph G is improperly vertex-colorable in n colors, but not in $n-1$ colors, such that each color class C induces a subgraph $G[C]$ with $\text{col}(G[C]) \leq k$, but every proper subgraph of G is colorable in this fashion in $n-1$ colors. Thus Mihók's concept "$(2, k-1)$-critical" is equivalent to the concept "$(k+1)$-critical w.r.t. col." The $(n, k-1)$-critical graphs were first studied for the case $k = 2$ by Bollobás and Harary [1975], Kronk and Mitchem [1975], and Mihók [1976, 1981a].

Question 4 is due to Borodin, who posed it in 1974. It has been investigated by A.V. Kostochka. The graphs in question form a subset of the graphs critical w.r.t. χ.

Finally a remark on similar questions for list coloring. The list-chromatic number $\chi_\ell(G)$ satisfies $\chi(G) \leq \chi_\ell(G) \leq \text{col}(G)$, hence each of the four problems given above may be split into two new problems, by replacing $\chi(G)$ and $\text{col}(G)$, respectively, by $\chi_\ell(G)$.

■ 4.2. COLORING OF DECOMPOSABLE GRAPHS

If the edges of G can be decomposed into two subsets, one inducing a subgraph H_1 of coloring number at most 2 (i.e., it is a forest), and the other inducing a subgraph H_2 of coloring number at most 3, can G be 5-colored? ■

The question is due to M. Tarsi, who conjectured that the answer is affirmative, and it appeared in Klein and Schönheim [1991]. The problem is similar in spirit to a question of N.W. Sauer raised at the International Conference on Combinatorics celebrating P. Erdős' eightieth birthday, held in Keszthely, Hungary, in July 1993: If the edge set of G can be partitioned into sets E_1 and E_2 such that the connected components of the subgraph induced by E_1 are "stars" (i.e., complete bipartite graphs

$K_{1,m}$ for $m \geq 1$) and the subgraph induced by E_2 is a forest, is it true that G is 3-colorable? Sauer's question was answered affirmatively by Stiebitz [1993]. Stiebitz has proved, more generally, that if G is the union of a forest and a bipartite graph B with bipartition (R, S), where every vertex in S has degree at most t in B, then G is $(t+2)$-choosable, that is, $\chi_\ell(G) \leq t+2$ [personal communication in 1994].

It is easy to find a 6-coloring of G by combining a 2-coloring of H_1 with a 3-coloring of H_2, since adjacent vertices in G must receive different colors in at least one of the two colorings. Asking if a 5-coloring is possible can be considered an interesting parallel to the 5-flow conjecture of W.T. Tutte (Problem 13.1). The decomposition of G into subgraphs H_1 and H_2 is dual, in a sense, to a decomposition used by P.D. Seymour to prove the 6-flow theorem (Theorem 33 in Chapter 1). Seymour proved that the edge set of any bridge-less graph can be partitioned into two subsets, one inducing a subgraph with a 2-flow, the other inducing a subgraph with a 3-flow, and these flows together easily combine into a 6-flow of the union (see also the discussion in Problem 13.2).

A far more general problem was suggested by Klein and Schönheim [1991, 1993] and Klein [1992]. G is called $(m_1, m_2, \ldots, m_k)$-composed if G is the union of k graphs $H_1, H_2, \ldots, H_k$, where H_i is m_i-degenerate (i.e., has coloring number at most $m_i + 1$) for $i = 1, 2, \ldots, k$, and $m_1 \leq m_2 \leq \cdots \leq m_k$. Define

$$v(m_1, m_2, \ldots, m_k) = m_1 + m_2 + \cdots + m_k + \left\lfloor \frac{1}{2} \cdot \left(1 + \sqrt{1 + 8 \cdot \textstyle\sum_{i<j} m_i m_j}\right) \right\rfloor.$$

Klein and Schönheim conjectured that every $(m_1, m_2, \ldots, m_k)$-composed graph can be $v(m_1, m_2, \ldots, m_k)$-colored. The motivation for this conjecture is a theorem by Klein and Schönheim [1991] for $k = 2$, and by Klein [1992] and Klein and Schönheim [1993] for general k, that the complete graph K_n is $(m_1, m_2, \ldots, m_k)$-composed if and only if $n \leq v(m_1, m_2, \ldots, m_k)$. The case $(m_1, m_2) = (1, 2)$ of the latter conjecture is the conjecture of Tarsi stated above.

4.3. COLOR-BOUND FAMILIES OF GRAPHS

Let $\mathcal{G}(H)$ denote the set of all graphs not containing the graph H as an induced subgraph. If T is a tree, does there exist some function $f_T(x)$ with $f_T(x) \longrightarrow \infty$ as $x \longrightarrow \infty$ so that

$$\chi(G) \geq f_T(\text{col}(G)) \qquad \text{for all} \qquad G \in \mathcal{G}(T),$$

where col(G) is the coloring number of G (i.e., the largest number k so that G has a subgraph of minimum degree $k-1$)? ■

The question was asked by P. Mihók [personal communication in 1993]. It was inspired by a question raised by Gyárfás [1975] and independently Sumner [1981] asking for an upper bound on $\chi(G)$ for $G \in \mathcal{G}(H)$, expressed as a function of $\omega(G)$, the size of a largest complete subgraph in G (see Problem 8.11).

Kierstead and Penrice [1990] proved that the answer to the question by Gyárfás and Sumner is affirmative when H is a forest consisting of disjoint trees $T_1, T_2, \ldots, T_n$, if and only if it is affirmative for each tree T_i. A proof of this was also presented by Sauer [1993]. Wagon [1980] had previously obtained this result for the case when each tree T_i is a complete graph K_2. The corresponding statement is not true for Mihók's problem, since forbidding the graph consisting of three vertices and only one edge as an induced subgraph, one obtains the class of all complete multipartite graphs, and these can have arbitrarily high coloring numbers for any fixed chromatic number.

For any graph H with at least one cycle, the question corresponding to Mihók's for the family of graphs $\mathcal{G}(H)$ has a negative answer, since there exist bipartite graphs of arbitrarily high vertex degrees (and hence arbitrarily high coloring number) and with shortest cycles of arbitrarily high length. However, it was shown by Gasparyan and Markosyan [1990] (see also Reed [1990]) that

$$\chi(G) \geq 1 + \mathrm{col}(G)/2$$

holds for any graph G without an induced even cycle.

If T is a tree with fewer than 4 vertices, the answer is clearly affirmative with $f_T(x) = x$.

■ 4.4. EDGE-DISJOINT PLACEMENTS

Let G_1 and G_2 be two graphs both on n vertices, and let $\Delta(G_1)$ and $\Delta(G_2)$ be their maximum degrees. Is it true that if

$$(\Delta(G_1) + 1) \cdot (\Delta(G_2) + 1) \leq n + 1,$$

then G_1 and G_2 can be embedded edge-disjointly in the same complete n-graph K_n? Or equivalently, if the condition is satisfied, is G_1 then isomorphic to a subgraph of the complement $\overline{G}_2$ of G_2? ■

The problem is due to Bollobás and Eldridge [1976, 1978] (see also Bollobás [1978c]) and independently to Catlin [1976, 1977]. The problem was studied by Catlin [1974, 1976, 1980] and by Sauer and Spencer [1978], who proved that the conclusion is true if

$$2 \cdot \Delta(G_1) \cdot \Delta(G_2) < n$$

(the first mention of this result seems to be a footnote in the paper by Catlin [1974]). It would be of considerable interest even to improve the factor 2 to a smaller value.

This problem does not have the form of a coloring problem; however, a coloring of a graph G can be thought of as an edge-disjoint placement of G and a graph with complete connected components. In particular, an affirmative answer would provide a very considerable extension of the deep coloring theorem (Theorem 13 in Chapter 1) of Hajnal and Szemerédi [1970], as noted by Bollobás [1978a, 1978c]. The theorem of Hajnal and Szemerédi states that a graph G of maximum degree $\Delta(G)$ and with $|V(G)| = k \cdot (\Delta(G) + 1)$ has a $(\Delta(G) + 1)$-coloring in which each color class has precisely k vertices.

■ 4.5. POWERS OF HAMILTON CYCLES

Let k be a positive integer, and let G be a graph with minimum degree $\delta(G)$ at least $k|V(G)|/(k + 1)$. Does G contain a kth power of a Hamilton cycle as a subgraph? That is, can the vertices of G be arranged in a cyclic order $x_1, x_2, \ldots, x_{|V(G)|}$, so that each x_i is adjacent in G to all the vertices $x_{i+1}, x_{i+2}, \ldots, x_{i+k}$ for $i = 1, 2, \ldots, |V(G)|$, adding subscripts modulo $|V(G)|$? ■

The conjecture that the answer is affirmative was made by Seymour [1974b]. The case $k = 1$ is a classical theorem of Dirac [1952c]: If $G \neq K_2$ and $\delta(G) \geq |V(G)|/2$, then G has a Hamilton cycle. The case $k = 2$ is an unsolved conjecture due to Pósa [1963].

Pósa's conjecture that if G satisfies $\delta(G) \geq 2|V(G)|/3$ then G contains the square of a Hamilton cycle, was studied by Faudree, Gould, Jacobson, and Schelp [1991] and by Fan and Häggkvist [1994]. Fan and Kierstead [1993] and J. Quintana [personal communication from H.A. Kierstead in 1994], proved the so far strongest known partial results, namely that "$\geq 2|V(G)|/3$" can be replaced by "$\geq (17|V(G)| + 9)/24$" and by "$\geq 7|V(G)|/10 + 2$," respectively. Fan [personal communication in 1993] reported the following result obtained by Kierstead: For any $\varepsilon > 0$ there is a number $c(\varepsilon)$ such that if

$$\delta(G) \geq \frac{2 + \varepsilon}{3} \cdot |V(G)| + c(\varepsilon),$$

then G contains the square of a Hamilton cycle. Moreover, Fan and Kierstead [personal communication in 1994] proved that if G has minimum degree $\delta(G)$ at least $(2|V(G)| - 1)/3$, then G contains the square of a Hamilton path.

Seymour's conjecture is not obviously connected to graph coloring. However, its truth would extend the theorem of Hajnal and Szemerédi [1970] (Theorem 13 in Chapter 1) that every graph G of maximum degree $\Delta(G)$ and with $k \cdot (\Delta(G) + 1)$ vertices has a $(\Delta(G) + 1)$-coloring in which every color class has size k. Consider the complement $\overline{G}$ of such a G with minimum degree

$$\delta(\overline{G}) = |V(G)| - 1 - \Delta(G) = \frac{k-1}{k} \cdot |V(\overline{G})|.$$

Seymour's conjecture would imply that $\overline{G}$ contains the $(k-1)$st power of a Hamilton cycle C. Thus G can be colored by giving any k consecutive vertices on C color 1, then the following k consecutive vertices on C color 2, and continuing like this until all vertices of G are colored.

■ 4.6. BROOKS' THEOREM FOR TRIANGLE-FREE GRAPHS

Find a best possible upper bound for the chromatic number $\chi(G)$ of a graph G in terms of the maximum degree $\Delta(G)$, provided that the graph is triangle-free or, more generally, provided that the graph contains no K_{r+1}. ■

This problem is due to Vizing [1968], who after mentioning Brooks' theorem said: "*Further investigations could be conducted taking into account a more exact relation between the maximum degree and the maximum size of a complete subgraph. Perhaps one should start with estimates of the chromatic number of a graph without triangles and given maximal degree for vertices.*" Another early reference is the paper by Grünbaum [1970a].

The theorem of Brooks [1941] (Theorem 11 in Chapter 1) states that if $3 \leq \Delta(G)$ and G contains no $K_{\Delta(G)+1}$, then $\chi(G) \leq \Delta(G)$. Borodin and Kostochka [1977], Catlin [1978a], and Lawrence [1978] independently proved that if $3 \leq r \leq \Delta(G)$ and G contains no K_{r+1}, then

$$\chi(G) \leq \frac{r}{r+1} \cdot (\Delta(G) + 2).$$

If G is triangle-free it does not contain a K_4, and therefore

$$\chi(G) \leq \frac{3}{4} \cdot (\Delta(G) + 2).$$

Catlin [1978a] remarked: "*However, in this case it would be natural to expect that the bound can be improved.*" For graphs G without 4-cycles, Catlin [1978b] proved that

$$\chi(G) \leq \frac{2}{3} \cdot (\Delta(G) + 3).$$

Kostochka [personal communication in 1987] proved in 1978[2] that this bound is indeed true also for triangle- free graphs. It has been proved by J.H. Kim for triangle-

[2]The proof was published by the Russian Academy of Sciences in Novosibirsk in 1982, but it is not generally accessible.

free graphs without 4-cycles, and by A. Johansson for triangle-free graphs in general, that

$$\max \frac{\chi(G)}{\Delta} \to 0 \quad \text{as} \quad \Delta \to \infty,$$

where the maximum is taken over all triangle-free graphs G with $\Delta(G) = \Delta$ [personal communication from R. Häggkvist in 1993]. The bound proved by Johansson is

$$\chi(G) \leq \frac{c \cdot \Delta(G)}{\log \Delta(G)}$$

for a constant $c > 0$. It follows from work by Kostochka and Mazurova [1977] that this bound is of best possible order of magnitude. They found for any given integer $\Delta > 3$ a triangle-free graph G for which $\Delta(G) = \Delta$ and

$$\chi(G) > \frac{\Delta}{2 \log \Delta}.$$

They in fact proved the much stronger result that such a G exists with any prescribed length of a shortest cycle. In contrast, Kostochka [1978a] has shown that for any graph G where all cycles are sufficiently long (depending on $\Delta = \Delta(G)$),

$$\chi(G) \leq \frac{\Delta}{2} + 2.$$

A. Johansson [personal communication in 1994] has suggested that if $r \geq 2$ and G contains no K_{r+1}, and if Δ is sufficiently large, then

$$\chi(G) \leq c(r) \cdot \frac{\Delta}{\log^{(r-1)} \Delta},$$

where $c(r)$ is a constant depending only on r, and $\log^{(r-1)}$ denotes the $(r-1)$ times iterated logarithm.

Catlin [1978b] managed to weaken the condition on G and still get good bounds on $\chi(G)$. If $r \geq 3$ and G contains no K_{r+2}^- (the complete $(r+2)$-graph minus one edge) as a subgraph, then

$$\chi(G) \leq \frac{r}{r+1} \cdot (\Delta(G) + 3).$$

A further improvement for triangle-free graphs of relatively high minimum degree has been obtained quite recently by G. Jin. If G is without triangles and satisfies

$$\delta(G) > \frac{10}{29} \cdot |V(G)|,$$

where $\delta(G)$ is the minimum degree of G, then G is 3-colorable [personal communication from R. Häggkvist in 1993].

4.7. GRAPHS WITHOUT LARGE COMPLETE SUBGRAPHS

Find the best possible lower bound in terms of k for the maximum degree $\Delta(G)$ of a k-chromatic graph G not containing a K_{k-1}. ■

The theorem of Brooks [1941] (Theorem 11 in Chapter 1) gives $\Delta(G) \geq k$. A conjecture of Lawrence [1978], also mentioned by Bollobás [1978a], suggests the bound $\Delta(G) \geq 2k - 5$. For $4 \leq k \leq 5$ this is true by Brooks' theorem. However, for $k \geq 8$ the conjecture is false, as shown by the counterexamples of Catlin [1979] to Hajós' conjecture (see also the paper by Krusenstjerna-Hafstrøm and Toft [1981]): If G consists of five disjoint copies of $K_{2\alpha+1}$ arranged in a cycle with neighboring ones completely joined, then G has chromatic number $5\alpha + 3$, does not contain a $K_{4\alpha+3}$, and is $(6\alpha + 2)$-regular. Thus the factor 2 in Lawrence's conjecture is too large; the most one can hope for is 6/5.

If instead of not containing a K_{k-1}, one imposes the weaker condition of not containing a K_k, then Brooks' bound $\Delta(G) \geq k$ is the best possible for $4 \leq k \leq 8$. However, Borodin and Kostochka [1977] conjectured that it can be improved for $k \geq 9$ (Problem 4.8).

Mozhan [1983] considered the stronger condition that G should contain no complete graph on more than $2\lfloor(k-1)/3\rfloor + 1$ vertices as a subgraph, and he proved that for $k \geq 9$ this implies that $\Delta(G) \geq k + 1$. Borodin [personal communication in 1993] has informed us that there exists an unpublished deposited manuscript by Mozhan, where it is proved even stronger, that if the largest complete subgraph of G has at most $k - 29$ vertices, the conclusion $\Delta(G) \geq k + 1$ also holds.

4.8. k-CHROMATIC GRAPHS OF MAXIMUM DEGREE k

Does there exist a k-chromatic graph without a K_k as a subgraph and of maximum degree k for any value of $k \geq 9$? ■

Borodin and Kostochka [1977] posed this problem and conjectured that the answer is no. Beutelspacher and Hering [1984] proved that if $\chi(G) = \Delta(G) = k$ and G does not contain a K_k, then $|V(G)| \geq 2k - 1$, and they showed that there are 13 critical graphs for which $|V(G)| = 2k - 1$. All of them are 8-colorable, and only one has chromatic number equal to 8, namely the 8-chromatic counterexample to Hajós' conjecture, due to Catlin [1979], consisting of five disjoint copies of K_3 in cyclic order with neighboring copies completely joined. Beutelspacher and Hering thus also (independently) arrived at the question if $\chi(G) = \Delta(G) = k$ and $G \not\supseteq K_k$ can occur only for finitely many k.

■ 4.9. TOTAL COLORING

Let G be any graph or multigraph on n vertices. The total chromatic number $\chi''(G)$ is the smallest number of colors needed to color all the elements of $V(G) \cup E(G)$ in such a way that no two adjacent or incident elements receive the same color. Let Δ be the maximum degree of G. If G is simple, does

$$\chi''(G) \leq \Delta + 2$$

hold? More generally, if μ denotes the multiplicity of G, does

$$\chi''(G) \leq \Delta + \mu + 1$$

hold? ■

Total coloring was introduced by Vizing [1964, 1965b] and independently by Behzad [1965] in his Ph.D. thesis. They both formulated the question above for simple graphs and conjectured that the answer is affirmative. The conjecture was mentioned in 1967 by A.A. Zykov in the problem session at a conference in Manebach, German Democratic Republic (see Zykov [1968b]). Also the conjecture that the answer to the second question above is affirmative is due to Behzad [1965] and independently to Vizing [1968].

An easy lower bound for the total chromatic number is $\chi''(G) \geq \Delta + 1$. Upper bounds were obtained by Behzad, Chartrand, and Cooper [1967], who proved the conjecture for bipartite multigraphs, by Kostochka [1977a, 1978b, 1979], and by Hind [1988, 1990]. Kostochka proved that

$$\chi''(G) \leq \left\lfloor \frac{3}{2}\Delta \right\rfloor$$

for any multigraph G. Initially, Kostochka proved this bound for all except finitely many values of Δ. Later he obtained proofs also for the exceptional values. Hind gave the bound

$$\chi''(G) \leq \chi'(G) + 2\left\lceil \sqrt{\chi(G)} \right\rceil,$$

where $\chi'(G)$ is the edge-chromatic number of G. Recently, Sánchez-Arroyo [1993] proved

$$\chi''(G) \leq \chi'(G) + \left\lfloor \frac{1}{3}\chi(G) \right\rfloor + 2,$$

giving a bound which is in some cases better than Hind's, that is, when the chromatic number of G is not too large, among such values all $\chi(G)$ satisfying $2 \leq \chi(G) \leq 23$.

Häggkvist and Chetwynd [1992] proved that if t satisfies $t! > n$, then

$$\chi''(G) \le \chi'(G) + t.$$

In particular, the inequality $\chi''(G) \le \Delta + \mu + t$ follows by the theorem of Vizing (Theorem 29 in Chapter 1), stating that $\chi'(G) \le \Delta + \mu$. An upper bound of $\chi''(G) \le \chi'(G) + t + 1$, again with $t! > n$ as above, was obtained independently by McDiarmid and Reed [1993]. The bound is better than Hind's bound when $\chi(G)$ is moderately large, say $\chi(G) \ge (\log n)^3$.

Hind [1992] proved, using the theorem of Hajnal and Szemerédi [1970] (Theorem 13 in Chapter 1), the bound

$$\chi''(G) \le \Delta + 2\left\lceil \frac{n}{\Delta} \right\rceil + 1,$$

which is better than other known bounds when n is large and Δ is large relative to n.

R. Häggkvist [personal communication in 1993] has informed us of a recent result obtained by A. G. Chetwynd and himself:

$$\chi''(G) \le \Delta + 18 \cdot \Delta^{1/3} \cdot \log(3\Delta).$$

The total chromatic number $\chi''(G)$ is closely related to the list-edge-chromatic number $\chi'_\ell(G)$, defined as follows. Let each edge xy of G be assigned a list $\Lambda(xy)$ of λ colors; then a Λ-coloring is an edge-coloring in which every edge is assigned a color from its list. The list-edge-chromatic number $\chi'_\ell(G)$ is the minimum number λ for which such a coloring exists for any given Λ with λ colors in each list. Clearly, $\chi'_\ell(G) \ge \chi'(G)$. The bound

$$\chi''(G) \le \chi'_\ell(G) + 2$$

is easily obtained: First color the vertices of G with at most $\chi'_\ell(G) + 2 > \Delta$ different colors. Now color each edge of G with a color chosen from the $\chi'_\ell(G)$ different colors not assigned to either vertex incident to it. The list-coloring conjecture (see Section 1.9), that every graph or multigraph G satisfies $\chi'_\ell(G) = \chi'(G)$, would imply, using Vizing's theorem (Theorem 29 of Chapter 1), that

$$\chi''(G) \le \Delta + \mu + 2$$

always holds.

The total coloring conjecture has been proved for graphs of sufficiently small maximum degree. It was proved for $\Delta = 3$ by Rosenfeld [1971] and independently by Vijayaditya [1971], and an algorithmic proof was presented by Yap [1989a]. For $\Delta = 4$ Kostochka [1977b] gave a proof of $\chi''(G) \le 6$. The case $\Delta = 5$ was settled in the doctoral thesis of Kostochka [1978b, 1979], who proved that $\chi''(G) \le \Delta + 2$ is valid for all multigraphs G with $\Delta \le 5$ (this proof was published in Russian, but

Kostochka has informed us that an English version is under preparation [personal communication in 1994]). Also, for simple graphs G of maximum degree satisfying $\Delta(G) \geq |V(G)| - 5$, the conjecture has been verified by Yap, Wang, and Zhang [1989, 1992].

For a simple graph G of minimum degree $\delta(G) \geq 5(|V(G)| + 1)/6$ the conjecture has been verified by Chetwynd, Hilton, and Zhao Cheng [1991], and also for even values of $|V(G)|$ when $\delta(G) \geq 3(|V(G)| - 1)/4$. Hilton and Hind [1993] improved these results by showing that the condition $\delta(G) \geq 3|V(G)|/4$ is sufficient for all values of $|V(G)|$. For (simple) complete r-partite graphs the conjecture was proved by Yap [1989a].

If G is a simple planar graph, Behzad and Vizing's total coloring conjecture has been proved with the exception of the cases $\Delta = 6$ or 7. For $\Delta \leq 5$ it follows by the results on graphs of small maximum degree mentioned above. For $\Delta \geq 14$ the total chromatic number of G is, in fact, equal to $\Delta + 1$, as proved by Borodin [1989b], who also added the question: Is there a smaller number that can replace 14? In the same paper Borodin proved the conjecture for G simple and planar with $\Delta \geq 9$. For $\Delta = 8$, Borodin [personal communication in 1993] has informed us that $\chi''(G) \leq 10$ can be proved from the four-color theorem (see Section 1.2) combined with a theorem of Vizing [1968] that the edge-chromatic number of a simple planar graph G with $\Delta \geq 8$ is $\chi'(G) = \Delta$. To prove this, first color the edges with colors $1, 2, \ldots, 8$ such that no two adjacent edges receive the same color, and color the vertices with colors $7, 8, 9, 10$ so that no two adjacent vertices receive the same color. Then recolor every edge e whose color is either 7 or 8 with one of the at least two colors from $\{7, 8, 9, 10\}$ not occurring on either vertex incident to e. That such a recoloring is possible is a special, and easy, case of the theorem of V.G. Vizing and independently P. Erdős, A.L. Rubin, and H. Taylor (Theorem 35 in Chapter 1) generalizing Brooks' theorem to list colorings: The list-chromatic number of any graph is at most as large as its maximum degree, except if the graph is complete or an odd cycle. This completes the proof for $\Delta = 8$. The argument can immediately be generalized, also for nonplanar multigraphs, to a proof of the statement that $\chi(G) \leq 4$ implies

$$\chi''(G) \leq \chi'(G) + 2.$$

The idea of the proof above for $\Delta = 8$ seems due to Yap [1989b], but a full proof may have been first presented by Andersen [1993]. Yap [1989b] noted that the remaining open cases $\Delta = 6$ and 7 of the total coloring conjecture for simple planar graphs would follow from Vizing's planar graph conjecture (see Problem 12.7) that if $\Delta \geq 6$, then $\chi'(G) = \Delta$. With $\Delta = 6$, Borodin [1989b] has obtained $\chi''(G) \leq 9$.

McDiarmid and Reed [1993] proved that the total coloring conjecture of Behzad and Vizing is in a very strong sense true for almost all simple graphs: There exists a constant $c < 1$ so that only a fraction $o(c^{n^2})$ of all simple graphs on n vertices are counterexamples to the conjecture having $\chi'' > \Delta + 2$.

The simple graphs G satisfying the total coloring conjecture fall into two classes. Type 1 graphs have $\chi''(G) = \Delta(G) + 1$, whereas type 2 graphs have $\chi''(G) = \Delta(G) + 2$. The question of distinguishing between the two types has been addressed by Chetwynd and Hilton [1988b] and Hilton [1990, 1991]. Chetwynd and Hilton

[1988b] proposed a conjecture (Conjecture 1 in their paper) to do so for graphs of high maximum degree; however, the conjecture fails already for $\Delta(G) = |V(G)| - 2$, as proved by Chen and Fu [1992]. (But perhaps the examples of Chen and Fu are the only counterexamples when $\Delta(G) > \frac{1}{2}|V(G)|$; this has very recently been proposed in a manuscript "Totally critical graphs and the conformability conjecture" by G.M. Hamilton, A.J.W. Hilton, and H.R. Hind.)

Sánchez-Arroyo [1989a, 1989b] proved that it is ***NP***-complete to decide for a given graph G if $\chi''(G) = \Delta(G) + 1$. McDiarmid and Sánchez-Arroyo [1994] proved that for every fixed $k \geq 3$ it is even ***NP***-complete to decide for a given k-regular bipartite graph G if $\chi''(G) = k + 1$.

Chetwynd [1990] gave a survey on total coloring, and Häggkvist and Chetwynd [1992] wrote a detailed survey on list coloring and total coloring.

4.10. EQUITABLE COLORING

The equitable chromatic number, $\chi_{\text{Eq}}(G)$, is defined as the least k for which G has a k-coloring in which the sizes of the color classes differ by at most one. If G is connected, is $\chi_{\text{Eq}}(G)$ at most the maximum degree $\Delta(G)$, unless G is an odd cycle or complete? ■

The question is due to Meyer [1973]. An affirmative answer would extend the famous theorem of Brooks [1941] (Theorem 11 in Chapter 1). Hajnal and Szemerédi [1970] proved that $\chi_{\text{Eq}}(G) \leq \Delta(G) + 1$ for all G (Theorem 13 in Chapter 1).

Lih and Wu [1991] have shown that the answer is affirmative for all connected bipartite graphs. Chen and Lih [1994] determined the equitable chromatic number of any tree. If the answer is negative in general, one might ask for a characterization of those G for which $\chi_{\text{Eq}}(G) = \Delta(G) + 1$.

4.11. ACYCLIC COLORING

The acyclic chromatic number, $a(G)$, is defined as the least k for which G has a usual k-coloring, so that, in addition, the union of any two color classes induces a subgraph of G without cycles. Find a best possible upper bound for $a(G)$ in terms of the maximum degree Δ of G. ■

The acyclic chromatic number was introduced by Grünbaum [1973] and was studied, in particular for graphs embedded on surfaces, by Albertson and Berman [1976] and Borodin [1979a, 1993a] (see Theorem 8 in Chapter 1), among several others. Grünbaum [1973] noted that it is easy to obtain bounds for $a(G)$ that are quadratic in Δ, and he asked if $a(G) \leq \Delta + 1$. Albertson and Berman [1976] reported that this was answered in the negative by P. Erdős, who proved by probabilistic methods the existence of graphs G with $a(G) \geq \Delta^{4/3-\varepsilon}$.

Albertson and Berman [1976] mentioned a conjecture of Erdős that $a(G) = o(\Delta^2)$. This conjecture was proved by Alon, McDiarmid, and Reed [1991], who showed probabilistically that there exists a constant $c_1 > 0$ such that $a(G) \leq c_1 \cdot \Delta^{4/3}$. On the other hand, they proved that there exists a constant $c_2 > 0$ such that there exist graphs of maximum degree Δ and acyclic chromatic number larger than $c_2 \cdot \Delta^{4/3} \cdot (\log \Delta)^{-1/3}$ for infinitely many Δ. Alon, McDiarmid, and Reed remarked that they would suspect the upper bound to be of best possible magnitude, but that no families of graphs G have been constructed for which $a(G)$ grows this rapidly as a function of Δ.

For specific values of Δ the problem does not seem very much studied. Clearly, $a(G) \leq 2$ if and only if G is a forest. Grünbaum [1973] proved that $a(G) \leq 4$ for $\Delta = 3$, and $a(G) \leq 6$ for $\Delta = 4$, and he remarked that the former bound is best possible, but that he expected that the latter could be improved. The bound $a(G) \leq 5$ for $\Delta = 4$ was indeed proved by Burstein [1979], who mentioned that this result had also been obtained by A.V. Kostochka.

Answering a question of Albertson and Berman [1976], Kostochka [1978b] proved in his doctoral thesis that it is an ***NP***-complete problem to decide if a given G satisfies $a(G) \leq 3$. In fact, the problem remains ***NP***-complete even if G is restricted to be planar and bipartite with $\text{col}(G) \leq 3$. Alon, McDiarmid, and Reed [1991] remarked that there is a simple polynomial algorithm for coloring G acyclically with Δ^2 colors, and that it would be interesting to find a polynomial algorithm for coloring acyclically with $o(\Delta^2)$ colors.

■ 4.12. MELNIKOV'S VALENCY-VARIETY PROBLEM

The valency-variety, or degree-variety, $w(G)$, of a graph G is the number of different degrees in G. Is

$$\chi(G) \geq \left\lceil \frac{\lfloor w(G)/2 \rfloor}{|V(G)| - w(G)} \right\rceil + 1?$$

■

This problem is due to L.S. Melnikov and was mentioned by Vizing [1968] and by Zykov [1968a]. The suggested lower bound would be best possible as shown by Melnikov. A best possible upper bound established by Nettleton [1960] and Dirac [1964b] is

$$\chi(G) \leq |V(G)| - \lfloor w(G)/2 \rfloor .$$

■ 4.13. INDUCED-ODD DEGREE SUBGRAPHS

Let $f(G)$ denote the largest cardinality of a set of vertices $W \subseteq V(G)$ such that the induced subgraph $G[W]$ has only odd vertex degrees. If G

is without isolated vertices, is

$$\chi(G) \geq \frac{|V(G)|}{f(G)} ?$$

■

The question was asked by Scott [1992], who proved that for any graph G without isolated vertices,

$$\chi(G) \geq \frac{|V(G)|}{2f(G)} .$$

The bound suggested above would be the best possible, as shown by the graph obtained from the complete 6-graph by removing the edges of a perfect matching.

■ 4.14. STRONG CHROMATIC NUMBER

If the number $|V(G)|$ of vertices of G is divisible by the number $k \geq 1$, then G is said to be strongly k-colorable if every partition of $V(G)$ into sets of size k allows a proper k-coloring of G such that each set of the partition contains a vertex from every color class. If $|V(G)|$ is not divisible by k, G is defined to be strongly k-colorable if the graph obtained by adding $k \cdot \lceil |V(G)|/k \rceil - |V(G)|$ new isolated vertices to G is strongly k-colorable. For any graph G, the strong chromatic number, $s\chi(G)$, of G is the smallest number k for which G is strongly k-colorable. What are the best possible bounds on $s\chi(G)$ in terms of the maximum degree $\Delta(G)$ of G? ■

The problem was introduced and studied by Alon [1988] and independently by Fellows [1990]. They both obtained exponential upper bounds on $s\chi(G)$ in terms of $\Delta(G)$. A much better bound was later established by Alon [1992], who showed that there exists a constant c such that $s\chi(G) \leq c \cdot \Delta(G)$, with a proof suggesting a rough estimate of $c \approx 10^8$. Alon [1988] studied the problem of finding a lower bound for $s\chi(G)$ and proved that for any $\Delta \geq 1$ there exists a graph G with $\Delta(G) = \Delta$ and $s\chi(G) > 3\lfloor \Delta/2 \rfloor$, hence implying that $c > 3/2$. Alon [1992] remarked that the actual value of c is probably much closer to 3/2 than to the estimated upper bound. The lower bound has, however, been improved by Fleischner and Stiebitz [1993] to $c \geq 2$.

A particularly well-studied special case is the graph C_{3n}, the cycle of length $3n$. The question if $s\chi(C_{3n}) = 3$ for all $n \geq 1$ was asked by P. Erdős at MIT in 1987 as an extension of a problem of D.Z. Du, D.F. Hsu, and F.K. Hwang (see West [1991a] and also Fleischner and Stiebitz [1992]). Fellows [1990] reported that the question was asked independently by J. Schönheim, and that I. Schur formulated the following equivalent conjecture many years ago: For any partition of the set $\mathbf{Z}$ of integers into triples, there is another partition (S_1, S_2, S_3) of $\mathbf{Z}$ such that each S_i contains a member of each triple but no consecutive pair of integers. (It has later transpired that the problem was incorrectly attributed to Schur, due to a misunderstanding in an oral communication in which Hsu was mentioned [personal communication from Fellows in 1994].) The problem was solved in the affirmative by Fleischner and Stiebitz [1992] using the deep theorem of N. Alon and M. Tarsi (Theorem 34 in Chapter 1); a proof was also obtained by Sachs [1993] using elementary methods.

For a cycle with $4n$ vertices it was remarked by Alon [1992] that C_{4n} can be shown strongly 4-colorable using the methods of Alon [1988, 1992] or Fellows [1990]. This was also proved by W. Fernandez de la Véga. However, a characterization of graphs G with $\Delta(G) = 2$ and $s\chi(G) = 4$ seems unknown. Constructions of Gallai [1963a] show that if G has a subgraph consisting of two disjoint cycles, one of length m and one of length $2m$, where $m \geq 3$, then $s\chi(G) > 3$. J. Huang [personal communication in 1993] observed that $s\chi(G) > 3$ if G contains a cycle of length 1 modulo 3. Fleischner and Stiebitz [1993] remarked that the best possible upper bound for $s\chi(G)$ for all G with $\Delta(G) = 2$ is not known.

BIBLIOGRAPHY

[1976] Albertson M.O. and D.M. Berman. The acyclic chromatic number. In: *Proc. 7th S–E Conference on Combinatorics, Graph Theory and Computing, Baton Rouge, 1976, Congr. Num., 17*, pages 51–60, 1976.

[1988] Alon N. The linear arboricity of a graph. *Israel J. Math.* **62**, 311–325, 1988.

[1992] Alon N. The strong chromatic number of a graph. *Random Structures Algorithms* **3**, 1–7, 1992.

[1991] Alon N., C.J.H. McDiarmid, and B.A. Reed. Acyclic coloring of graphs. *Random Structures Algorithms* **2**, 277–288, 1991.

[1993] Andersen L. *Total colouring of simple graphs* (in Danish). Master's thesis, University of Aalborg, 1993.

[1965] Behzad M. *Graphs and their chromatic numbers*. Ph.D. thesis, Michigan State University, 1965.

[1967] Behzad M., G. Chartrand, and J.K. Cooper Jr. The colour numbers of complete graphs. *J. London Math. Soc.* **42**, 225–228, 1967.

[1984] Beutelspacher A. and P.R. Hering. Minimal graphs for which the chromatic number equals the maximal degree. *Ars Combin.* **18**, 201–216, 1984.

[1978a] Bollobás B. *Extremal Graph Theory*. Academic Press, 1978.

[1978c] Bollobás B. Problem. In: J.-C. Bermond, J.-C. Fournier, M. Las Vergnas, and D. Sotteau, editors, *Problèmes Combinatoires et Théorie des Graphes*, page 437. Édition Centre National Recherche Scientifique, 1978.

[1976] Bollobás B. and S.E. Eldridge. Problem. In: *Proc. Fifth British Combinatorial Conference Aberdeen, 1975*, page 690. Utilitas Mathematics Publishing Company, 1976.

[1978] Bollobás B. and S.E. Eldridge. Packings of graphs and applications to computational complexity. *J. Combin. Theory Ser. B* **25**, 105–124, 1978.

[1975] Bollobás B. and F. Harary. Point-arboricity critical graphs exist. *J. London Math. Soc. (2)* **12**, 97–102, 1975.

[1979a] Borodin O.V. On acyclic colorings of planar graphs. *Discrete Math.* **25**, 211–236, 1979.

[1989b] Borodin O.V. On the total coloring of planar graphs. *J. Reine Angew. Math.* **394**, 180–185, 1989.

[1993a] Borodin O.V. Four problems on plane graphs raised by Branko Grünbaum. *Contemporary Math.* **147**, 149–156, 1993.

[1977] Borodin O.V. and A.V. Kostochka. On an upper bound of a graph's chromatic number, depending on the graph's degree and density. *J. Combin. Theory Ser. B* **23**, 247–250, 1977.

[1941] Brooks R.L. On colouring the nodes of a network. *Proc. Cambridge Phil. Soc.* **37**, 194–197, 1941.

[1979] Burstein M.I. Every 4–valent graph has an acyclic 5-coloring (in Russian). *Soobšč. Akad. Nauk Gruzin. SSR* **93**, 21–24, 1979 (Georgian and English summaries).

[1974] Catlin P.A. Subgraphs of graphs I. *Discrete Math.* **10**, 225–233, 1974.

[1976] Catlin P.A. *Embedding subgraphs and coloring graphs under extremal degree conditions.* Ph.D. thesis, Ohio State University, 1976.

[1977] Catlin P.A. Embedding subgraphs under extremal degree conditions. In: *Proc. 8th S–E Conference on Combinatorics, Graph Theory and Computing, Baton Rouge, 1977, Congr. Num., 19*, pages 139–145, 1977.

[1978a] Catlin P.A. A bound on the chromatic number of a graph. *Discrete Math.* **22**, 81–83, 1978.

[1978b] Catlin P.A. Another bound on the chromatic number of a graph. *Discrete Math.* **24**, 1–6, 1978.

[1979] Catlin P.A. Hajós' graph-coloring conjecture: variations and counterexamples. *J. Combin. Theory Ser. B* **26**, 268–274, 1979.

[1980] Catlin P.A. On the Hajnal–Szemerédi theorem on disjoint cliques. *Utilitas Math.* **17**, 163–177, 1980.

[1992] Chen B.-L. and H.-L. Fu. Total colorings of graphs of order $2n$ having maximum degree $2n - 2^*$. *Graphs Combin.* **8**, 119–123, 1992.

[1994] Chen B.-L. and K.-W. Lih. Equitable coloring of trees. *J. Combin. Theory Ser. B.* **61**, 83–87, 1994.

[1990] Chetwynd A.G. Total colorings of graphs. In: R. Nelson and R.J. Wilson, editors, *Graph Colourings*, pages 65–78. Longman, 1990.

[1988b] Chetwynd A.G. and A.J.W. Hilton. Some refinements of the total chromatic number conjecture. In: *Proc. 19th S–E Conference on Combinatorics, Graph Theory and Computing, Boca Raton, Congr. Num., 66*, pages 195–216, 1988.

[1991] Chetwynd A.G., A.J.W. Hilton, and Zhao Cheng. The total chromatic number of graphs of high minimum degree. *J. London Math. Soc. (2)* **44**, 193–202, 1991.

[1979] Christen C.A. and S.M. Selkow. Some perfect coloring properties of graphs. *J. Combin. Theory Ser. B* **27**, 49–59, 1979.

[1952c] Dirac G.A. Some theorems on abstract graphs. *Proc. London Math. Soc. (3)* **2**, 69–81, 1952.

[1964b] Dirac G.A. Valency-variety and chromatic number of abstract graphs. *Wiss. Z. Martin-Luther-Univ. Halle–Wittenberg Math.-Natur. Reihe* **13**, 59–64, 1964.

[1966] Erdös P. and A. Hajnal. On chromatic number of graphs and set-systems. *Acta Math. Acad. Sci. Hungar.* **17**, 61–99, 1966.

[1994] Fan G. and R. Häggkvist. The square of a Hamiltonian cycle. *SIAM J. Discrete Math.* **7**, 203–212, 1994.

[1993] Fan G. and H.A. Kierstead. The square of paths and cycles. Manuscript, 1993.

[1991] Faudree R.J., R.J. Gould, M.S. Jacobson, and R.H. Schelp. On a problem of Paul Seymour. In: V.R. Kulli, editor, *Recent Advances in Graph Theory*, pages 197–215. Vishwa International Publishers, 1991.

[1990] Fellows M.R. Transversals of vertex partitions in graphs. *SIAM J. Discrete Math.* **3**, 206–215, 1990.

[1969] Finck H.-J. and H. Sachs. Über eine von H.S. Wilf angegebene Schranke für die chromatische Zahl endlicher Graphen. *Math. Nachr.* **39**, 373–386, 1969.

[1992] Fleischner H. and M. Stiebitz. A solution to a colouring problem of P. Erdös. *Discrete Math.* **101**, 39–48, 1992.

[1993] Fleischner H. and M. Stiebitz. Some remarks on the cycle plus triangles problem. Manuscript, May 1993.

[1963a] Gallai T. Kritische Graphen I. *Publ. Math. Inst. Hungar. Acad. Sci.* **8**, 165–192, 1963.

[1976] Garey M.R., D.S. Johnson, and L.J. Stockmeyer. Some simplified ***NP***-complete graph problems. *Theoret. Comput. Sci.* **1**, 237–267, 1976.

[1990] Gasparyan G.S. and S.E. Markosyan. ω-perfect graphs (in Russian). *Kibernetika (Kiev)* no.2, 8–11, 1990. English translation in *Cybernetics* **26**, 152–156.

[1970a] Grünbaum B. A problem in graph coloring. *Amer. Math. Monthly* **77**, 1088–1092, 1970.

[1973] Grünbaum B. Acyclic colorings of planar graphs. *Israel J. Math.* **14**, 390–408, 1973.

[1975] Gyárfás A. On Ramsey covering numbers. In: A. Hajnal, R. Rado, and V.T. Sós, editors, *Infinite and Finite Sets*, volume 10 of *Colloquia Mathematica Societatis János Bolyai*, pages 801–816. North-Holland, 1975.

[1992] Häggkvist R. and A.G. Chetwynd. Some upper bounds on the total and list chromatic numbers of multigraphs. *J. Graph Theory* **16**, 503–516, 1992.

[1970] Hajnal A. and E. Szemerédi. Proof of a conjecture of Erdős. In: P. Erdős, A. Rényi, and V.T. Sós, editors, *Combinatorial Theory and Its Applications, Vol. II*, volume 4 of *Colloquia Mathematica Societatis János Bolyai*, pages 601–623. North-Holland, 1970.

[1967] Halin R. Unterteilungen vollständiger Graphen in Graphen mit unendlicher chromatischer Zahl. *Abh. Math. Sem. Univ. Hamburg* **31**, 156–165, 1967.

[1990] Hilton A.J.W. A total-chromatic number analogue of Plantholt's theorem. *Discrete Math.* **79**, 169–175, 1989/90.

[1991] Hilton A.J.W. The total chromatic number of nearly complete bipartite graphs. *J. Combin. Theory Ser. B* **52**, 9–19, 1991.

[1993] Hilton A.J.W. and H.R. Hind. The total chromatic number of graphs having large maximum degree. *Discrete Math.* **117**, 127–140, 1993.

[1988] Hind H.R. *Restricted edge-colourings*. Ph.D. thesis, Peterhouse College, Cambridge, 1988.

[1990] Hind H.R. An improved bound for the total chromatic number of a graph. *Graphs Combin.* **6**, 153–159, 1990.

[1992] Hind H.R. An upper bound for the total chromatic number of dense graphs. *J. Graph Theory* **16**, 197–203, 1992.

[1981] Holyer I. The ***NP***-completeness of edge-coloring. *SIAM J. Comput.* **10**, 718–720, 1981.

[1990] Kierstead H.A. and S.G. Penrice. Recent results on a conjecture of Gyárfás. In: *Proc. 21st S–E Conference on Combinatorics, Graph Theory and Computing, Baton Rouge, 1990, Congr. Num., 79*, pages 182–186, 1990.

[1992] Klein R. *On* $(m_1, m_2, \ldots, m_k)$*-composed graphs*. Ph.D. Thesis, Tel Aviv University, 1992.

[1991] Klein R. and J. Schönheim. Decomposition of K_n into two graphs, respectively m_1- and m_2-degenerate, and colorability of graphs having such a decomposition. Technical Report 91/107 of the series Sonderforschungsbereich 343, Bielefeld University, 1991.

[1993] Klein R. and J. Schönheim. Decomposition of K_n into degenerate graphs. In: T.H. Ku and H.P. Yap, editors, *Combinatorics and Graph Theory. Proc. Spring School and International Conference on Combinatorics, Hefei, China, 6–27 April 1992*, pages 141–155. World Scientific Publishing Co., 1993.

[1977a] Kostochka A.V. An analogue of Shannon's estimate for complete colorings (in Russian). *Metody Diskret. Analiz.* **30**, 13–22, 1977.

[1977b] Kostochka A.V. The total coloring of a multigraph with maximal degree 4. *Discrete Math.* **17**, 161–163, 1977.

[1978a] Kostochka A.V. Degree, girth and chromatic number. In: A. Hajnal and V.T. Sós, editors, *Combinatorics*, volume 18 of *Colloquia Mathematica Societatis János Bolyai*, pages 679–696. North-Holland, 1978.

[1978b] Kostochka A.V. *Upper bounds of chromatic functions of graphs* (in Russian). Doctoral Thesis, Novosibirsk, 1978.

[1979] Kostochka A.V. Exact upper bound for the total chromatic number of a graph (in Russian). In: *Proc. 24th Int. Wiss. Koll., Tech. Hochsch. Ilmenau, 1979*, pages 33–36, 1979.

[1977] Kostochka A.V. and N.P. Mazurova. An inequality in the theory of graph coloring (in Russian). *Metody Diskret. Analiz.* **30**, 23–29, 1977.

[1975] Kronk H.V. and J. Mitchem. Critical point-arboritic graphs. *J. London Math. Soc. (2)* **9**, 459–466, 1975.

[1981] Krusenstjerna-Hafstrøm U. and B. Toft. Some remarks on Hadwiger's conjecture and its relations to a conjecture of Lovász. In: G. Chartrand, Y. Alavi, D.L. Goldsmith, L. Lesniak-Foster, and D.R. Lick, editors, *The Theory and Applications of Graphs: Proc. 4th International Graph Theory Conference, Kalamazoo, 1980*, pages 449–459. Wiley, 1981.

[1978] Lawrence J. Covering the vertex set of a graph with subgraphs of smaller degree. *Discrete Math.* **21**, 61–68, 1978.

[1970] Lick D.R. and A.T. White. *k*-degenerate graphs. *Canad. J. Math.* **22**, 1082–1096, 1970.

[1991] Lih K.-W. and P.-L. Wu. On equitable coloring of bipartite graphs. Technical report, Institute of Mathematics, Academia Sinica, Taipei, 1991.

[1993] McDiarmid C.J.H. and B.A. Reed. On total colourings of graphs. *J. Combin. Theory Ser. B* **57**, 122–130, 1993.

[1994] McDiarmid C.J.H. and A. Sánchez-Arroyo. Total colouring regular bipartite graphs is ***NP***-hard. *Discrete Math.* **124**, 155–162, 1994.

[1968] Matula D.W. A min-max theorem for graphs with application to graph coloring. *SIAM Rev.* **10**, 481–482, 1968.

[1973] Meyer W. Equitable coloring. *Amer. Math. Monthly* **80**, 920–922, 1973.

[1976] Mihók P. On the point-arboricity critical graphs. In: *Graphs, Hypergraphs and Block Systems*, pages 155–161. Zielona Góra, 1976.

[1981a] Mihók P. On the structure of the point arboricity critical graphs. *Math. Slovaca* **31**, 101–106, 1981.

[1981b] Mihók P. On graphs critical with respect to vertex partition numbers. *Discrete Math.* **37**, 123–126, 1981.

[1983] Mozhan N.N. The chromatic number of graphs with a density not exceeding two-thirds of the maximal degree (in Russian). *Metody Diskret. Analiz.* **39**, 52–65, 1983.

[1960] Nettleton R.E. Some generalized theorems on connectivity. *Canad. J. Math.* **12**, 546–554, 1960.

[1963] Pósa L. On the circuits of finite graphs. *Magyar Tud. Akad. Mat. Kutató Int. Közl.* **8**, 355–361, 1963.

[1990] Reed B.A. Perfection, parity, planarity and packing paths. In: R. Kannan and W.R. Pulleyblank, editors, *Integer Programming and Combinatorial Optimization*, pages 407–419. University of Waterloo Press, 1990.

[1971] Rosenfeld M. On the total coloring of certain graphs. *Israel J. Math.* **9**, 396–402, 1971.

[1993] Sachs H. Elementary proof of the cycle-plus-triangles theorem. In: D. Miklós, V.T. Sós, and T. Szőnyi, editors, *Combinatorics: Paul Erdős Is Eighty*, volume 1 of *Bolyai Society Mathematical Studies*, pages 347–359. János Bolyai Mathematical Society, 1993. Republished by GERAD, Montréal, 1994.

[1989a] Sánchez-Arroyo A. *Total colourings and complexity*. Master's thesis, University of Oxford, 1989.

[1989b] Sánchez-Arroyo A. Determining the total colouring number is ***NP***-hard. *Discrete Math.* **78**, 315–319, 1989.

[1993] Sánchez-Arroyo A. A new upper bound for total colourings of graphs. Manuscript, 1993.

[1993] Sauer N.W. Vertex partition problems. In: D. Miklós, V.T. Sós, and T. Szőnyi, editors, *Combinatorics: Paul Erdős Is Eighty*, volume 1 of *Bolyai Society Mathematical Studies*, pages 361–377. János Bolyai Mathematical Society, 1993.

[1978] Sauer N.W. and J.H. Spencer. Edge-disjoint placements of graphs. *J. Combin. Theory Ser. B* **25**, 295–302, 1978.

[1992] Scott A.D. Large induced subgraphs with all degrees odd. *Combin. Probab. Comput.* **1**, 335–349, 1992.

[1974b] Seymour P.D. Problem 3. In: T.P. McDonough and V.C. Mavron, editors, *Combinatorics*, page 201. Cambridge University Press, 1974.

[1976] Simões-Pereira J.M.S. A survey of k-degenerate graphs. *Graph Theory Newslett.* **5**(6), 1–7, 1976.

[1993] Stiebitz M. The forest plus star colouring problem. Technical report, Institute of Mathematics, Technical University of Ilmenau, 1993. To appear in *Discrete Math.*

[1981] Sumner D.P. Subtrees of a graph and chromatic number. In: G. Chartrand, Y. Alavi, D.L. Goldsmith, L. Lesniak-Foster, and D.R. Lick, editors, *The Theory and Applications of Graphs: Proc. 4th International Graph Theory Conference, Kalamazoo, 1980*, pages 557–576. Wiley 1981.

[1968] Szekeres G. and H.S. Wilf. An inequality for the chromatic number of a graph. *J. Combin. Theory* **4**, 1–3, 1968.

[1971] Vijayaditya N. On total chromatic number of a graph. *J. London Math. Soc. (2)* **3**, 405–408, 1971.

[1964] Vizing V.G. On an estimate of the chromatic class of a p-graph (in Russian). *Metody Diskret. Analiz.* **3**, 25–30, 1964.

[1965b] Vizing V.G. The chromatic class of a multigraph (in Russian). *Kibernetika (Kiev)* no.3, 29–39, 1965. English translation in *Cybernetics* **1**, 32–41.

[1968] Vizing V.G. Some unsolved problems in graph theory (in Russian). *Uspekhi Mat. Nauk* **23**, 117–134, 1968. English translation in *Russian Math. Surveys* **23**, 125–141.

[1980] Wagon S. A bound on the chromatic number of graphs without certain induced subgraphs. *J. Combin. Theory Ser. B* **29**, 345–346, 1980.

[1991a] West D.B. Open problems. *SIAM J. Discrete Math. Newslett.* **1**(3), 9–12, 1991.

[1989a] Yap H.P. Total colourings of graphs. *Bull. London Math. Soc.* **21**, 159–163, 1989.

[1989b] Yap H.P. Total-colourings of graphs. Manuscript, 1989.

[1989] Yap H.P., J.F. Wang, and Z.F. Zhang. Total chromatic number of graphs of high degree. *J. Austral. Math. Soc. Ser. A* **47**, 445–452, 1989.

[1992] Yap H.P., J.F. Wang, and Z.F. Zhang. Total chromatic number of graphs of high degree II. *J. Austral. Math. Soc. Ser. A* **53**, 219–228, 1992.

[1968a] Zykov A.A. Problem 11. In: H. Sachs, H.-J. Voss, and H. Walther, editors, *Beiträge zur Graphentheorie vorgetragen auf dem Internationalen Kolloquium in Manebach DDR vom 9.–12. Mai 1967*, page 228. B.G. Teubner, 1968.

[1968b] Zykov A.A. Problem 12. In: H. Sachs, H.-J. Voss, and H. Walther, editors, *Beiträge zur Graphentheorie vorgetragen auf dem Internationalen Kolloquium in Manebach DDR vom 9.–12. Mai 1967*, page 228. B.G. Teubner, 1968.

5

Critical Graphs

■ 5.1. CRITICAL GRAPHS WITH MANY EDGES

Let $F_k(n)$ denote the maximum number of edges possible in a k-critical graph ($k \geq 4$) on n vertices.

Does

$$\lim_{n \to \infty} \frac{F_k(n)}{n^2}$$

exist for all $k \geq 4$?

Is $F_6(n) = n^2/4 + n$ for $n \equiv 2$ modulo 4?

Does there exist a constant $c > 1/16$ such that $F_4(n) \geq c \cdot n^2$ for infinitely many values of n? ■

Critical graphs were first defined and used by G.A. Dirac when he studied for his Ph.D. at the University of London under the supervision of R. Rado. In 1949 he told P. Erdős about his new concept, and Erdős' first reaction was to ask for the maximum number of edges in critical graphs (see Erdős [1985, 1989] and Toft [1970a]). Dirac immediately observed that a 6-critical graph with many edges can be obtained by completely joining two disjoint odd cycles of the same length (see Dirac [1952a]). If G has n vertices, the number of edges is $n^2/4 + n$. To this day no better 6-critical example has been obtained for any value of n, nor has it been proved that the graphs constructed by Dirac are the best possible (except for small values of n). We conjecture that the graphs of Dirac are not the best possible. If it is required only that $\chi(G - x) < \chi(G) = 6$ for all vertices x of G (i.e., G is 6-vertex-critical), then it is known that there are such graphs on n vertices and at least $3n^2/10$ edges (see Toft [1978]).

The bounds $F_4(n) > n^2/16$ for all values of n with $n \geq 4$ and $n \neq 5$, and $F_5(n) > 4n^2/31$ for all values of n with $n \geq 5$ and $n \neq 6$, were shown constructively by Toft [1970a]. Again, no better constructions have been found. However, it seems likely that the constant $4/31$ for 5-critical graphs can be improved—for 5-vertex-critical graphs it can be improved to $1/4$; this can be obtained from two odd cycles of equal lengths using the result of Toft [1978] that a graph is the complement of a vertex-critical graph if and only if all its blocks are. In the 4-critical graphs with many edges constructed by Toft [1970a] most edges are concentrated in bipartite subgraphs, but V. Rödl and Stiebitz [1987b] proved that this need not be so.

The corresponding question for k-critical r-uniform hypergraphs was studied by Toft [1975a], who proved that the correct order of magnitude is $O(n^r)$ for $k \geq 4$ and gave for $k = 3$ the lower bound $c \cdot n^{r-1}$, where $c > 0$ is a constant. Lovász [1976] subsequently gave an interesting linear algebra proof that the correct order of magnitude is indeed $O(n^{r-1})$ for $k = 3$.

5.2. MINIMUM DEGREE OF 4- AND 5-CRITICAL GRAPHS

Let $\delta_k(n)$ be the maximum possible minimum degree in a k-critical graph on n vertices. What is the order of magnitude of $\delta_4(n)$? In particular does there exist a constant $c > 0$ such that $\delta_4(n) \geq cn$? What is the order of magnitude of $\delta_5(n)$? Do r-regular 4-critical graphs exist for all $r \geq 3$? ■

Simonovits [1972] and Toft [1972b] independently proved that $\delta_4(n) \geq cn^{1/3}$ by construction. This may be the best possible order of magnitude. The "in particular"-question above is due to P. Erdős (see Erdős [1985, 1989]). Erdős posed the question about $\delta_4(n)$ when G.A. Dirac in 1949 presented him with examples of 6-critical graphs obtained by completely joining two odd cycles of the same length, showing $\delta_6(n) \geq n/2$. Perhaps one should first try to look at the problem of determining $\delta_k(n)$ for $k = 5$—it might be easier than for $k = 4$.

Erdős [1989] asked the question about the existence of r-regular 4-critical graphs and he conjectured that the answer is affirmative. He also remarked that no such graph seems known for any value of $r \geq 6$. By the theorem of R.L. Brooks (Theorem 11 in Chapter 1), the only 3-regular 4-critical graph is the complete graph K_4. Gallai [1963a] constructed the first examples of 4-regular 4-critical graphs, in fact an infinite family of these. Later constructions were due to Koester [1985b, 1990] (of planar examples) and Youngs [1992].

A triangle-free edge-transitive 4-regular 4-critical graph G was discovered by Jensen and Royle [1989]. The vertices of G are numbered $1, 2, \ldots, 13$ with ij an edge of G if and only if $i - j \equiv 1, 5, 8$, or 12 modulo 13. G has the curious property of being embeddable on the torus such that the toroidal dual graph G^* is isomorphic to G. G is not embeddable on the Klein bottle. Five-regular 4-critical graphs seem more scarce. One example has been found by T.R. Jensen. The vertices are the integers $1, 2, \ldots, 24$. The edge set consists of the pairs of vertices with difference 5 or 12, the pairs of even vertices with difference 2, and the pairs of odd vertices with difference 4. (All differences are taken modulo 24.)

5.3. CRITICAL GRAPHS WITH FEW EDGES

Let $f_k(n)$ denote the minimum number of edges possible in a k-critical graph ($k \geq 4$) on n vertices ($n \geq k+2$). What is the value of $f_k(n)$? For $3 \leq s \leq k$, let $R(k,s)$ denote the largest integer such that every k-critical graph G without a complete s-graph as a subgraph satisfies

$$2 \cdot |E(G)| \geq (k-1) \cdot |V(G)| + R(k,s).$$

Determine $R(k,s)$. ■

The natural first question asks for an extension of the theorem of R.L. Brooks (Theorem 11 in Chapter 1), which implies that

$$2f_k(n) \geq (k-1)n + 1.$$

Dirac [1957c] proved (Theorem 16 in Chapter 1)

$$2f_k(n) \geq (k-1)n + (k-3).$$

Dirac's proof was rather long. Shorter and more elegant proofs were found by Kronk and Mitchem [1972a] and Weinstein [1975]. Dirac [1974] also gave a complete description of the extremal cases.

Gallai [1963a] proved that the second term tends to infinity with n:

$$2f_k(n) \geq (k-1)n + \frac{k-3}{k^2-3}n$$

(Theorem 17 in Chapter 1). Moreover, Gallai [1963b] determined the exact value of $f_k(n)$ for $n \leq 2k-1$. Ore [1967] suggested the following: If G is k-critical on n vertices with a minimum number of edges, then a k-critical graph obtained from G and K_k by the construction of Hajós [1961] (see Section 1.4) is perhaps a k-critical graph on $n+k-1$ vertices with a minimum number of edges? This would imply that $f_k(n+k-1) = f_k(n) + \frac{1}{2}k(k-1) - 1$.

For $k = 4$, the suggestion of Ore together with $f_4(4) = 6, f_4(6) = 10, f_4(7) = 11$, and $f_4(8) = 13$ (see Toft [1974b]) would imply equality in

$$f_4(n) \leq \left\lfloor \frac{5}{3}n \right\rfloor.$$

In general, the values of $f_k(n)$ for $n \leq 2k$ and the suggestion of Ore imply a suggested value of $f_k(n)$ for all n. However, $f_k(2k)$ is not known for $k \geq 5$.

The same question may be asked for k-critical graphs with no K_ℓ as a subgraph, ℓ fixed, $3 \leq \ell < k$. At the International Conference on Combinatorics in Keszthely, Hungary, July 1993, J. Schönheim asked this question in general, and he also asked in particular: Is it true that a 10-critical graph on n vertices without K_9 as subgraph has more than $5n - 11$ edges?

The fact that $(\lim_{n\to\infty} f_k(n)/n)$ exists was mentioned by Liu [1976], but the argument goes back, via H.L. Abbott and L. Moser, to Fekete [1923]. We shall present the argument in the case $k = 4$. We let $f(n)$ denote $f_4(n)$.

By Hajós' construction and the preceding remarks,

$$f(n+m) \leq f(n) + f(m+1)\ (n \geq 6, m \geq 5) \tag{5.1}$$

and

$$f(n) \leq \frac{5}{3}n\ (n \geq 6). \tag{5.2}$$

Let $\varepsilon > 0$ be given, and let

$$\alpha = \liminf \frac{f(n)}{n} \leq \limsup \frac{f(n)}{n} = \beta.$$

There is a value $N \geq 5/(3\varepsilon)$ for which

$$\frac{f(N)}{N} \leq \alpha + \varepsilon.$$

Fix such an $N \geq 6$ and let n be some suitably large integer. Write $n = Nq + r$, where we may assume that

$$\max\left\{6, \frac{5(6+N)}{3\varepsilon N}\right\} \leq q \text{ and } 6 \leq r < 6 + N. \tag{5.3}$$

Then by repeated applications of 5.1 we have

$$\begin{aligned} f(n) &= f(Nq + r) \\ &\leq f(N) + f(N(q-1) + r + 1) \\ &\leq f(N) + f(N) + f(N(q-2) + r + 2) \\ &\ \vdots \\ &\leq f(N) \cdot q + f(r + q) \\ &\leq f(N) \cdot q + f(q) + f(r + 1). \end{aligned}$$

Hence by 5.2 and 5.3

$$\begin{aligned}\frac{f(n)}{n} &\leq \frac{f(N)\cdot q + f(q) + f(r+1)}{Nq+r} \\ &\leq \frac{f(N)\cdot q}{Nq} + \frac{\frac{5}{3}q}{Nq} + \frac{\frac{5}{3}(6+N)}{Nq} \\ &\leq (\alpha+\varepsilon)+\varepsilon+\varepsilon \\ &= \alpha + 3\varepsilon.\end{aligned}$$

Since this holds for all $\varepsilon > 0$ and all sufficiently large n, we conclude that $\beta \leq \alpha$, hence $\alpha = \beta$, and the desired limit exists.

The conjectured value of $(\lim_{n\to\infty} f_4(n)/n)$ is $5/3$.

As mentioned above, Dirac [1957c] proved that $R(k,k) = k - 3$ (Theorem 16 in Chapter 1), and moreover proved the generalization $R(k,s) \geq (k-3) + (k-s)$. Weinstein [1975] proved that $R(k,s) \geq (k-3) + 2(k-s)$, and that for fixed s the growth of $R(k,s)$ is in fact at least quadratic in k.

■ 5.4. FOUR-CRITICAL AMENABLE GRAPHS

Let G be a graph of chromatic number $\chi(G) = k > 1$. G is called amenable if for every nonconstant mapping $f : V(G) \rightarrow \{1,2,\ldots,k\}$ there exists a proper vertex coloring $\varphi : V(G) \rightarrow \{1,2,\ldots,k\}$ so that $\varphi(x) \neq f(x)$ for all $x \in V(G)$. Determine the minimum number, n_4, of vertices in a 4-critical nonamenable graph. ■

T. Gallai asked if all 4-critical graphs are amenable (see Toft [1970a]). However, a construction due to V. Rödl and Stiebitz [1987b] gave, as pointed out by B. Toft, examples of 4-critical nonamenable graphs. The smallest such example shows that $n_4 \leq 50$.

Brown, Kelly, Schönheim, and Woodrow [1990] proved the so far best known lower bound $n_4 \geq 10$. The obvious generalization of determining the number n_k for $k \neq 4$ has also been studied. For $k = 3$, Kelly and Kelly [1954] observed that all odd cycles are amenable, hence n_3 does not exist. B.Aa. Sørensen and Toft [1974a] proved that $n_k \leq 11k - 24$ for all $k \geq 5$.

■ 5.5. FOUR-CRITICAL DEGREE 5 PROBLEM

Let G be a 4-critical graph with a vertex x of degree 5 and let N be the set of neighbors of x in G. Let $f : N \rightarrow \{1,2,3\}$ be a nonconstant mapping. Does there exist a 3-coloring φ of $G - x$ with colors 1, 2, and 3, such that $\varphi(y) \neq f(y)$ for all y in N? ■

This question was raised by Toft [1974a], who proved that the answer is always YES for x of degree less than 5 and sometimes NO for x of degree greater than 5. For general $k \geq 5$ the answer to the corresponding question is YES for x of degree $2k - 3$ or less, and NO for x of degree $2k - 2$ or more (see Toft [1974a]); hence the only remaining unanswered case is the one above.

■ 5.6. LARGE CRITICAL SUBGRAPHS OF CRITICAL GRAPHS

For $k \geq 4$, does a large k-critical graph necessarily contain a large $(k-1)$-critical subgraph? ■

This question was asked by J. Nešetřil and V. Rödl at the International Colloquium on Finite and Infinite Sets in Keszthely in Hungary in 1973 celebrating the sixtieth birthday of P. Erdős. For $k = 4$ the answer is affirmative: Kelly and Kelly [1954] proved that every large k-critical graph contains a long cycle, and since a k-critical graph is 2-connected with odd cycles, there has also to be a long odd cycle. An alternative argument was given by Voss [1977, 1991].

For $k \geq 5$ the question is unsettled. Stiebitz [1987b] proved that if every $(k-1)$-critical subgraph of a k-critical graph G is as small as possible (i.e., is a K_{k-1}), then $G = K_k$.

■ 5.7. CRITICAL SUBGRAPH COVERING A 2-PATH

Let P be a path of length 2 in a k-critical graph ($k \geq 4$). Is there a $(k-1)$-critical subgraph containing P? ■

This question is due to Toft [1974b], who proved that for each pair of edges e_1 and e_2 in a k-critical graph there is a $(k-1)$-critical subgraph containing e_1 but not containing e_2.

An affirmative answer to the question for $k = 4$ follows from a theorem of Dirac [1964a], and was also obtained by Wessel [1981]. If the answer to the question is negative for some value of $k \geq 5$, then by the construction of Hajós [1961] (see Section 1.4) it is possible to obtain arbitrarily large k-critical graphs where all $(k-1)$-critical subgraphs are of bounded size. Hence the answer to the question of J. Nešetřil and V. Rödl above (Problem 5.6) would also be negative.

For k-vertex-critical graphs Toft [1974b] observed that for any two vertices x and y there is always an induced $(k-1)$-vertex-critical subgraph containing x but not y. An example of a 4-critical graph containing two vertices x and y not joined by any induced path of odd length, and thus not both contained in the same induced odd cycle, has been found by S. Hougardy [personal communication in 1993]. This negatively answers a question posed by Duchet and Meyniel [1989] if it is true that two distinct vertices in a k-critical graph are always joined by some induced path of odd length.

■ 5.8. NONINDUCED CRITICAL SUBGRAPHS

Let G be a k-critical graph ($k \geq 4$). If G is not a complete graph, does there exist a $(k-1)$-critical noninduced subgraph of G? ■

Stiebitz [1985] asked this question and proved the weaker result that if every $(k-1)$-critical subgraph of G is a copy of the complete graph K_{k-1}, then G is itself the complete graph K_k.

Although not difficult to prove, Stiebitz' weaker result is of interest for two different reasons. First, it shows that if all $(k-1)$-critical subgraphs of a k-critical graph are as small as possible ($k \geq 4$), then G itself is as small as possible. This is of interest in connection with the problem by J. Nešetřil and V. Rödl on the existence of large $(k-1)$-critical subgraphs of k-critical graphs (Problem 5.6). Second, it implies that the nonperfect graphs for which all proper subgraphs are perfect are precisely the odd cycles of length at least 5 (compare with the strong perfect graph conjecture as stated in Section 1.6). Thus the problem asks for generalizations of these statements.

■ 5.9. NUMBER OF CRITICAL SUBGRAPHS

If G is a k-critical graph on n vertices, is it true that G contains n distinct $(k-1)$-critical subgraphs? ■

The question was asked by T. Gallai [personal communication in 1984], and it was mentioned by Stiebitz [1985], who proved that G contains at least $\log_2 n$ distinct $(k-1)$-critical subgraphs.

The problem has a natural generalization to vertex-critical graphs: If G is k-vertex-critical, does G contain n distinct induced subgraphs that are $(k-1)$-vertex-critical? We note that an affirmative answer to this question would solve Gallai's problem above. Abbott and Zhou [1993b] improved Stiebitz' lower bound by showing that if G is k-vertex-critical, then G contains at least $(n(k-1)!)^{1/(k-1)}$ subgraphs that are $(k-1)$-vertex-critical. The results were proved by using an observation of Toft [1974b], that for any two vertices x and y of a k-vertex-critical graph there is an induced $(k-1)$-vertex-critical subgraph containing x but not y. It follows by a simple argument that G has at least $\log_2 n$ induced $(k-1)$-vertex-critical subgraphs. Abbott and Zhou [1993b] refined this argument and obtained their stronger result.

T. Gallai conjectured that every k-critical graph G on n vertices contains at most n copies of K_{k-1} as subgraphs. This conjecture was settled in the affirmative by Stiebitz [1985] for $k = 4$ and by Abbott and Zhou [1992] for all $k \geq 5$. Moreover, Abbott and Zhou [1992] proved that if G contains exactly n copies of K_{k-1}, where $k \geq 4$, then $G = K_k$, and if G is 4-critical with $n \geq 6$ and contains exactly $n-1$ triangles, then G is an odd wheel (i.e., a vertex completely joined to an odd cycle). Abbott and Zhou suggested that perhaps every k-critical graph on $n \geq k+2$ vertices contains at most $n-k+3$ complete $(k-1)$-graphs ($k \geq 5$). An odd wheel completely joined to a complete graph shows that this would be best possible. For $k \leq 7$ the suggestion of Abbott and Zhou was confirmed by Su [1994a, 1994b],who conjectured (and proved for $k \leq 7$) that every k-critical graph on $n \geq k+2$ vertices has an edge contained in at most one K_{k-1}.

■ 5.10. SUBGRAPHS OF CRITICAL GRAPHS

If H is a subgraph of some k-critical graph ($k \geq 4$), is it then true that H is a subgraph of a k-critical graph G, where $|V(G)| \leq c_k \cdot |V(H)|$ for some constant c_k depending only on k? ■

This question was posed by Stiebitz [1987b]. Greenwell and Lovász [1974] characterized the possible subgraphs of k-critical graphs, and this was done independently by Müller [1975, 1979]: A graph H is a proper subgraph of some k-critical graph if and only if H is $(k-1)$-colorable and for every edge $xy \in E(H)$ the graph $H - xy$ is $(k-1)$-colorable with x and y colored the same. In particular, any $(k-2)$-colorable graph can be used as H, and for such H there is a k-critical G such that $H \subseteq G$ and

$$|V(G)| \leq 2 \cdot |V(H)| + d_k$$

for a constant d_k depending only on k, as observed by Toft [1974a].

■ 5.11. MINIMAL CIRCUMFERENCE OF CRITICAL GRAPHS

The circumference of a graph G is the maximum length of a cycle in G. Let

$$L_k(n) = \min_G(\text{circumference of } G)$$

where the minimum is taken over all k-critical graphs G with at least n vertices. What is the order of magnitude of $L_k(n)$? ■

$L_k(n)$ is a nondecreasing function of n. Kelly and Kelly [1954] proved that, in fact, $L_k(n) \longrightarrow \infty$ as $n \longrightarrow \infty$. Concerning an upper bound, Kelly and Kelly [1954] proved that $L_4(n) \leq c \cdot (\log n)^2$. After subsequent improvements by Dirac [1955] and Read [1957], Gallai [1963a] obtained the best result so far, namely

$$L_k(n) < \frac{2(k-1)}{\log(k-2)} \cdot \log n$$

for $k \geq 4$ and infinitely many values of n.

■ 5.12. THE ERDŐS–LOVÁSZ TIHANY PROBLEM

Let G be a k-chromatic graph containing no K_k, and let a and b be natural numbers with $a, b \geq 2$ and $a + b = k + 1$. Does there exist a pair of disjoint subgraphs of G of chromatic numbers a and b, respectively? ■

The case $k = 5$ and $a = b = 3$ was asked by Erdős [1968] and generalized by L. Lovász in this way.

The question of Erdős ($k = 5$ and $a = b = 3$) was settled in the affirmative by Brown and Jung [1969], and the case $(k, a, b) = (4, 2, 3)$ can be dealt with likewise. The case $(k, a, b) = (5, 2, 4)$ has been solved by Mozhan [1986] and independently by Stiebitz [1987a, 1988]. Moreover, Stiebitz [1988] solved the cases $(k, a, b) = (6, 3, 4)$ and $(7, 3, 5)$.

For the case $a = 2$ the problem can be formulated as follows: A k-chromatic graph G is "double-critical" if for every edge xy of G the graph $G - x - y$ is $(k - 2)$-colorable. Is K_k the only double-critical graph? This formulation resembles Hadwiger's conjecture, and some (not very deep) relations between the two were exhibited in Krusenstjerna-Hafstrøm and Toft [1981].

■ 5.13. PARTIAL JOINS OF CRITICAL GRAPHS

Let two disjoint graphs G_1 and G_2 be k_1- and k_2-critical, respectively. A partial join of G_1 and G_2 is obtained from the union of G_1 and G_2 by adding a set of new edges each joining a vertex of G_1 to a vertex of G_2. Characterize the critical graphs that are partial joins of smaller critical graphs. In particular, do so when G_1 and G_2 are both odd cycles, or when G_1 is complete. ■

For complete joins, where all vertices of G_1 are joined to all vertices of G_2, the graph G obtained is $(k_1 + k_2)$-critical as observed by Dirac [1952a]. The questions on partial joins were asked by T. Gallai [personal communication in 1969], who knew examples of critical graphs being partial noncomplete joins of smaller critical graphs (see the description of Gallai's graphs given by Krusenstjerna-Hafstrøm and Toft [1981]). Gallai also observed that every critical partial join of two complete graphs must be a complete join—this is equivalent to the fact that the complement of every bipartite graph is a perfect graph!

Other examples of critical partial joins were presented by Youngs [1992]. Stiebitz and Wessel [1993] characterized the critical partial joins of a complete graph and an odd cycle.

■ 5.14. VERTEX-CRITICAL GRAPHS WITHOUT CRITICAL EDGES

Is there some positive function f so that for every $k \geq 4$ there exists a graph G satisfying

(i) $\chi(G) = k$,
(ii) $\chi(G - x) = k - 1$ for all $x \in V(G)$ and
(iii) $\chi(G - A) = k$ for all $A \subseteq E(G)$ with $|A| \leq f(n)$

where $n = |V(G)|$? If there is such an f, how fast can it grow? ■

This question was formulated by Erdős [1989]. At the time when this problem was raised, there were no known graphs G of any chromatic number k satisfying (i)–(iii) even with $f(n) = 1$, that is, vertex-critical graphs without critical edges. For $k \leq 3$, clearly no such graph exists. A 5-vertex-critical graph without critical edges was given by Brown [1992], who observed that the following graph due to Chvátal, Graham, Perold, and Whitesides [1979] will do: The vertices of G are the integers modulo 17 with i and j joined by an edge if and only if $i - j \in \{2, 6, 7, 8, 9, 10, 11, 15\}$ (addition modulo 17). Similarly constructed examples have been found by T.R. Jensen; in particular 5-vertex-critical 8-regular examples on 21 and 25 vertices, a 6-vertex-critical 10-regular example on 21 vertices, and an 8-vertex-critical 14-regular example on 29 vertices.

It remains open whether any examples exist for $k = 4$. It seems unknown if there exist examples of k-vertex-critical graphs without critical edges for infinitely many k, or examples of 5-vertex-critical graphs on n vertices and without critical edges for infinitely many n.

BIBLIOGRAPHY

[1992] Abbott H.L. and B. Zhou. On a conjecture of Gallai concerning complete subgraphs of k-critical graphs. *Discrete Math.* **100**, 223–228, 1992.

[1993b] Abbott H.L. and B. Zhou. $(k - 1)$-critical subgraphs of k-critical graphs. Manuscript, 1993. Submitted to *Combinatorica*.

[1992] Brown J.I. A vertex critical graph without critical edges. *Discrete Math.* **102**, 99–102, 1992.

[1990] Brown J.I., D. Kelly, J. Schönheim, and R.E. Woodrow. Graph coloring satisfying restraints. *Discrete Math.* **80**, 123–143, 1990.

[1969] Brown W.G. and H.A. Jung. On odd circuits in chromatic graphs. *Acta Math. Acad. Sci. Hungar.* **20**, 129–134, 1969.

[1979] Chvátal V., R.L. Graham, A.F. Perold, and S.H. Whitesides. Combinatorial designs related to the strong perfect graph conjecture. *Discrete Math.* **26**, 83–92, 1979.

[1952a] Dirac G.A. A property of 4-chromatic graphs and some remarks on critical graphs. *J. London Math. Soc.* **27**, 85–92, 1952.

[1955] Dirac G.A. Circuits in critical graphs. *Monatsh. Math.* **59**, 178–187, 1955.

[1957c] Dirac G.A. A theorem of R.L. Brooks and a conjecture of H. Hadwiger. *Proc. London Math. Soc. (3)* **7**, 161–195, 1957.

[1964a] Dirac G.A. On the structure of 5- and 6-chromatic abstract graphs. *J. Reine Angew. Math.* **214/215**, 43–52, 1964.

[1974] Dirac G.A. The number of edges in critical graphs. *J. Reine Angew. Math.* **268/269**, 150–164, 1974.

[1989] Duchet P. and H. Meyniel. Problem 6. In: L.D. Andersen, I.T. Jakobsen, C. Thomassen, B. Toft, and P.D. Vestergaard, editors, *Graph Theory in Memory of G.A. Dirac*, volume 41 of *Annals of Discrete Mathematics*, page 516. North-Holland, 1989.

[1968] Erdős P. Problem 2. In: *Theory of Graphs: Proc. Colloquium, Tihany, Hungary, 1966*, page 361. Academic Press, 1968.

[1985] Erdős P. Problems and results on chromatic numbers in finite and infinite graphs. In: Y. Alavi, G. Chartrand, L. Lesniak, D.R. Lick, and C.E. Wall, editors, *Graph Theory with Applications to Algorithms and Computer Science*, pages 201–213. Wiley, 1985.

[1989] Erdős P. On some aspects of my work with Gabriel Dirac. In: L.D. Andersen, I.T. Jakobsen, C. Thomassen, B. Toft, and P.D. Vestergaard, editors, *Graph Theory in Memory of G.A. Dirac*, volume 41 of *Annals of Discrete Mathematics*, pages 111–116. North-Holland, 1989.

[1923] Fekete M. Über die Verteilung der Wurzeln bei gewissen algebraischen Gleichungen mit ganzzahligen Koefficienten. *Math. Z.* **17**, 228–249, 1923.

[1963a] Gallai T. Kritische Graphen I. *Publ. Math. Inst. Hungar. Acad. Sci.* **8**, 165–192, 1963.

[1963b] Gallai T. Kritische Graphen II. *Publ. Math. Inst. Hungar. Acad. Sci.* **8**, 373–395, 1963.

[1974] Greenwell D. and L. Lovász. Applications of product coloring. *Acta Math. Acad. Sci. Hungar.* **25**, 335–340, 1974.

[1961] Hajós G. Über eine Konstruktion nicht n-färbbarer Graphen. *Wiss. Z. Martin-Luther-Univ. Halle–Wittenberg Math.-Natur. Reihe* **10**, 116–117, 1961.

[1989] Jensen T.R. and G.F. Royle. Small graphs of chromatic number 5: a computer search. Manuscript, 1989. To appear in *J. Graph Theory* **19**, 1995.

[1954] Kelly J.B. and L.M. Kelly. Paths and circuits in critical graphs. *Amer. J. Math.* **76**, 786–792, 1954.

[1985b] Koester G. Note to a problem of T. Gallai and G.A. Dirac. *Combinatorica* **5**, 227–228, 1985.

[1990] Koester G. 4-critical 4-valent planar graphs constructed with crowns. *Math. Scand.* **67**, 15–22, 1990.

[1972a] Kronk H.V. and J. Mitchem. On Dirac's generalization of Brooks' theorem. *Canad. J. Math.* **24**, 805–807, 1972.

[1981] Krusenstjerna-Hafstrøm U. and B. Toft. Some remarks on Hadwiger's conjecture and its relations to a conjecture of Lovász. In: G. Chartrand, Y. Alavi, D.L. Goldsmith, L. Lesniak-Foster, and D.R. Lick, editors, *The Theory and Applications of Graphs. Proc. 4th International Graph Theory Conference, Kalamazoo, 1980*, pages 449–459. Wiley, 1981.

[1976] Liu A. *Some results on hypergraphs.* Ph.D. thesis, University of Alberta, Canada, 1976.

[1976] Lovász L. Chromatic number of hypergraphs and linear algebra. *Studia Sci. Math. Hungar.* **11**, 113–114, 1976.

[1986] Mozhan N.N. On doubly critical graphs with chromatic number five (in Russian). Technical Report 14, Institute of Technology, Omsk, 1986.

[1975] Müller V. On colorable critical and uniquely colorable critical graphs. In: M. Fiedler, editor, *Recent Advances in Graph Theory: Proc. Symposium, Prague, June 1974*, pages 385–386. Academia Praha, 1975.

[1979] Müller V. On colorings of graphs without short cycles. *Discrete Math.* **26**, 165–176, 1979.

[1967] Ore O. *The Four-Color Problem.* Academic Press, 1967.

[1957] Read R.C. Maximal circuits in critical graphs. *J. London Math. Soc.* **32**, 456–462, 1957.

[1972] Simonovits M. On colour-critical graphs. *Studia Sci. Math. Hungar.* **7**, 67–81, 1972.

[1985] Stiebitz M. *Beiträge zur Theorie der färbungskritischen Graphen.* Dissertation zu Erlangung des akademischen Grades Dr.sc.nat., Technische Hochscule Ilmenau, 1985.

[1987a] Stiebitz M. K_5 is the only double-critical 5-chromatic graph. *Discrete Math.* **64**, 91–93, 1987.

[1987b] Stiebitz M. Subgraphs of colour-critical graphs. *Combinatorica* **7**, 303–312, 1987.

[1988] Stiebitz M. On k-critical n-chromatic graphs. In: A. Hajnal, L. Lovász, and V.T. Sós, editors, *Combinatorics*, volume 52 of *Colloquia Mathematica Societatis János Bolyai*, pages 509–514. North-Holland, 1988.

[1993] Stiebitz M. and W. Wessel. On colouring partial joins of a complete graph and a cycle. *Math. Nachr.* **163**, 109–116, 1993.

[1994a] Su X.-Y. On complete subgraphs of color-critical graphs. Manuscript, 1994. To appear in *Discrete Math.*

[1994b] Su X.-Y. Complete $(k-1)$-subgraphs of k-critical graphs. Manuscript, 1994.

[1970a] Toft B. On the maximal number of edges of critical k-chromatic graphs. *Studia Sci. Math. Hungar.* **5**, 461–470, 1970.

[1972b] Toft B. Two theorems on critical 4-chromatic graphs. *Studia Sci. Math. Hungar.* **7**, 83–89, 1972.

[1974a] Toft B. Color-critical graphs and hypergraphs. *J. Combin. Theory Ser. B* **16**, 145–161, 1974.

[1974b] Toft B. On critical subgraphs of colour-critical graphs. *Discrete Math.* **7**, 377–392, 1974.

[1975a] Toft B. On colour-critical hypergraphs. In: A. Hajnal, R. Rado, and V.T. Sós, editors, *Infinite and Finite Sets*, volume 10 of *Colloquia Mathematica Societatis János Bolyai*, pages 1445–1457. North-Holland, 1975.

[1978] Toft B. An investigation of colour-critical graphs with complements of low connectivity. In: B. Bollobás, editor, *Advances in Graph Theory*, volume 3 of *Annals of Discrete Mathematics*, pages 279–287. North-Holland, 1978.

[1977] Voss H.-J. Graphs with prescribed maximal subgraphs and critical chromatic graphs. *Comment. Math. Univ. Carolinae* **18**, 129–142, 1977.

[1991] Voss H.-J. *Cycles and Bridges in Graphs.* Deutscher Verlag der Wissenschaften, Kluwer Academic Publishers, 1991.

[1975] Weinstein J. Excess in critical graphs. *J. Combin. Theory Ser. B* **18**, 24–31, 1975.

[1981] Wessel W. Critical lines, critical graphs and odd cycles. Technical report, Akademie der Wissenschaften der DDR, Institut für Mathematik, 1981.

[1992] Youngs D.A. Gallai's problem on Dirac's construction. *Discrete Math.* **101**, 343–350, 1992.

6

The Conjectures of Hadwiger and Hajós

■ 6.1. HADWIGER'S CONJECTURE

Let G be a k-chromatic graph. Is it true that by successive contractions of edges, G can be contracted into a graph containing K_k as a subgraph?

Alternatively, is it true that G contains k disjoint connected subgraphs $G_1, G_2, \ldots, G_k$ such that for all i and j, where $1 \leq i < j \leq k$, there is an edge xy in G with $x \in V(G_i)$ and $y \in V(G_j)$?

A third formulation: Let $G \geq H$ mean that H is a minor of G (i.e., that H can be obtained from G by successive deletions of vertices and edges and/or contraction of edges). Let $\mathcal{G}$ be a minor-closed class of graphs (i.e., if $G \in \mathcal{G}$ and $G \geq H$, then $H \in \mathcal{G}$). Is it true that the maximum chromatic number of graphs in a minor-closed class $\mathcal{G}$ is equal to the size of a maximum complete graph in $\mathcal{G}$?

A fourth formulation: A k-chromatic graph G is contraction-critical if ($G \geq H$ and $G \neq H$) implies that $\chi(H) < k$. Is the only k-chromatic contraction-critical graph the complete k-graph K_k? ■

That the answer to each of these equivalent questions is affirmative was conjectured by Hadwiger [1943] and proved by him for $k = 3$. For $k = 4$ it was proved by Dirac [1952a]. For $k = 5$ an affirmative answer implies the four-color theorem. Wagner [1937] characterized the maximal graphs not contractible into K_5 and observed that they are all 4-colorable provided that all planar graphs are (Theorem 7 in Chapter 1). Thus Hadwiger's conjecture for $k = 5$ and the four-color theorem are equivalent. For short proofs of Wagner's theorem and the equivalence theorem, see Halin [1964] and Young [1971], respectively. The book by Ore [1967] contains a detailed chapter on Hadwiger's conjecture.

When Dirac [1952a] proved Hadwiger's conjecture for $k = 4$, he actually proved the stronger statement that any graph of minimum degree at least 3 can be contracted into a graph containing K_4 as a subgraph. Thus any graph G without such a contraction (usually such a graph is referred to as a "series-parallel" graph) has coloring number satisfying $\operatorname{col}(G) \leq 3$ and there is a simple sequential algorithm for 3-coloring G (see Theorem 12 in Chapter 1, and also the survey paper by Matula, Marble, and Isaacson [1972]). More recently, Seymour [1990] has proved that the following "almost greedy" algorithm also k-colors any series-parallel graph G for any $k \geq 3$ (a similar algorithm is due to Brelaz [1979]):

> Initially, none of the vertices of G are colored. At each step of the algorithm, some vertices have been colored, and some remain uncolored. Among the uncolored vertices, one is chosen such that the number of different colors received by its already colored neighbours in G is maximum. This vertex is then colored with any color $1, 2, \ldots, k$ not already received by any of its neighbours, and the algorithm proceeds to the next step. When all vertices of G have been colored, or when the algorithm fails in being able to assign a color to some vertex, it stops.

As indicated by Seymour, for $k > 2$ there is a bipartite graph that the algorithm may fail to k-color. He remarked that the series-parallel graphs form a surprisingly large class of graphs for which this simple algorithm always succeeds.

Dirac [1964a] proved for $k = 6$ that a 6-chromatic graph G can be contracted into a graph containing a K_6^-, a complete 6-graph with one edge missing. Mayer [1989b, 1992] initiated an attack on Hadwiger's conjecture for $k = 6$ by extending the proof methods used to solve the four-color problem (see Theorem 5 in Chapter 1). Mayer obtained a number of partial results, but not a complete proof. Robertson, Seymour, and Thomas [1993a] recently obtained a proof of Hadwiger's conjecture for $k = 6$. The result can be considered an extension of the result by Wagner [1937] (Theorem 7 in Chapter 1) that the truth of the conjecture for $k = 5$ is equivalent to the four-color theorem. Robertson, Seymour, and Thomas [1993a] showed that any minimum contraction-critical 6-chromatic graph, different from the complete graph K_6, is an "apex graph" (i.e., it has a vertex whose removal leaves a planar graph). As a consequence, the four-color theorem implies Hadwiger's conjecture for $k = 6$.

A graph is said to be "discatenable" if it can be embedded in $\mathbf{R}^3$ such that any two disjoint cycles are not concatenated (i.e., they are contained in disjoint topological balls). Sachs [1983] asked for the maximum chromatic number of discatenable graphs. As a consequence of the theorem of Robertson, Seymour, and Thomas [1993a] the statement that all discatenable graphs are 5-colorable is equivalent to the four-color theorem. For characterizations of discatenable graphs, see Robertson, Seymour, and Thomas [1993b].

Dirac [1964a] observed that contraction-critical k-chromatic graphs different from K_k have minimum degree at least k. Mayer [1989a] proved the Brooks-type theorem that not all vertices can have degree k, and this was extended by Stiebitz and Toft [1993] to the result that there must be at least $k - 4$ vertices of degree at least $k + 1$. Stiebitz and Toft also proved that no four vertices of degree k induce a K_4^- (i.e., K_4 minus an edge), thus giving an abstract extension of the planar graph diamond theorem of Birkhoff [1913] that no four vertices of degree 5 in a minimum counterexample to the four-color theorem induce a K_4^-.

Jakobsen [1970, 1971] proved for $k = 7, 8$, and 9 that a k-chromatic graph can be contracted into a graph containing a K_7^{--}, K_7^{-}, and K_7, respectively. Wagner [1964] proved the existence of a function $f(k)$ such that any $f(k)$-chromatic graph can be contracted into a graph containing a K_k. A strong result of this type follows from an extremal result of Mader [1968a]: If a graph G satisfies

$$|E(G)| \geq 8 \cdot \left\lceil \frac{k \log k}{\log 2} \right\rceil \cdot |V(G)|,$$

then G can be contracted into a graph containing a K_k. It follows that every graph of chromatic number $16 \cdot \lceil k \log k / \log 2 \rceil$ can be contracted into a graph containing a K_k. Kostochka [1982] improved Mader's extremal result, replacing $8 \cdot \lceil k \log k / \log 2 \rceil$ by $c \cdot k \cdot \sqrt{\log k / \log 2}$. Moreover, Kostochka proved that this order of magnitude is best possible. This was also noted by Fernandez de la Végа [1983] as a consequence of the calculations of Bollobás, Catlin, and Erdős [1980]. Thomason [1984] estimated a value of $c \simeq 2.68$ for k sufficiently large.

Connectivity properties of contraction-critical k-chromatic graphs have been found by Dirac [1960], Mader [1968b], and Toft [1972a]. Bollobás, Catlin, and Erdős [1980] and independently Kostochka [1982] have proved that Hadwiger's conjecture is true for almost all graphs. By joining G completely to a large complete graph, Hadwiger's conjecture holds for the graph H obtained, if and only if it holds for G. Using this very simple observation it follows that to prove Hadwiger's conjecture for all k it suffices to prove it for all k and graphs H on at most ck vertices for any fixed $c > 1$.

Vizing [1968] asked if Hadwiger's conjecture can be proved at least for line graphs of multigraphs. For line graphs of simple graphs it is not difficult to prove Hadwiger's conjecture using the theorem of V.G. Vizing (Theorem 29 in Chapter 1).

Concerning algorithms related to the general Hadwiger conjecture, N. Robertson reported (at the 11th British Combinatorial Conference at Goldsmith's College, University of London, in July 1987) on recent developments in his joint work with P.D. Seymour on graph minors. They had been able to describe an algorithm that for fixed k accepts a graph G and in low-degree polynomial time finds either

- **(a)** a K_{k+1} minor of G,
- **(b)** a k-coloring of G, or
- **(c)** a minor H of G with no K_{k+1} as a minor and no k-coloring (i.e., a counterexample to Hadwiger's conjecture).

■ 6.2. HAJÓS' CONJECTURE

Let G be a k-chromatic graph with $k = 5$ or 6. Does G contain a subdivision of K_k? (A subdivision is obtained by replacing edges by internally disjoint paths.) ■

This question is attributed to G. Hajós. For $k \leq 4$ the answer to the corresponding question is YES, as proved by Dirac [1952a]. For $k \geq 7$ the answer is NO, as pointed out by Catlin [1979]. The counterexamples of Catlin consist of five disjoint complete graphs G_1, G_2, G_3, G_4, and G_5, with G_5 and G_1 completely joined, and G_i and G_{i+1} completely joined for $i = 1$, 2, 3, and 4. Let $C(t)$ denote such a graph, with all the graphs G_i being complete t-graphs. Then $C(3)$ is a counterexample to Hajós' conjecture for $k = 8$.

The graphs described above were known prior to the paper of Catlin [1979]; they were in fact discovered and investigated by T. Gallai [personal communication in 1969] (see Krusenstjerna-Hafstrøm and Toft [1981] for an account of Gallai's results), and independently by Borodin and Kostochka [1977], who conjectured that all k-chromatic graphs of maximum degree k contain K_k as a subgraph when $k \geq 9$ (see Problem 4.8). The reason for assuming that $k \geq 9$ was Borodin's and Kostochka's knowledge of $C(3)$ as an 8-chromatic counterexample to their conjecture [personal communication from Borodin in 1993]. Also, the Catlin graphs are line graphs of so-called ring multigraphs (cycles with multiple edges). The edge-chromatic numbers of ring multigraphs (i.e., the usual chromatic numbers of their line graphs) were given in Section 4.1 of the book by Ore [1967], where they were attributed to B.L. Rothschild and J. Stemple.

Thus the main achievement of Catlin [1979] was not the discovery of the explicit graphs providing counterexamples to Hajós' conjecture, but rather, that he lead us out of a blind alley thinking that the conjectures of Hadwiger and Hajós were basically of the same nature, as had been the common belief prior to his work. The two conjectures are indeed very different. In fact, Erdős and Fajtlowicz [1981] proved that the statement "G contains a subdivision of the complete $\chi(G)$-graph" is false for almost all graphs (see also the sharp estimate of $\chi(G)/\omega_{top}(G)$ for random graphs by Bollobás and Catlin [1981], where $\omega_{top}(G)$ is the largest k for which G contains a subdivision of K_k).

Mader [1972] proved the extremal theorem that if $|E(G)| \geq 2^{k-1} \cdot |V(G)|$ holds, then G contains a subdivision of K_k. This implies that if G is $(2^k + 1)$-chromatic, then G contains a subdivision of K_k. For small values of k stronger results are known: Pelikán [1969] proved from an extremal result for K_5^- (i.e., K_5 minus an edge), that every 5-chromatic graph contains a subdivision of K_5^-. Thomassen [1974] proved that if $|V(G)| \geq 5$ and $|E(G)| \geq 4|V(G)| - 10$, then G contains a subdivision of K_5, implying that every 9-chromatic graph contains a subdivision of K_5. Dirac [1964c] had conjectured that $|E(G)| \geq 4|V(G)| - 10$ can be replaced by $|E(G)| \geq 3|V(G)| - 5$, but this remains an open problem.

6.3. THE (m, n)- AND $[m, n]$-CONJECTURES

A graph is said to have property P_m if it contains no subdivision of the complete graph K_{m+1} and no subdivision of the complete bipartite graph $K_{\lfloor (m+2)/2 \rfloor, \lceil (m+2)/2 \rceil}$. (A subdivision is obtained by replacing edges by internally disjoint paths.)

Let $\chi_n(G)$ denote the minimum number of colors in an improper coloring of the vertices of G in which each color class induces a subgraph having property P_n. For which values of m and n ($m \geq n \geq 1$) is it true that all graphs G having property P_m satisfy $\chi_n(G) \leq m - n + 1$?

Let $\chi_n'(G)$ denote the minimum number of classes in a partition of the edges of G for which each class of edges forms the edge set of a subgraph having property P_n. For which values of m and n ($m \geq n \geq 2$) is it true that all graphs G having property P_m satisfy $\chi_n'(G) \leq m - n + 1$?

What are the answers to the corresponding questions in terms of contractions rather than subdivisions? That is, in terms of the property Q_m that no K_{m+1} and no $K_{\lfloor (m+2)/2 \rfloor, \lceil (m+2)/2 \rceil}$ can be obtained by successive deletions of vertices and edges and/or contraction of edges, rather than in terms of the property P_m. ■

The conjectures that the answers to the first two questions are always affirmative, called the (m, n)-conjecture for vertex partitions and the $[m, n]$-conjecture for edge partitions, were posed by Chartrand, Geller, and Hedetniemi [1971] in an interesting paper, where they pointed out that many seemingly unrelated concepts in graph theory are closely related.

For $m = 1, 2, 3$, and 4 the properties P_m are "without edges," "without cycles," "outerplanar," and "planar," respectively. Thus χ_1 is the usual chromatic number, χ_2 the vertex arboricity, χ_3 the vertex outer thickness, and χ_4 the vertex thickness. Moreover, χ_2' is the arboricity, χ_3' the outer thickness, and χ_4' the thickness.

Chartrand, Geller, and Hedetniemi [1971] proved the (m, n)-conjecture for all ordered pairs (m, n) with $1 \leq n \leq m \leq 4$, except for the case $(m, n) = (4, 1)$, which is the four-color theorem (see Section 1.2). They also proved the $[m, n]$-conjecture for all ordered pairs $[m, n]$ with $2 \leq n \leq m \leq 4$, except the case $[m, n] = [4, 3]$. The $[4, 3]$-conjecture was recently proved by Heath [1993], who also obtained a polynomial algorithm to divide the edges of a planar graph into two outerplanar subgraphs. This is perhaps the deepest achievement in this area so far (not taking the four-color theorem into account). Finding the proof of this result was very difficult and required considerable effort [personal communication from Heath in 1993]. Heath was motivated by applications to stack and queue layouts in computer science, and he developed his proof without knowing about the paper of Chartrand, Geller, and Hedetniemi [1971].

The (m, n)- and $[m, n]$-conjectures were mentioned in the excellent survey by Woodall [1990]. (In this survey it was, however, mistakenly stated that the $[4, 3]$-conjecture had been solved by Chartrand, Geller, and Hedetniemi [1971].) Woodall [1990] remarked that perhaps the corresponding conjectures in terms of contractions, rather than subdivisions, are more likely to be true.

Indeed, it follows from arguments as those used by Erdős and Fajtlowicz [1981] that for all fixed values of n the (m, n)-conjecture is false for almost all graphs. Moreover, Hanson and Toft [1994] have proved that the type of graphs used by

Catlin [1979] to give counterexamples to Hajós' conjecture (see Problem 6.2) also give counterexamples to the (m, n)-conjecture for $n = 1$ and m sufficiently large. They asked if the smallest such counterexample is the graph $C(11)$ consisting of five disjoint complete 11-graphs G_1, G_2, G_3, G_4, and G_5, with G_5 completely joined to G_1, and G_i completely joined to G_{i+1}, $i = 1, 2, 3$, and 4. This graph seems to have property P_{27}, but $\chi(C(11)) = 28 > m - n + 1 = 27$. However, Hanson and Toft [1994] proved that $C(t)$ satisfies property P_m for $m = 2t + 2\lfloor\sqrt{t+4}\rfloor$. Thus $\chi_1(C(t)) \geq 5t/2 > m - n + 1 = m$ for $t \geq 29$. This shows constructively that the (m, n)-conjecture fails for $n = 1$ and m sufficiently large.

The case $n = 1$ of the (m, n)-contraction conjecture of Woodall [1990], that all graphs G having property Q_m satisfy $\chi_n(G) \leq m - n + 1$, may be regarded as a weaker version of Hadwiger's conjecture (see Problem 6.1). The case $(4, 1)$ of the (m, n)-contraction conjecture is precisely the four-color theorem. Hanson and Toft [1994] noted that, interestingly, the next two cases $(5, 1)$ and $(6, 1)$ seem simpler. They can be proved without resorting to the four-color theorem, using a result of Mader [1968b] that a k-chromatic graph with $k = 6$ or 7 can be contracted either into a K_k or into a k-connected graph.

■ 6.4. HADWIGER DEGREE OF A GRAPH

Let G be any graph. The Hadwiger degree $\mu(G)$ is defined as the smallest nonnegative integer such that there exists a graph $H \neq G$ with $\chi(H) \geq \chi(G)$ obtained from a subgraph G' of G by contracting $\mu(G)$ mutually disjoint, nontrivial, connected subgraphs of G' into equally many single vertices. If no such integer exists, define $\mu(G) = \infty$.

By definition, G is critical if and only if $\mu(G) \geq 1$, and G is contraction-critical if and only if $\mu(G) = \infty$. Is it true that if $\chi(G) \leq 7$ and G is not complete, then $\mu(G) \leq 1$? ■

The concept of Hadwiger degree was introduced and studied by Krusenstjerna-Hafstrøm and Toft [1981] in their remarks to the conjecture of Hadwiger [1943]. A possible formulation of Hadwiger's conjecture is the following:

Conjecture. *If $\chi(G) = k$ and $G \neq K_k$, then $\mu(G) \leq k$.*

Krusenstjerna-Hafstrøm and Toft proved the affirmative answer to the question for 4-chromatic graphs; this follows from the theorem of Krusenstjerna-Hafstrøm and Toft [1980], which states that any such graph G, where $G \neq K_4$, contains an odd cycle C without diagonals so that $G - V(C)$ is a connected graph with at least 2 vertices. Hence:

Theorem. *If $\chi(G) = 4$ and $G \neq K_4$, then $\mu(G) \leq 1$.*

They remarked that this strengthening of Hadwiger's conjecture for $k = 4$ does not follow directly from the truth of G. Hajós' conjecture for $k = 4$ (first proved by Dirac [1952a]) that every 4-chromatic graph contains a subdivision of K_4 as a subgraph.

Krusenstjerna-Hafstrøm and Toft [1981] showed that for any $n > 1$ there exists a graph G with Hadwiger degree $\mu(G) = n$. Each of their examples satisfies $\chi(G) = 5\mu(G) - 2$, thus every known example of a graph with Hadwiger degree at least 2 has chromatic number at least 8.

6.5. GRAPHS WITHOUT ODD-K_5

A graph H is called an odd-K_k if the following hold:

(a) $V(H)$ can be partitioned into k subsets $V_1, V_2, \ldots, V_k$ each of which induces a tree in H.

(b) For each pair $i, j \in \{1, 2, \ldots, k\}$ with $i \neq j$ there exists exactly one edge e_{ij} from V_i to V_j.

(c) Each cycle C in H satisfies $|E(C)| \equiv |E(C) \cap \widetilde{E}| \pmod 2$, where

$$\widetilde{E} = \{e_{ij} : i, j = 1, 2, \ldots, k; i \neq j\}.$$

If G does not contain an odd-K_5 as a subgraph, is G 4-colorable? ■

The question was raised by A.M.H. Gerards at the Fourth Bellairs Workshop on Combinatorial Optimization held in March 1993 in Holetown, Barbados. He conjectured that the answer is affirmative. The statement implies the four-color theorem but is not implied by it directly.

Gerards [personal communication in 1993] and P.D. Seymour asked more generally: Is it true that every k-chromatic graph contains an odd-K_k as a subgraph? This would be substantially stronger than the conjecture of Hadwiger [1943]. An affirmative answer for $k = 4$ was given by Catlin [1979].

6.6. SCHEME CONJECTURE

A k-scheme in a graph G consists of a subset $X = \{x_1, x_2, \ldots, x_k\}$ of k different vertices of G together with $\binom{k}{2}$ paths P_{ij} $(1 \leq i < j \leq k)$ in G, where P_{ij} joins x_i and x_j and where P_{ij} and P_{hl} are disjoint whenever i, j, h, l are all different. Does every k-chromatic graph contain a k-scheme? ■

The question was raised by P. Duchet and H. Meyniel at the Silver Jubilee Conference on Combinatorics and Optimization at the University of Waterloo in 1982 (see Duchet [1984b]).

If Hadwiger's conjecture is true, then the scheme conjecture is also true. Thus the scheme conjecture can be viewed as a weakened form of Hadwiger's conjecture. Conversely, for $k = 5$ the scheme conjecture implies the four-color theorem, as explained by Duchet [1984b], hence it also implies Hadwiger's conjecture for $k = 5$ by the equivalence theorem of Wagner [1937] (Theorem 7 in Chapter 1). The scheme conjecture for $k = 5$, Hadwiger's conjecture for $k = 5$, and the four-color theorem are thus all equivalent.

■ 6.7. CHROMATIC 4-SCHEMES

Let G be a 4-chromatic graph. Does there exist a 4-coloring f of G and four vertices x_1, x_2, x_3, x_4 in G such that $f(x_i) = i$ for $i = 1, 2, 3, 4$ and such that x_i and x_j are joined by a path P_{ij} in G whose vertices are alternately colored i and j for $1 \leq i < j \leq 4$? Such a 4-coloring is called a chromatic 4-scheme. ■

This question is due to Duchet [1984b], who proved that the corresponding statement is not true for k-chromatic graphs, for $k \geq 5$, when generalizing the definition of a chromatic 4-scheme in the obvious way to a chromatic k-scheme for $k \geq 5$. For $k \leq 3$ the corresponding statement for any k-chromatic graph is true.

For the more restricted class of k-critical graphs, the statement that any such graph allows a chromatic k-scheme is false for each value of k for which the k-chromatic counterexample exhibited by Duchet [1984b] is also k-critical. Duchet's examples were first published by Catlin [1979], and it follows from results of T. Gallai (see Krusenstjerna-Hafstrøm and Toft [1981]) that these graphs are k-critical (disregarding the special cases of complete graphs and odd cycles) if and only if $k \geq 8$ for k even, or $k \geq 9$ and not of the form $2^\alpha + 3$ or $3 \cdot 2^\alpha + 3$ for k odd. We do not know any k-critical counterexamples for other values of k. In particular, we do not know if such examples exist for $k = 4, 5, 6, 7$, or 13.

A further weakening of the question for general $k \geq 4$ was conjectured by Duchet and H. Meyniel (see Duchet [1984b]): Is it true that every k-chromatic graph G contains a subgraph G' such that G' has a chromatic k-scheme? G' does not have to be k-chromatic. For $k = 5$ even this statement implies the four-color theorem (see Problem 6.6).

■ 6.8. ODD SUBDIVISIONS OF K_4

Does every 4-chromatic graph contain a subdivision of K_4 in which each of the six paths replacing the edges of K_4 has odd length? ■

This question was raised by Toft [1975b]. Since then it has appeared elsewhere (e.g. in Thomassen [1988]) and it was mentioned by P.D. Seymour at the Graph Minors Workshop in Seattle, 1991 (see West [1991b]).

Catlin [1979] proved that a 4-chromatic graph has a subdivision in which each of the four triangles of the K_4 is transformed into an odd cycle. This requires at least two of the six paths to be odd. Krusenstjerna-Hafstrøm and Toft [1980] proved that at least the three paths corresponding to the edges of a K_3 subgraph of K_4 can be chosen to be odd; moreover, Thomassen and Toft [1981] proved that three edges of any spanning tree of K_4 can be left undivided, corresponding to paths of length one. It seems unknown if four of the six paths replacing the edges of K_4 can be chosen to be odd. Recently, Jensen and Shepherd [1993] answered Toft's question affirmatively for line graphs and for 4-critical graphs of minimum degree three.

■ 6.9. NONSEPARATING ODD CYCLES IN 4-CRITICAL GRAPHS

Is it true that every 4-critical graph G contains an odd cycle C without diagonals such that $G - V(C)$ is connected and has at most two end blocks? ■

The question was asked by Krusenstjerna-Hafstrøm and Toft [1980], who proved that there exists a C such that $G - V(C)$ is connected. Moreover, they gave an example showing that there does not always exist a C such that $G - V(C)$ is 2-connected.

If true, the statement above would imply that every 4-chromatic graph contains a special subdivision of K_4 in which the three edges of any spanning tree of K_4 are left undivided and at least one of the remaining three edges is replaced by an odd path (see also Problem 6.8).

■ 6.10. MINIMAL EDGE CUTS IN CONTRACTION-CRITICAL GRAPHS

Let G be a contraction-critical k-chromatic graph ($k \geq 5$), that is, $\chi(G) = k$, and if H is any proper minor of G, then $\chi(H) < k$ (see also Problem 6.1). Let E be a minimal edge cut of G with G_1 and G_2 the two connected components of $G - E$, and let $n \leq k/2$. If $|E| \leq nk - n^2$, is it then true that $|V(G_i)| \leq n - 1$ for either $i = 1$ or $i = 2$ except precisely when $|E| = nk - n^2$ and $G = K_k$? ■

The question is due to Toft [1972a], who proved that the answer is affirmative for $1 \leq n \leq 3$. Note that if Hadwiger's conjecture is true (i.e., every k-chromatic graph can be contracted into a graph containing K_k as a subgraph, see Problem 6.1), then $G = K_k$ and the answer is affirmative.

■ 6.11. KOSTOCHKA'S CONJECTURE ON HADWIGER NUMBER

Let G be any graph on n vertices. The Hadwiger number $\eta(G)$ is defined as the maximum number of vertices in a complete graph into which a subgraph of G can be contracted. If G has Hadwiger number k, is it true that

$$\eta(\overline{G}) \geq \left(\frac{k+1}{2k}\right) \cdot n - c_k$$

where $\overline{G}$ denotes the complement of G and c_k is a constant depending only on k? ■

This was conjectured by Kostochka [1989] (see also Stiebitz [1992]). The lower bound is the best possible, as can be seen from the graphs consisting of disjoint unions of complete graphs K_k.

The conjecture of Hadwiger [1943] may be formulated in terms of the Hadwiger number: Is it true that $\chi(G) \leq \eta(G)$? Hadwiger's conjecture would imply the conjecture due to Duchet and Meyniel [1982], that the following question has an affirmative answer: If $\alpha(G)$ denotes the maximum size of an independent set of vertices in G, is it true that

$$\alpha(G) \cdot \eta(G) \geq n?$$

Kostochka [1989] noted that his conjecture would imply Duchet and Meyniel's conjecture.

Nordhaus and Gaddum [1956] proved the following inequality relating the chromatic number of G to the chromatic number of G's complement $\overline{G}$:

$$\chi(G) + \chi(\overline{G}) \leq n + 1.$$

Zelinka [1976] asked if the similar inequality holds for the Hadwiger number, that is, is it true that

$$\eta(G) + \eta(\overline{G}) \leq n + 1?$$

This question was answered in the negative by Kostochka [1981]. However, Kostochka did prove, for $n \geq 5$, the best possible bound

$$\eta(G) + \eta(\overline{G}) \leq \left\lfloor \frac{6n}{5} \right\rfloor.$$

Stiebitz [1992] considered the general problem of finding lower and upper bounds for $\eta(\overline{G})$ when $\eta(G)$ is fixed. He settled Kostochka's conjecture for $k \leq 3$ by proving that if $2 \leq k \leq 3$, then

$$\eta(\overline{G}) \geq \left(\frac{k+1}{2k}\right) \cdot n - \frac{k+1}{2}.$$

For $k = 2$ this inequality was first proved by Kostochka [1989].

BIBLIOGRAPHY

[1913] Birkhoff G.D. The reducibility of maps. *Amer. J. Math.* **35**, 115–128, 1913.

[1981] Bollobás B. and P.A. Catlin. Topological cliques of random graphs. *J. Combin. Theory Ser. B* **30**, 224–227, 1981.

[1980] Bollobás B., P.A. Catlin, and P. Erdős. Hadwiger's conjecture is true for almost every graph. *European J. Combin.* **1**, 195–199, 1980.

[1977] Borodin O.V. and A.V. Kostochka. On an upper bound of a graph's chromatic number, depending on the graph's degree and density. *J. Combin. Theory Ser. B* **23**, 247–250, 1977.

[1979] Brelaz D. New methods to color the vertices of a graph. *Comm. ACM* **22**, 251–256, 1979.

[1979] Catlin P.A. Hajós' graph-coloring conjecture: variations and counterexamples. *J. Combin. Theory Ser. B* **26**, 268–274, 1979.

[1971] Chartrand G., D.P. Geller, and S.T. Hedetniemi. Graphs with forbidden subgraphs. *J. Combin. Theory* **10**, 12–41, 1971.

[1952a] Dirac G.A. A property of 4-chromatic graphs and some remarks on critical graphs. *J. London Math. Soc.* **27**, 85–92, 1952.

[1960] Dirac G.A. Trennende Knotenpunktmengen und Reduzibilität abstrakter Graphen mit Anwendung auf das Vierfarbenproblem. *J. Reine Angew. Math.* **204**, 116–131, 1960.

[1964a] Dirac G.A. On the structure of 5- and 6-chromatic abstract graphs. *J. Reine Angew. Math.* **214/215**, 43–52, 1964.

[1964c] Dirac G.A. Homomorphism theorems for graphs. *Math. Ann.* **153**, 69–80, 1964.

[1984b] Duchet P. Schemes of graphs. In: B. Bollobás, editor, *Graph Theory and Combinatorics*, pages 145–153. Academic Press, 1984.

[1982] Duchet P. and H. Meyniel. On Hadwiger's number and the stability number. In: B. Bollobás, editor, *Graph Theory*, volume 13 of *Annals of Discrete Mathematics*, pages 71–74. North-Holland, 1982.

[1981] Erdős P. and S. Fajtlowicz. On the conjecture of Hajós. *Combinatorica* **1**, 141–143, 1981.

[1983] Fernandez de la Véga W. On the maximum density of graphs which have no subcontractions to K^s. *Discrete Math.* **46**, 109–110, 1983.

[1943] Hadwiger H. Über eine Klassifikation der Streckenkomplexe. *Vierteljahrsch. Naturforsch. Ges. Zürich* **88**, 133–142, 1943.

[1964] Halin R. Über einen Satz von K. Wagner zum Vierfarbenproblem. *Math. Ann.* **153**, 47–62, 1964.

[1994] Hanson D. and B. Toft. The (m, n)-conjecture is false. *Bull. Inst. Combin. Appl.* **11**, 59–66, 1994.

[1993] Heath L.S. Edge coloring planar graphs with two outerplanar subgraphs. Manuscript, 1993. Submitted to *J. Graph Theory*.

[1970] Jakobsen I.T. *On certain homomorphism-properties of graphs with applications to the conjecture of Hadwiger*. Ph.D. thesis, University of London, 1970. *Various Publ. Ser.* **15** (Aarhus University).

[1971] Jakobsen I.T. A homomorphism theorem with an application to the conjecture of Hadwiger. *Studia Sci. Math. Hungar.* **6**, 151–160, 1971.

[1993] Jensen T.R. and F.B. Shepherd. Note on a conjecture of Toft. Manuscript, 1993. To appear in *Combinatorica*.

[1981] Kostochka A.V. On Hadwiger numbers of a graph and its complement. In: A. Hajnal, L. Lovász, and V.T. Sós, editors, *Finite and Infinite Sets*, volume 37 of *Colloquia Mathematica Societatis János Bolyai*, pages 537–545. North-Holland, 1981.

[1982] Kostochka A.V. The minimum Hadwiger number for graphs with a given mean degree of vertices (in Russian). *Metody Diskret. Analiz.* **38**, 37–58, 1982.

[1989] Kostochka A.V. A lower bound for the product of the Hadwiger number of a graph and its complement (in Russian). *Combin. Anal.* **8**, 50–62, 1989.

[1980] Krusenstjerna-Hafstrøm U. and B. Toft. Special subdivisions of K_4 and 4-chromatic graphs. *Monatsh. Math.* **89**, 101–110, 1980.

[1981] Krusenstjerna-Hafstrøm U. and B. Toft. Some remarks on Hadwiger's conjecture and its relations to a conjecture of Lovász. In: G. Chartrand, Y. Alavi, D.L. Goldsmith, L. Lesniak-Foster, and D.R. Lick, editors, *The Theory and Applications of Graphs: Proc. 4th International Graph Theory Conference, Kalamazoo, 1980*, pages 449–459. Wiley, 1981.

[1968a] Mader W. Homomorphiesätze für Graphen. *Math. Ann.* **178**, 154–168, 1968.

[1968b] Mader W. Über trennende Eckenmengen in homomorphiekritischen Graphen. *Math. Ann.* **175**, 243–252, 1968.

[1972] Mader W. Hinreichende Bedingungen für die Existens von Teilgraphen, die zu einem vollständigen Graphen homöomorph sind. *Math. Nachr.* **53**, 145–150, 1972.

[1972] Matula D.W., G. Marble, and J.D. Isaacson. Graph coloring algorithms. In: R.C. Read, editor, *Graph Theory and Computing*, pages 109–122. Academic Press, 1972.

[1989a] Mayer J. Conjecture de Hadwiger: Un graphe k-chromatique contraction-critique n'est pas k-régulier. In: L.D. Andersen, I.T. Jakobsen, C. Thomassen, B. Toft, and P.D. Vestergaard, editors, *Graph Theory in Memory of G.A. Dirac*, volume 41 of *Annals of Discrete Mathematics*, pages 341–345. North-Holland, 1989.

[1989b] Mayer J. Hadwiger's conjecture ($k = 6$): neighbour configurations of 6-vertices in contraction-critical graphs. *Discrete Math.* **74**, 137–148, 1989.

[1992] Mayer J. Conjecture de Hadwiger: ($k = 6$). II. Réductions de sommets de degré 6 dans les graphes 6-chromatiques contraction-critiques. *Discrete Math.* **101**, 213–222, 1992.

[1956] Nordhaus E.A. and J.W. Gaddum. On complementary graphs. *Amer. Math. Monthly* **63**, 175–177, 1956.

[1967] Ore O. *The Four-Color Problem*. Academic Press, 1967.

[1969] Pelikán J. Valency conditions for the existence of certain subgraphs. In: P. Erdős and G. Katona, editors, *Theory of Graphs, Proc. Colloquium, Tihany, Hungary, September 1966*. Academic Press, 1969.

[1993a] Robertson N., P.D. Seymour, and R. Thomas. Hadwiger's conjecture for K_6-free graphs. *Combinatorica* **13**, 279–361, 1993.

[1993b] Robertson N., P.D. Seymour, and R. Thomas. Linkless embeddings of graphs in 3-space. *Bull. Amer. Math. Soc.* **28**, 84–89, 1993.

[1983] Sachs H. On a spatial analogue of Kuratowski's theorem on planar graphs: an open problem. In: M. Borowiecki, J.W. Kennedy, and M.M. Sysło, editors, *Graph Theory, Łagów, 1981*, volume 1018 of *Lecture Notes in Mathematics*, pages 230–241. Springer-Verlag, 1983.

[1990] Seymour P.D. Colouring series–parallel graphs. *Combinatorica* **10**, 379–392, 1990.

[1992] Stiebitz M. On Hadwiger's number: a problem of the Nordhaus–Gaddum type. *Discrete Math.* **101**, 307–317, 1992.

[1993] Stiebitz M. and B. Toft. An abstract generalization of a map reduction theorem of Birkhoff. Technical report, Odense University, 1993. Submitted to *J. Combin. Theory Ser. B*.

[1984] Thomason A.G. An extremal function for contractions of graphs. *Math. Proc. Cambridge Philos. Soc.* **95**, 261–265, 1984.

[1974] Thomassen C. Some homomorphism properties of graphs. *Math. Nachr.* **64**, 119–133, 1974.

[1988] Thomassen C. Paths, circuits and subdivisions. In: L.W. Beineke and R.J. Wilson, editors, *Selected Topics in Graph Theory*, volume 3, pages 97–131. Academic Press, 1988.

[1981] Thomassen C. and B. Toft. Nonseparating induced cycles in graphs. *J. Combin. Theory Ser. B* **31**, 199–224, 1981.

[1972a] Toft B. On separating sets of edges in contraction-critical graphs. *Math. Ann.* **196**, 129–147, 1972.

[1975b] Toft B. Problem 10. In: M. Fiedler, editor, *Recent Advances in Graph Theory: Proc. Symposium, Prague, June 1974*, pages 543–544. Academia Praha, 1975.

[1968] Vizing V.G. Some unsolved problems in graph theory (in Russian). *Uspekhi Mat. Nauk* **23**, 117–134, 1968. English translation in *Russian Math. Surveys* **23**, 125–141.

[1937] Wagner K. Über eine Eigenschaft der ebenen Komplexe. *Math. Ann.* **114**, 570–590, 1937.

[1964] Wagner K. Beweis eine Abschwächung der Hadwiger–Vermutung. *Math. Ann.* **153**, 139–141, 1964.

[1991b] West D.B. Open problems. *SIAM J. Discrete Math. Newslett.* **2**(1), 10–12, 1991.

[1990] Woodall D.R. Improper colourings of graphs. In: R. Nelson and R.J. Wilson, editors, *Graph Colourings*, volume 218 of *Pitman Research Notes in Mathematics Series*, pages 45–63. Longman, 1990.

[1971] Young H.P. A quick proof of Wagner's equivalence theorem. *J. London Math. Soc. (2)* **3**, 661–664, 1971.

[1976] Zelinka B. Hadwiger numbers of finite graphs. *Math. Slovaca* **26**, 23–30, 1976.

7

Sparse Graphs

■ 7.1. BLANCHE DESCARTES' TRIANGLE-FREE GRAPHS

The graphs G_k ($k \geq 3$) of Blanche Descartes are defined as follows. G_3 is an odd cycle. G_{k+1} has a set X of $k(|V(G_k)| - 1) + 1$ independent vertices and $\binom{|X|}{|V(G_k)|}$ mutually disjoint copies of G_k (all disjoint from X as well). Each copy of G_k corresponds to a unique subset of X of size $|V(G_k)|$ and the copy and the subset are joined in G_{k+1} by $|V(G_k)|$ disjoint edges. Are the graphs G_{k+1} thus obtained $(k + 1)$-critical? ■

If G_3 is the 7-cycle, then each G_k has minimum cycle length at least 6 and moreover satisfies $\chi(G_k) \geq k$. This is the famous construction of Descartes [1947, 1948b, 1954] of triangle-free k-chromatic graphs (see Theorem 21 in Chapter 1). After presenting the argument for $\chi(G_k) \geq k$, Descartes wrote: "*This does not prove that G_k is k-chromatic, but if it is not we can obtain a k-chromatic graph from it by deleting some vertices and their incident edges.*" It seems likely, however, that G_k is in fact k-critical. This question was mentioned by Toft [1970b]. For the other early constructions of triangle-free k-chromatic graphs by Zykov [1949] and Mycielski [1955], the criticality problem was solved by Schäuble [1969].

It is easy to see that G_4 is 4-critical. Suppose that G_k is k-critical for some $k \geq 4$. Then G_{k+1} is $(k + 1)$-critical if the following question has a positive answer. Let $f : V(G_k) \rightarrow \{1, 2, \ldots, k\}$ be a nonconstant mapping. Does there exist a k-coloring φ of G_k with colors $1, 2, \ldots, k$ such that $\varphi(x) \neq f(x)$ for all vertices $x \in V(G_k)$? The answer for k-critical graphs in general is NO, as shown by B.Aa. Sørensen for $k \geq 5$ (see Toft [1974a]) and by V. Rödl and Stiebitz [1987b] for the case $k = 4$, even with f taking only three different values. T. Gallai observed that the answer for k-critical graphs in general is YES when f takes only two different values [personal communication in 1969] (see Toft [1974a]).

■ 7.2. GRÜNBAUM'S GIRTH PROBLEM

Does there exist a k-chromatic k-regular graph without triangles for $k = 5$ and $k = 6$? ■

Grünbaum [1970a] conjectured more generally that for $k \geq 3$ and $g \geq 4$ there is a k-chromatic k-regular graph without cycles of length less than g, that is, a graph of girth at least g.

The conjecture of Grünbaum was disproved for $k \geq 7$ and $g \geq 4$ independently at about the same time by Borodin and Kostochka [1977], Catlin [1978a], and Lawrence [1978] (see Section 1.10 and Problem 4.6). For the cases $(k, g) = (4, 4)$ and $(k, g) = (4, 5)$ positive answers were given by Chvátal [1970b] and Grünbaum [1970a]. Kostochka [1978a] proved that there is no 5-chromatic graph of maximum degree five with shortest cycle of length at least 35. Also, for 6-chromatic graphs Grünbaum's conjecture fails for g large enough. We do not know what the situation is for 4-chromatic graphs.

■ 7.3. SMALLEST TRIANGLE-FREE k-CHROMATIC GRAPHS

Let $n(k)$ denote the smallest possible number of vertices in a k-chromatic graph without triangles. What is the order of magnitude of $n(k)$? Is $n(k+1) \leq 2n(k)$ for all $k \geq 4$? What is the value of $n(k)$ for $k = 6$? ■

The problem is due to Erdős [1967].

The constructions of Descartes [1954], Kelly and Kelly [1954], Mycielski [1955], and Zykov [1949] all give at least exponential upper bounds for $n(k)$, the best being Mycielski's construction showing that $n(k) \leq 3 \cdot 2^{k-2} - 1$. For $k = 2, 3,$ and 4, in fact, equality holds: $n(2) = 2$, $n(3) = 5$, and $n(4) = 11$. For $k = 5$ Grinstead, Katinsky, and Van Stone [1989] showed that $21 \leq n(5) \leq 22$, and Jensen and Royle [1989] found that, in fact, $n(5) = 22$. From this and Mycielski's construction it follows that $n(6) \leq 2n(5) + 1 = 45$. We do not know if $n(k+1) \leq 2n(k)$ for all $k \geq 5$.

Erdős was the first to give polynomial upper bounds. A geometric construction of Erdős [1958] shows $n(k) \leq c \cdot k^{50}$, where c is constant, as explained by Sachs [1969]. The probabilistic method of Erdős [1959, 1961] gives even better bounds. The lower bound $\lfloor (\log n)/(4 \log k) \rfloor$ for the maximal girth of a graph on n vertices of chromatic number at least k (see Theorem 22 in Chapter 1 and the discussion following it) implies that $n(k) \leq k^{16}$. In fact, Erdős [1961] proved by probabilistic arguments the existence of a constant c_1 such that for all $n \geq 2$ there is a triangle-free graph G on n vertices with $\alpha(G) < c_1 \cdot \sqrt{n} \cdot \log n$, where $\alpha(G)$ is the size of a largest independent set of vertices in G. This G has chromatic number k, where

$$k \geq n/\alpha(G)$$

and thus

$$k > \frac{\sqrt{n}}{c_1 \cdot \log n} \geq \frac{\sqrt{n(k)}}{c_1 \cdot \log n(k)}.$$

Therefore,

$$\begin{aligned} n(k) &\leq c_1^2 \cdot k^2 \cdot \log^2 n(k) \\ &\leq c_1^2 \cdot k^2 \cdot \log^2(k^{16}). \end{aligned}$$

It follows that there is a constant c_2 such that for all k

$$n(k) \leq c_2 \cdot (k \cdot \log k)^2.$$

As far as a lower bound for $n(k)$ is concerned, using the Ramsey result in the classical paper by Erdős and Szekeres [1935] that $r(3, \ell) \leq \ell(\ell + 1)/2$ (i.e., that any graph on $\ell(\ell + 1)/2$ vertices either has a triangle or ℓ independent vertices), it follows that $n(k) \geq k(k - 1)/2 + 1$. (Explanation: A triangle-free graph G on $k(k - 1)/2$ vertices has a set X_1 of $k - 1$ independent vertices. Color all vertices in X_1 by the color 1. The triangle-free graph $G - X_1$ has $(k - 1)(k - 2)/2$ vertices and a set X_2 of $k - 2$ independent vertices. Color all vertices in X_2 by the color 2. Continue like this to obtain a $(k - 1)$-coloring.)

The bound on the Ramsey number $r(3, \ell)$ was improved by Graver and Yackel [1968] and by Ajtai, Komlós, and Szemerédi [1980], who proved that there is a positive constant c_3 such that any triangle-free graph G on n vertices has at least $c_3 \cdot \sqrt{n \log n}$ independent vertices. This implies, as we shall indicate below, the lower bound in

$$c_5 \cdot k^2 \cdot \log k \leq n(k) \leq c_2 \cdot (k \cdot \log k)^2. \qquad (*)$$

Thus the order of magnitude of $n(k)$ is almost determined.

Let us explain how the lower bound of $(*)$ can be obtained. We do this in some more detail than in the interesting paper by Erdős and Hajnal [1985] (note that this paper contains a misprint in the formulation of $(*)$). We begin by making the following general observation.

Proposition. *Let $\mathcal{G}$ be a class of graphs that is closed under taking induced subgraphs. Suppose that $\alpha(G) \geq f(|V(G)|)$ for all $G \in \mathcal{G}$, where $f : [2, \infty) \to \mathbf{R}_+$ is a positive, nondecreasing, continuous function. Then*

$$\chi(G) \leq 2 + \int_2^{|V(G)|} \frac{1}{f(x)} dx$$

holds for every $G \in \mathcal{G}$ with $|V(G)| \geq 2$.

Proof. The proof is by induction over the number of vertices $n \geq 2$. For $n = 2$ the result is obviously true. Assume that it is true for graphs with fewer than n vertices and let $|V(G)| = n$. Let X be an independent subset of $V(G)$ of size $\alpha(G)$. Then if $|X| \geq n - 1$, G must clearly be bipartite. Otherwise, $G - X$ has at least 2 vertices, and by induction hypothesis

$$\begin{aligned}
\int_2^n \frac{1}{f(x)} dx &= \int_2^{n-\alpha(G)} \frac{1}{f(x)} dx + \int_{n-\alpha(G)}^n \frac{1}{f(x)} dx \\
&\geq \chi(G - X) - 2 + \int_{n-\alpha(G)}^n \frac{1}{f(n)} dx \\
&= \chi(G - X) - 2 + \alpha(G) \cdot \frac{1}{f(n)} \\
&\geq \chi(G - X) - 2 + 1 \\
&\geq \chi(G) - 1 - 2 + 1 \\
&= \chi(G) - 2.
\end{aligned}$$

This proves the result. Observe that what we do here is to use a greedy coloring of G: color as many vertices as possible with color 1, then as many as possible with color 2, etc. ■

With $\mathcal{G}$ the class of triangle-free graphs and $f(x) = c_3 \cdot \sqrt{x \log x}$ it follows for a triangle-free k-chromatic graph on n vertices that

$$\begin{aligned}
k &\leq 2 + \int_2^n \frac{1}{c_3\sqrt{x \log x}} dx \\
&\leq c_4 \cdot \sqrt{n}/\sqrt{\log n}
\end{aligned}$$

and hence that

$$\begin{aligned}
n &\geq c_5 \cdot k^2 \cdot \log n \\
&\geq c_5 \cdot k^2 \cdot \log k,
\end{aligned}$$

which gives the lower bound of (*).

Similar to asking for a smallest possible number of vertices in a triangle-free k-chromatic graph, one can ask for the smallest possible number $n_g(k)$ of vertices in a k-chromatic graph without cycles of length less than g. Even for $k = 4$ and $g = 5$ the answer is unknown. A 4-critical example showing $n_5(4) \leq 21$ can be described as follows: The vertices are the integers $1, 2, \ldots, 21$ with i and j joined by an edge if and only if either $i + j = 0$ modulo 3 and $i - j = \pm 1$ modulo 7, or $i + j = 1$ modulo 3 and $i - j = \pm 1, \pm 2$, or ± 3 modulo 21 [personal communication from D.A. Youngs in 1991].

7.4. LARGE BIPARTITE SUBGRAPHS OF TRIANGLE-FREE GRAPHS

Let G be a triangle-free graph on n vertices. Is it possible to obtain a bipartite graph from G by deleting at most $n^2/25$ edges? ■

The question was asked by Erdős [1990a], who conjectured that the answer is affirmative. That the number $n^2/25$ would be the best possible can be seen from a graph whose vertex set is the union of 5 disjoint classes $U_1, \ldots, U_5$ all of the same size m, such that U_i is completely joined by m^2 edges to U_{i+1} (adding modulo 5) for $i = 1, \ldots, 5$.

Erdős, Faudree, Pach, and Spencer [1988] proved that every triangle-free graph G can be made bipartite by deleting at most $n^2/18 + n/2$ edges, and they also proved that $n^2/25$ edges are sufficient if G satisfies $|E(G)| \geq n^2/5$. Erdős, Győri, and Simonovits [1991] proved the stronger result that if L is any fixed 3-chromatic graph, and G is a graph having at least $n^2/5 - o(n^2)$ edges and not containing L as a subgraph, then G can be made bipartite by removing at most $n^2/25 + o(n^2)$ edges.

Using a "greedy" algorithm, it is easy to find a subset of at most $\frac{1}{2}|E(G)|$ edges of any graph G whose removal leaves a bipartite subgraph. Hence the conjecture by Erdős remains open only in the interval

$$\frac{2n^2}{25} < |E(G)| < \frac{n^2}{5}.$$

The similar problem for triangle-free regular graphs has also been studied (see Locke [1990] for results and references).

If G has no complete 4-graph K_4 as a subgraph, it has been conjectured that G can be made bipartite by deleting at most $(1/9 + o(1))n^2$ edges. This would be best possible, as can be seen from complete tripartite graphs with the same number of vertices in each class. This conjecture was mentioned by Erdős, Faudree, Pach, and Spencer [1988].

7.5. SPARSE SUBGRAPHS

If G has a large chromatic number (depending on g and k), must G contain a subgraph with chromatic number at least k in which each cycle has length at least g? ■

This question is due to A. Hajnal and Erdős [1969, 1971].

Rödl [1977] proved that for given k and n there exists a number $\varphi(k, n)$, such that if G is a graph with chromatic number at least $\varphi(k, n)$, then G either contains a complete n-graph or G has a triangle-free k-chromatic subgraph. This implies an affirmative answer to the question for $g = 4$, by the existence theorems for triangle-free k-chromatic graphs for all k (Zykov [1949], Descartes [1954], Kelly and Kelly [1954], and Mycielski [1955]).

■ 7.6. NUMBER OF ODD CYCLE LENGTHS

Let $C_O(G)$ denote the set of odd cycle lengths in a graph G:

$$C_O(G) = \{2m + 1 : G \text{ contains a cycle of length } 2m + 1, \text{ for } m \geq 1\}.$$

If $|C_O(G)| = k$, is it true that

$$\chi(G) \leq 2k + 2,$$

with equality if and only if G contains a complete graph on $2k + 2$ vertices as a subgraph? ■

This was conjectured by B. Bollobás and P. Erdős (see Erdős [1990b]). Erdős [1990b] reported that for $k = 1$ the conjecture is true, as observed by Bollobás and S. Shelah.

An elegant argument by Neumann-Lara [1982] shows that every edge in a $(2k + 3)$-critical graph G is contained in a cycle of length 1 modulo $2m$ for all m in the interval $1 \leq m \leq k + 1$.

■ 7.7. MAXIMUM GIRTH OF k-CHROMATIC GRAPHS

Let $g_k(n)$, where $n \geq k \geq 4$, be the maximum girth of k-chromatic graphs on n vertices (i.e., there exists a k-chromatic graph on n vertices in which all cycles have length at least $g_k(n)$, and $g_k(n)$ is the largest such number).

Does

$$\lim_{n \to \infty} \frac{g_k(n)}{\log n}$$

exist? ■

This question is due to Erdős [1959, 1962], who proved that for $k \geq 4$ (see Section 1.5)

$$\frac{\log n}{4 \log k} \leq g_k(n) \leq \frac{2 \log n}{\log(k - 2)} + 2.$$

The upper bound for $g_k(n)$ is easy to obtain by considering the distance classes with respect to any fixed vertex of a graph with the given three parameters. For $k = 3$ one has that $g_3(n)$ equals n for n odd, and $n - 1$ for n even.

Kostochka [1988] considered a very related problem: Let $k_g(n)$, where $n \geq g \geq 4$, be the maximum chromatic number of graphs on n vertices in which all cycles have length at least g. For g fixed, what is the asymptotic behavior of $k_g(n)$ as $n \longrightarrow \infty$? Kostochka proved that

$$\frac{n^{1/(g-1)}}{\log n} \leq k_g(n) \leq n^{2/(g-2)},$$

where the upper bound is again derived from a simple argument involving distance classes. Note that the lower bound for $k_g(n)$ can be translated to a new lower bound for $g_k(n)$, which is asymptotically worse than the bound obtained by Erdős but improves it when n is small.

7.8. MAXIMUM RATIO χ/ω

Let $\omega(G)$ denote the number of vertices of a largest complete subgraph of G. Does the limit

$$\lim_{n\to\infty} \max \left\{ \left(\frac{\chi(G)}{\omega(G)} \right) \Big/ \left(\frac{n}{\log^2 n} \right) : |V(G)| = n \right\}$$

exist? ■

This question is due to Erdős [1967], who proved the existence of constants c_1 and c_2 such that

$$\frac{c_1 \cdot n}{\log^2 n} \leq \max \left(\frac{\chi(G)}{\omega(G)} \right) \leq \frac{c_2 \cdot n}{\log^2 n}.$$

7.9. CHROMATIC NUMBER OF SPARSE RANDOM GRAPHS

For $c > 0$ constant, let $G_c(n)$ be a random graph on n vertices and with edge probability c/n; hence $G_c(n)$ has $\sim cn/2$ edges. What is the largest integer $k = f(c)$ for which the probability that $G_c(n)$ has chromatic number exactly k has a nonzero limit as $n \longrightarrow \infty$? ■

The question was asked by P. Erdős in the appendix, written by him, to the book by Alon and Spencer [1992]. For $0 < c \leq 1$ it is known that $f(c) = 3$. This was proved by Łuczak and Wierman [1989].

Erdős and Spencer [1974] (p. 96) gave as an exercise to prove that for every $\varepsilon > 0$ there exists $c(\varepsilon)$ such that $\forall c > c(\varepsilon)$:

$$\Pr\left(\frac{c}{2\log c}(1-\varepsilon) \leq \chi(G_c(n)) \leq \frac{c}{\log c}(1+\varepsilon) \right) \longrightarrow 1 \quad \text{as} \quad n \longrightarrow \infty.$$

Thus the behavior of $f(c)$ as $c \to \infty$ is roughly determined. Similar results, especially for edge probabilities $c(n)/n$, where $c(n)$ is a function of n with $c(n) \to \infty$ and $c(n)/n \to 0$ as $n \to \infty$, can be found in Bollobás [1985].

BIBLIOGRAPHY

[1980] Ajtai M., J. Komlós, and E. Szemerédi. A note on Ramsey numbers. *J. Combin. Theory Ser. A* **29**, 354–360, 1980.

[1992] Alon N. and J.H. Spencer. *The Probabilistic Method.* Wiley, 1992. (With an appendix on open problems by P. Erdős).

[1985] Bollobás B. *Random Graphs.* Academic Press, 1985.

[1977] Borodin O.V. and A.V. Kostochka. On an upper bound of a graph's chromatic number, depending on the graph's degree and density. *J. Combin. Theory Ser. B* **23**, 247–250, 1977.

[1978a] Catlin P.A. A bound on the chromatic number of a graph. *Discrete Math.* **22**, 81–83, 1978.

[1970b] Chvátal V. The smallest triangle-free 4-chromatic 4–regular graph. *J. Combin. Theory* **9**, 93–94, 1970.

[1947] Descartes B. A three-colour problem. *Eureka* **9**, 1947.

[1948b] Descartes B. Solutions to problems in Eureka No. 9. *Eureka* **10**, 1948.

[1954] Descartes B. Solution to advanced problem No. 4526. *Amer. Math. Monthly* **61**, 532, 1954.

[1958] Erdős P. Remarks on a theorem of Ramsey. *Bull. Res. Council Israel. Sect. F* **7**, 21–24, 1957-58.

[1959] Erdős P. Graph theory and probability. *Canad. J. Math.* **11**, 34–38, 1959.

[1961] Erdős P. Graph theory and probability II. *Canad. J. Math.* **13**, 346–352, 1961.

[1962] Erdős P. On circuits and subgraphs of chromatic graphs. *Mathematika* **9**, 170–175, 1962.

[1967] Erdős P. Some remarks on chromatic graphs. *Colloq. Math.* **16**, 253–256, 1967.

[1969] Erdős P. Problems and results in chromatic graph theory. In: F. Harary, editor, *Proof Techniques in Graph Theory*, pages 27–35. Academic Press, 1969.

[1971] Erdős P. Some unsolved problems in graph theory and combinatorial analysis. In: D.J.A. Welsh, editor, *Combinatorial Mathematics and Its Applications (Proc. Conf., Oxford, 1969)*, pages 97–109. Academic Press, 1971.

[1990a] Erdős P. On some of my favourite problems in graph theory and block designs. In: M. Gionfriddo, editor, *Le Matematiche*, volume XLV, pages 61–74, 1990.

[1990b] Erdős P. Some of my favourite unsolved problems. In: A. Baker, B. Bollobás, and A. Hajnal, editors, *A Tribute to Paul Erdős*, pages 467–479. Cambridge University Press, 1990.

[1988] Erdős P., R.J. Faudree, J. Pach, and J.H. Spencer. How to make a graph bipartite. *J. Combin. Theory Ser. B* **45**, 86–98, 1988.

[1991] Erdős P., E. Győri, and M. Simonovits. How many edges should be deleted to make a triangle-free graph bipartite? In: *Sets, Graphs and Numbers*, volume 60 of *Colloquia Mathematica Societatis János Bolyai*, pages 239–263. North-Holland, 1991.

[1985] Erdős P. and A. Hajnal. Chromatic number of finite and infinite graphs and hypergraphs. *Discrete Math.* **53**, 281–285, 1985.

[1974] Erdős P. and J.H. Spencer. *Probabilistic Methods in Combinatorics.* Academic Press, 1974.

[1935] Erdős P. and G. Szekeres. A combinatorial problem in geometry. *Compositio Math.* **2**, 463–470, 1935.

[1968] Graver J.E. and J. Yackel. Some graph theoretic results associated with Ramsey's theorem. *J. Combin. Theory* **4**, 125–175, 1968.

[1989] Grinstead C.M., M. Katinsky, and D. Van Stone. On minimal triangle-free 5-chromatic graphs. *J. Combin. Math. Combin. Comput.* **6**, 189–193, 1989.

[1970a] Grünbaum B. A problem in graph coloring. *Amer. Math. Monthly* **77**, 1088–1092, 1970.

[1989] Jensen T.R. and G.F. Royle. Small graphs of chromatic number 5: a computer search. Manuscript, 1989. To appear in *J. Graph Theory* **19**, 1995.

[1954] Kelly J.B. and L.M. Kelly. Paths and circuits in critical graphs. *Amer. J. Math.* **76**, 786–792, 1954.

[1978a] Kostochka A.V. Degree, girth and chromatic number. In: A. Hajnal and V.T. Sós, editors, *Combinatorics*, volume 18 of *Colloquia Mathematica Societatis János Bolyai*, pages 679–696. North-Holland, 1978.

[1988] Kostochka A.V. Upper bounds on the chromatic number of graphs (in Russian). *Trudy Inst. Mat. (Novosibirsk)* **10**, 204–226, 265, 1988.

[1978] Lawrence J. Covering the vertex set of a graph with subgraphs of smaller degree. *Discrete Math.* **21**, 61–68, 1978.

[1990] Locke S.C. A note on bipartite subgraphs of triangle-free regular graphs. *J. Graph Theory* **14**, 181–185, 1990.

[1989] Łuczak T. and J.C. Wierman. The chromatic number of random graphs at the double-jump threshold. *Combinatorica* **9**, 39–49, 1989.

[1955] Mycielski J. Sur le coloriage des graphes. *Colloq. Math.* **3**, 161–162, 1955.

[1982] Neumann-Lara V. The dichromatic number of a digraph. *J. Combin. Theory Ser. B* **33**, 265–270, 1982.

[1977] Rödl V. On the chromatic number of subgraphs of a given graph. *Proc. Amer. Math. Soc.* **64**, 370–371, 1977.

[1969] Sachs H. Finite graphs (investigations and generalizations concerning the construction of finite graphs having given chromatic number and no triangles). In: W.T. Tutte, editor, *Recent Progress in Combinatorics*, pages 175–184. Academic Press, 1969.

[1969] Schäuble M. Bemerkungen zur Konstruktion dreikreisfreier k-chromatischer Graphen. *Wiss. Z. Tech. Hochsch. Ilmenau* **15**, 59–63, 1969.

[1987b] Stiebitz M. Subgraphs of colour-critical graphs. *Combinatorica* **7**, 303–312, 1987.

[1970b] Toft B. *Some contributions to the theory of colour-critical graphs*. Ph.D. thesis, University of London, 1970. *Various Publ. Ser.* **14** (Aarhus University).

[1974a] Toft B. Color-critical graphs and hypergraphs. *J. Combin. Theory Ser. B* **16**, 145–161, 1974.

[1949] Zykov A.A. On some problems of linear complexes (in Russian). *Mat. Sbornik N.S.* **24**, 163–188, 1949. English translation in *Amer. Math. Soc. Transl.* **79**, 1952. Reissued in *Translation Series 1* **7**, *Algebraic Topology*, 418–449 (American Mathematical Society 1962).

8

Perfect Graphs

■ 8.1. STRONG PERFECT GRAPH CONJECTURE

A graph G is perfect if every induced subgraph H of G has chromatic number equal to the size of a maximum complete subgraph of H,

$$\chi(H) = \omega(H) \text{ for every induced subgraph } H \text{ of } G.$$

It is easy to see that a perfect graph is a Berge graph (i.e., neither G nor its complement $\overline{G}$ contains an odd cycle of length at least 5 as an induced subgraph). Is it true that G is perfect if and only if G is a Berge graph? ■

The question and the conjecture that the answer is affirmative is due to Berge [1963a, 1966] (and it was also discovered by P.C. Gilmore). For more details on the background of the problem, see Section 1.6.

The corollary of the conjecture, now known as the perfect graph theorem, that G is perfect if and only if G's complement $\overline{G}$ is perfect, was proved by Lovász [1972a, 1972b] (Theorem 24 in Chapter 1), who proved two different extensions of this statement. Concerning one of the proofs, Lovász [1972b] wrote: *"It should be pointed out that thus the proof consists of two steps and the more difficult second step was first carried out by Fulkerson."* The contributions of D.R. Fulkerson are contained and described in Fulkerson [1970, 1971, 1972, 1973].

Extending the perfect graph theorem, Lovász [1972a] proved (Theorem 25 in Chapter 1) that a graph G is perfect if and only if every induced subgraph H of G satisfies

$$\alpha(H) \cdot \omega(H) \geq |V(H)|.$$

Thus to prove the strong perfect graph conjecture it suffices to prove that if G is a Berge graph, then $\alpha(G) \cdot \omega(G) \geq |V(G)|$. In particular, $\max\{\alpha(G), \omega(G)\}$ would be at least $|V(G)|^{1/2}$. It seems not even known if there is a positive constant ε such that $\max\{\alpha(G), \omega(G)\} \geq |V(G)|^{\varepsilon}$. This problem is due to L. Lovász (see Erdős and Hajnal [1989]).

Several other equivalent versions of the strong perfect graph conjecture were given by Sachs [1970], Wessel [1977] and Chvátal [1984a], among others. The draft paper by Chvátal [1993] contains a wealth of interesting problems and results of many different types concerning perfect graphs and Berge graphs, and it is updated with recent new additions by several contributors.

A version of the strong perfect graph conjecture in terms of vertex-critical graphs is due to Wessel [1977] (see also Toft [1985]):

Conjecture. *Let G be a vertex-critical k-chromatic graph ($k \geq 4$) with the property that for all $k' < k$ every induced vertex-critical k'-chromatic subgraph of G is the complete graph $K_{k'}$. Then either $G = K_k$ or $G = \overline{C_{2k-1}}$.*

This version of the conjecture was proved for $k = 4$ by Tucker [1977a]. Moreover, Stiebitz [1987b] noted the truth of the partly stronger and partly weaker version where the condition "for all $k' < k$ every induced vertex-critical k'-chromatic subgraph of G is the complete graph $K_{k'}$" is replaced by "every vertex-critical $(k-1)$-chromatic subgraph of G is the complete graph K_{k-1}."

Parthasarathy and Ravindra [1976] proved the strong perfect graph conjecture for graphs not containing, as an induced subgraph, a 3-star $K_{1,3}$. Hence to prove, in the formulation given above, that either $G = K_k$ or $G = \overline{C_{2k-1}}$, it is sufficient to prove that G does not have a $K_{1,3}$ as an induced subgraph.

Many classes of perfect graphs are known (see Berge [1975], Golumbic [1980] or Lovász [1983a]). Meyniel [1976] and independently Karapetyan and Markosyan [1976] proved a conjecture by Olaru [1972], that G is perfect if every odd cycle of G of length at least 5 has at least 2 diagonals. Such a graph G is often called a "Meyniel graph." Hoàng [1987] proved that a graph G is a Meyniel graph if and only if G is so-called "very strongly perfect"; that is, for every induced subgraph H of G, each vertex of H belongs to an independent set of vertices in H that intersects the vertex set of every maximal complete subgraph of H.

A recent probabilistic result by Prömel and Steger [1992] shows that almost all Berge graphs are perfect.

8.2. MARKOSYAN'S PERFECT GRAPH PROBLEMS

An edge xy of a graph G is said to be α-critical if $\alpha(G - xy) > \alpha(G)$, where $\alpha(H)$ is the size of a maximum independent set of vertices in H. Let C be the set of vertices of a connected component of the subgraph of G formed by all α-critical edges. The subgraph of G induced by C is called an α-critical component of G. (Note that not all edges of the α-critical components are necessarily α-critical.)

If G is a Berge graph (i.e., G contains neither odd cycles of length greater than or equal to 5 nor their complements as induced subgraphs), is it then true that all α-critical components of G are complete?

If all α-critical components of G and of all induced subgraphs of G are complete, is it then true that G is perfect? ■

The two questions were raised by Markosyan [1981] (see also *Math. Reviews* **87k**:05082). As explained below, the statement that both questions have positive answers is equivalent to the strong perfect graph conjecture (Problem 8.1).

In a sense Markosyan's problems break the strong perfect graph conjecture into two separate parts. This can be done using any property P for which (1) all perfect graphs have P, and (2) all graphs having P are Berge graphs. Then the strong perfect graph conjecture, that all Berge graphs are perfect, implies that (a) all Berge graphs have P, and (b) all graphs having P are perfect. On the other hand, (a) and (b) together imply the strong perfect graph conjecture.

Markosyan's problems are (a) and (b) above with P the property that the α-critical components of G and of its induced subgraphs are all complete. This P satisfies (2) as is easily seen. To see that it satisfies (1), let G be perfect. Then the complement of G is perfect and G has a covering of all its vertices with $\alpha(G)$ disjoint complete subgraphs. For any α-critical edge xy the two vertices x and y are contained in the same one of the $\alpha(G)$ complete subgraphs, implying that all α-critical components are complete.

■ 8.3. BOLD CONJECTURE

A pair of vertices (x, y) of a graph G is an ω-critical pair if $\omega(G + xy) > \omega(G)$, where $G + xy$ denotes the graph obtained by adding the edge xy to G and $\omega(H)$ is the size of a maximum complete subgraph in H. Let $\Omega(G)$ be the graph on the vertex set $V(G)$ whose edges correspond to the ω-critical pairs of G. Prove or disprove the following conjecture: If G is a uniquely $\chi(G)$-colorable perfect graph, then there exists some color class C (w.r.t. the unique coloring of G with $\chi(G)$ colors) such that the induced subgraph $\Omega(G)[C]$ is connected. ■

G. Bacsó raised this conjecture, naming it the "bold conjecture," at the International Conference on Combinatorics dedicated to P. Erdős on his eightieth birthday and held in July 1993 in Keszthely, Hungary.

The conjecture implies the strong perfect graph conjecture (Problem 8.1). The implication is not immediate, but Bacsó indicated that it can be deduced from linear algebra results of Padberg [1974], implying that if G is a minimal imperfect graph, then for every vertex $v \in V(G)$ the graph $G - v$ is uniquely colorable with $\chi(G) - 1$ colors. This statement was first explicitly stated and shown by Tucker [1977b] and later in a more general setting by Bland, Huang, and Trotter [1979]. These proofs were also based on linear algebra arguments, and no purely graph-theoretic proof seems known (see also Problem 8.8).

■ 8.4. RASPAIL (SHORT-CHORDED) GRAPHS

A graph G is called a Raspail graph (or a short-chorded graph) if every odd cycle of length at least 5 has a short diagonal, that is, an edge of the graph joining vertices of distance 2 apart on the cycle. Are Raspail graphs perfect? ■

These graphs were first studied around 1975–76 by C. Berge, P. Duchet, M. Las Vergnas, and H. Meyniel at the address 54 Boulevard Raspail in Paris. V. Chvátal therefore suggested the name Raspail graph (see Sun [1991]).

A proof that every Raspail graph is perfect would generalize many known results, since the class of Raspail graphs contains many of the known classes of perfect graphs—on the other hand, it seems not contained in any of the well-known classes (see the paper by Lubiw [1991] for details). Partial results have been obtained by Lubiw [1991] and Sun [1991].

As explained by Lubiw [1991], the complements of Raspail graphs are exactly the graphs in which every odd cycle of length at least 5 has a "long diagonal," that is, an edge of the graph joining two vertices as far apart as possible on the cycle. In particular, Raspail graphs are special Berge graphs, and thus the truth of the strong perfect graph conjecture (Problem 8.1) would imply that all Raspail graphs are perfect.

As a more restricted open problem, Lubiw [1991] asked for a proof that a Raspail graph is perfect if its complement is also a Raspail graph.

■ 8.5. "SEMISTRONG" PERFECT GRAPH CONJECTURE

Let the edges of a complete graph be (improperly) colored with colors red, green, and black in such a way that the red edges induce a subgraph that is perfect, as do the green edges. Suppose that the subgraph induced by the black edges is not perfect. Does there exist a K_4 subgraph with its six edges colored so that each of the colors red and green induces a simple path of length at least 1, and the color black induces a simple path of length at least 2? ■

Extending the perfect graph theorem of Lovász [1972a, 1972b] (Theorem 24 in Chapter 1) that the complement of a perfect graph is perfect, Cameron, Edmonds, and Lovász [1986] proved that if the edges of a complete graph are colored with three colors so that no triangle gets three different colors (such a coloring is sometimes called a "Gallai partition," e.g., in the paper by Körner, Simonyi, and Tuza [1992]), and if two of these colors each induces a perfect graph, then so does the third color. They made the conjecture that the answer to the question above is affirmative, in a different formulation, as a possible strengthening of Lovász's theorem.

Cameron, Edmonds, and Lovász observed that the strong perfect graph conjecture would imply their conjecture, and that this conjecture in turn would imply

Lovász's perfect graph theorem. A different conjecture of V. Chvátal having the same "semistrong" property was proved by B. Reed (see Section 1.6). Assigning colors to the edges of a complete graph subject to restrictions similar to those considered by Cameron, Edmonds, and Lovász has applications in algebraic logic (see Tuza [1991]). The result of Cameron, Edmonds, and Lovász has been generalized by Körner, Simonyi, and Tuza [1992] in terms of additivity of a functional called "graph entropy." The generalization is based on an information-theoretic characterization of perfect graphs obtained by Csiszár, Körner, Lovász, Marton, and Simonyi [1990]. (The connections to algebraic logic and information theory were pointed out to us by L. Lovász [personal communication in 1993].)

■ 8.6. HOÀNG'S CONJECTURE ON 2-COLORING EDGES

The "leaves" of a path are the two edges incident to the vertices of degree one in the path. Is it true that a graph G is perfect if the edges of G can be (improperly) colored with two colors so that the two leaves of every induced path on four vertices receive different colors? ■

The question is due to Hoàng [to appear] (it also appeared in *Math. Reviews* **92i**:05097). It is easy to see that a graph with an (improper) 2-coloring of the edges as described does not contain an odd cycle of length at least 5 nor its complement as an induced subgraph; that is, it must be a Berge graph. Hence the strong perfect graph conjecture would imply that the answer to the question is affirmative.

Similarly, Chvátal [1993] has asked if it is true that a graph G is perfect if there exists an orientation of G in which the two leaves of any induced path on four vertices point in different directions on the path (such a G is called an "opposition graph"). Partial results were obtained by Olariu [1988].

■ 8.7. NEIGHBORHOOD PERFECT GRAPHS

A neighborhood in a graph G is a subgraph induced by a vertex and its neighbors in G. The maximum number of edges in G, no two of which belong to the same neighborhood, is denoted by $\alpha_N(G)$. The minimum number of neighborhoods containing all edges of G is denoted by $\rho_N(G)$. If $\alpha_N(G') = \rho_N(G')$ for every induced subgraph G' of G, then G is called neighborhood perfect. Is it true that every neighborhood perfect graph is also perfect? ■

This question is due to Lehel and Tuza [1986], who observed that neighborhood perfect graphs are in fact Berge graphs (but the converse is not true). Thus the question can be considered a special case of the strong perfect graph conjecture (Problem 8.1).

No polynomial algorithm for recognizing a neighborhood perfect graph seems known, and neither does a structural characterization of this class of graphs.

8.8. MONSTERS

A minimal imperfect graph can be defined as a graph G satisfying

(a) $\chi(G) > \omega(G)$, and
(b) for every proper subset X of $V(G)$ the induced subgraph $G[X]$ satisfies $\chi(G[X]) = \omega(G[X])$.

A minimal imperfect graph with a largest complete subgraph of size $\omega(G) \geq 3$ and with a largest independent set of vertices of size $\alpha(G) \geq 3$ is called a "monster."

Describe properties of monsters. For example:

(1) Obtain lower bounds on the size of monsters.
(2) Prove that a monster has two complete subgraphs of size $\omega(G)$ with $\omega(G) - 1$ vertices in common.
(3) If C is a complete subgraph of G of size $\omega(G)$, prove that $G - C$ is uniquely $(\chi(G) - 1)$-colorable.
(4) Prove that a monster has no odd pair of vertices (i.e., it has no pair of nonadjacent vertices such that every induced path between them has odd length).
(5) Prove for all vertices x of a monster G that $G - x$ has an even pair of nonadjacent vertices. ■

The perfect graph theorem (Theorem 24 in Chapter 1) implies that the complement of a minimal imperfect graph is also minimal imperfect. A minimal imperfect graph G with $\omega(G) = 2$ is an odd cycle, and with $\alpha(G) = 2$ an odd-cycle complement. Thus monsters are precisely minimal counterexamples to the strong perfect graph conjecture (Problem 8.1). According to Chvátal [1993], the term "monster" was suggested by P. Duchet.

By Lovász's extended perfect graph theorem (Theorem 25 in Chapter 1), minimal imperfect graphs G with $|V(G)| = n$, $\omega(G) = \omega$, and $\alpha(G) = \alpha$ satisfy

(c) $n = 1 + \alpha\omega$,
(d) for all vertices x of G the subgraph $G - x$ has an ω-coloring with ω color classes of size α, and a covering with α disjoint complete ω-graphs.

Graphs satisfying (c) and (d) are called (α, ω)-graphs. Many examples of such graphs were presented by Chvátal, Graham, Perold, and Whitesides [1979].

The following properties were first obtained for minimal imperfect graphs by Padberg [1974], then for (α, ω)-graphs in general by Bland, Huang, and Trotter [1979].

(e) G has exactly n complete subgraphs of size ω and exactly n independent sets of size α,

(f) every vertex of G is contained in exactly ω complete subgraphs of size ω and in exactly α independent sets of size α,

(g) every complete subgraph of size ω is disjoint from exactly one independent set of size α, and vice versa.

Moreover, Tucker [1977b] and Bland, Huang, and Trotter [1979] proved for (α, ω)-graphs that

(h) For all vertices of G the graph $G - x$ is uniquely ω-colorable, and it has a unique cover with α disjoint complete subgraphs.

The proofs of (e)–(h), both for minimal imperfect graphs and for (α, ω)-graphs in general, were based on linear algebra arguments. Can more direct graph-theoretical proofs be given? The question is due to J. Fonlupt and mentioned by Chvátal [1993].

Many other properties of minimal imperfect graphs have been established by Chvátal [1985] and others (see Chvátal [1993]). Questions (1)–(5) above can all be found in the very interesting list of unsolved problems by Chvátal [1993], containing many additional questions.

Lam, Swiercz, Thiel, and Regener [1979] established that monsters have at least 21 vertices, and Chvátal [1993] reported that V.A. Gurvich and V.M. Udalov have increased this number to 25. Question (2) is due to Fonlupt and Sebő [1990] and question (3) to Sebő [1992], who proved that question (3) is equivalent to the strong perfect graph conjecture (Problem 8.1). Meyniel [1987] proved that no minimal imperfect graph contains an even pair. Question (4) was posed by Chvátal [1993], and question (5) by S. Hougardy.

■ 8.9. SQUARE-FREE BERGE GRAPHS

Let G be a graph whose induced subgraphs are neither odd cycles of length at least 5 nor the complements of odd cycles of length at least 5 nor 4-cycles. Is it true that G is perfect? ■

Gyárfás [1987] reported that the question is due to J. Lehel, who noted that the answer is known to be affirmative when "4-cycle" is replaced by "H" for any graph H on 4 vertices, except $H = 2K_2 = \overline{C_4}$, the complement of a 4-cycle. This was proved for $H = K_4$ by Tucker [1977a], for $H = K_4^-$ (K_4 minus an edge) by Parthasarathy and Ravindra [1979], and Tucker [1987] (who pointed out a mistake in the proof by Parthasarathy and Ravindra [1979]), for $H = K_{1,3}$ by Parthasarathy and Ravindra [1976], for $H = P_4$ (the path with 4 vertices) by Seinsche [1974], and for the graph consisting of a K_3 and a K_2 with a vertex in common, it follows from a theorem of Meyniel [1976] and, independently, Karapetyan and Markosyan [1976]. For the remaining 4-vertex graphs it follows from the perfect graph theorem by Lovász [1972a, 1972b] (Theorem 24 in Chapter 1), which states that G is perfect if and only if the complement $\overline{G}$ is perfect.

In Lehel's question as stated above, the condition that G must not contain an induced subgraph that is the complement of an odd cycle is superfluous. This follows from the observations that the complement of a 5-cycle is a 5-cycle and that the complement of an odd cycle of length at least 7 contains C_4 as an induced subgraph.

Wagon [1980] proved that if G is any graph without $\overline{C_4}$ as an induced subgraph, then $\chi(G) \leq \omega(G) \cdot (\omega(G) + 1)/2$. By the perfect graph theorem, the question by Lehel has an affirmative answer if and only if any G containing neither $\overline{C_4}$ nor the complement of an odd cycle of length at least 5 as an induced subgraph satisfies $\chi(G) = \omega(G)$.

If one cannot prove that when G contains neither C_4 nor any C_{2m+1}, for $m \geq 2$, as an induced subgraph, then $\chi(G) = \omega(G)$, is it at least possible to prove $\chi(G) \leq f(\omega(G))$ for some linear function f? Is there any function f for which it is possible to prove this?

■ 8.10. WEAKENED STRONG PERFECT GRAPH CONJECTURE

Does there exist some function f such that for every Berge graph G, that is, for every graph G without an odd cycle of length at least 5 or the complement of such a cycle as an induced subgraph,

$$\chi(G) \leq f(\omega(G)),$$

where $\omega(G)$ denotes the maximum size of a complete subgraph in G? ■

The strong perfect graph conjecture is equivalent to the statement that the best possible f one could hope for, namely $f(x) = x$, is a valid answer to this question. The much weaker conjecture, that some such f exists, was proposed by Gyárfás [1987], who added that it is surprising that even this is not known.

When asking this question, Gyárfás suggested that one could try to prove one of the following stronger conjectures:

Conjecture 1. *There exists a function f_1 such that for every graph G without any induced odd cycle of length at least 5 as a subgraph, $\chi(G) \leq f_1(\omega(G))$.*

Conjecture 2. *For every $m \geq 2$ there exists a function f_m such that for every graph G without any induced odd cycle of length at least $2m + 1$ as a subgraph, $\chi(G) \leq f_m(\omega(G))$.*

A weaker version of Conjecture 2 was also suggested by Gyárfás:

Conjecture 3. *For every $l \geq 4$ there exists a function g_l such that for every graph G without a cycle of length at least l as an induced subgraph, $\chi(G) \leq g_l(\omega(G))$.*

Conjecture 3 holds for $l = 4$ with $g_4(x) = x$, and this is equivalent to the statement that triangulated (also called "chordal" or "rigid circuit") graphs are perfect (Hajnal and Surányi [1958], Berge [1961], Dirac [1961]). For $l \geq 5$ the conjecture is open.

■ 8.11. GYÁRFÁS' FORBIDDEN SUBGRAPH CONJECTURE

Let H be any graph and let $\mathcal{G}(H)$ denote the set of all graphs not containing H as an induced subgraph. If F is a forest, does there exist a function f_F, such that $\chi(G) \leq f_F(\omega(G))$ for all $G \in \mathcal{G}(F)$, where $\omega(G)$ denotes the size of a maximum complete subgraph of G? ■

The question was posed by Gyárfás [1975], and independently by Sumner [1981]. A family $\mathcal{G}$ of graphs is called χ-bound if there exists a function f such that

$$\chi(G) \leq f(\omega(G)) \quad \text{for all} \quad G \in \mathcal{G}.$$

P. Erdős proved the existence, for all $g > 0$ and sufficiently large $k > 0$, of graphs $G_{g,k}$ with $\chi(G_{g,k}) = k$ and no cycles of length $< g$ (Theorem 22 in Chapter 1). Hence if $\mathcal{G}(H)$ is a χ-bound family, it follows that H is a forest. Gyárfás [1975] conjectured that the converse is also true.

Kierstead and Penrice [1990] proved that if F is a forest, then $\mathcal{G}(F)$ is χ-bound if and only if $\mathcal{G}(T)$ is χ-bound for every connected component T of F. This was also proved by Sauer [1993].

Gyárfás [1987] proved that for a star S on $n \geq 3$ vertices (i.e., S is the graph $K_{1,n-1}$ consisting of one vertex joined to $n-1$ independent vertices), $\mathcal{G}(S)$ is a χ-bound family and f_S satisfies

$$\frac{r(n-1, x+1) - 1}{n-2} \leq f_S(x) \leq r(n-1, x),$$

where r is the Ramsey function: $r(p,q)$ is the smallest number such that every graph on $r(p,q)$ vertices either contains a complete graph on p vertices or an independent set of q vertices. Gyárfás [1987] also proved that if P_n denotes the path on n vertices for $n \geq 2$, then $\mathcal{G}(P_n)$ is χ-bound, and

$$f_{P_n}(x) \leq (n-1)^{x-1}.$$

A graph B is called a "broom" if B can be obtained from a path and a star by identifying an end-vertex of the path with a vertex of the star. Gyárfás [1987] remarked that it is possible to extend the results above on stars and paths and prove that $\mathcal{G}(B)$ is χ-bound for any broom B. It has been proved by Kierstead and Penrice [1994] that $\mathcal{G}(T)$ is χ-bound for every tree T of radius at most two (i.e., there is a vertex in T of distance at most two from every other vertex). For triangle-free graphs, the positive answers for brooms and radius two trees were first obtained by Gyárfás, Szemerédi, and Tuza [1980].

A generalization of Gyárfás' problem is to ask, given a collection of graphs G_i for $1 \leq i \leq m$, if there is an upper bound for the chromatic numbers of the family of graphs $\mathcal{G}(G_1, G_2, \ldots, G_m)$ without any of the given graphs as an induced subgraph. For example, the conjecture of Gyárfás and Sumner says that $\mathcal{G}(K_n, T)$ has this property for any number n and any tree T. Kierstead and Rödl [1993] proved that the family $\mathcal{G}(K_n, K_{n,n}, T)$ is χ-bound for any n and any tree T, where $K_{n,n}$ denotes the complete bipartite graph with n vertices on each side. As an approximation to Gyárfás and Sumner's conjecture, Sauer [1993] suggested a problem of showing that there is an odd cycle C such that $\mathcal{G}(K_n, T, C)$ is χ-bound, assuming that $\mathcal{G}(K_n, T)$ is not χ-bound.

■ 8.12. QUASIPERFECT GRAPHS

Let $\vec{G}$ be an orientation of a graph G, where an edge e of G may correspond to two directed edges of $\vec{G}$ joining the same pair of vertices as e, one in each direction. A subset $S \subseteq V(G)$ is called a kernel of $\vec{G}$ if

(a) S is independent in G, and
(b) for every vertex x of G with $x \notin S$, there is a directed edge $\vec{xy}$ in $\vec{G}$ such that $y \in S$.

An orientation $\vec{G}$ of G is called "normal" (or sometimes "admissible") if every complete subgraph of G has a kernel in $\vec{G}$ (necessarily consisting of a single vertex).

G is "quasiperfect" (sometimes called "nearly perfect" or "solvable") if for every normal orientation of G, all induced subgraphs have kernels.

Conjecture 1. *Every perfect graph is quasiperfect.*

Conjecture 2. *Every quasiperfect graph is perfect.* ■

In May 1987 at the Nato Advanced Research Workshop on Cycles and Rays held in Montréal, C. Berge mentioned that Conjecture 1 had been raised by Berge and Duchet in 1979, and that Conjecture 2 had been raised by Berge and Duchet in 1982. Conjecture 1 was also mentioned by Berge [1985] (page 325), and it appeared in Berge and Duchet [1988a]. Conjecture 2 appeared in print in Berge and Duchet [1988b].

A quasiperfect graph is a Berge graph, since there is no kernel for the (normal) orientation of an odd cycle into a directed cycle, and similarly for the complement of an odd cycle. Hence the truth of the strong perfect graph conjecture (Problem 8.1) would imply Conjecture 2. Conversely, if it could be proved that a Berge graph is always quasiperfect, then Conjecture 2 would imply the strong perfect graph conjecture.

A weaker question than the conjunction of Conjecture 1 and Conjecture 2 was also mentioned by Berge in 1987: Is it true that the complement of a quasiperfect graph is quasiperfect? This would follow from Conjecture 1 and Conjecture 2 by the perfect graph theorem of Lovász [1972a, 1972b] (Theorem 24 in Chapter 1), that G is perfect if and only if $\overline{G}$ is perfect.

Kernels in directed graphs were studied by König [1936] (using the term "Punktbasis zweiter Art"), by von Neumann and Morgenstern [1944], and by Richardson [1953], who proved that every directed graph without odd directed cycles has a kernel.

For graphs G such that every odd cycle $(x_1, x_2, \ldots, x_{2k+1}, x_1)$ $(k \geq 2)$ has two diagonals $x_i x_{i+2}$ and $x_{i+1} x_{i+3}$ (adding subscripts modulo $2k + 1$) Conjecture 1 is true, as proved by Duchet and Meyniel [1983]. Conjecture 1 was proved by Maffray [1984, 1986] for graphs G such that every cycle in G of length at least 4 has a diagonal (i.e., G is triangulated, also called "chordal" or "rigid circuit"), or every odd cycle of G of length at least 5 contains two noncrossing diagonals (sometimes such a graph is called a "Gallai graph"). Blidia [1986] proved Conjecture 1 for parity graphs (a graph is a parity graph if it allows an orientation such that for every pair of vertices x, y, where $x \neq y$, all vertex-minimal directed paths from x to y have lengths of the same parity). Champetier [1989] proved Conjecture 1 for comparability graphs (the vertices of such a graph are the elements of a partially ordered set, with two of them adjacent in the graph if and only if they are comparable). A similar result for a more special class of perfect graphs was obtained by Blidia and Engel [1992].

Maffray [1992] proved that Conjecture 1 and Conjecture 2 are both true for line graphs. The line graph $L(G)$ of a graph G is the graph with vertex set $V(L(G)) = E(G)$ and two vertices of $L(G)$ joined by an edge of $L(G)$ if and only if they are adjacent edges in G. Hence a line graph is perfect if and only if it is quasiperfect. Perfect line graphs have been completely characterized by Trotter [1977], who proved that $L(G)$ is perfect if and only if G has no odd cycles of length at least 5.

It seems unknown if the perfect graphs by Meyniel [1976], in which every odd cycle of length at least 5 has at least two diagonals, are quasiperfect. It seems also unknown if the perfect "weakly triangulated" graphs by Hayward [1985] are quasiperfect; these are the graphs with the property that no induced subgraph is a cycle of length at least 5 (regardless of its parity) or the complement of such a cycle. Blidia, Duchet, and Maffray [1993] proved that when joining an independent set of vertices completely to a quasiperfect graph the resulting graph is again quasiperfect.

■ 8.13. PERFECT GRAPH RECOGNITION

Is there a polynomial algorithm to decide if any given graph G is perfect? That is, does there exist a polynomial algorithm to determine for any G

whether every induced subgraph H of G satisfies $\chi(H) = \omega(H)$, where $\omega(H)$ is the maximum size of a complete subgraph in H?

Is it an ***NP***-problem to decide whether a given graph G is perfect? That is, if G is perfect, does there exist a certificate of this fact that can be checked in polynomial time in the size of G? ■

The problem of polynomial recognition of perfect graphs has been studied by a number of people (see, e.g., Lovász [1983a], Whitesides [1984], Grötschel, Lovász, and Schrijver [1988], and Tuza [1990]). Chvátal, Lenhart, and Sbihi [1990] specifically addressed the second question "Are perfect graphs in ***NP***?" Many algorithmic questions for perfect graphs were answered by Grötschel, Lovász, and Schrijver [1981]; among other results they proved that the chromatic number of a perfect graph can be determined in polynomial time.

Based on the results of Lovász [1972a] and Padberg [1974], it can be proved that the decision problem "is G a perfect graph?" belongs to the class ***co-NP*** (see Lovász [1983a]); that is, if G is not perfect, then there exists a certificate of this fact that can be checked in polynomial time in the size of G (see Garey and Johnson [1979] for the precise definitions of ***NP*** and ***co-NP***).

Assuming the strong perfect graph conjecture is true, then proving that a graph G is perfect is equivalent to demonstrating that both G and its complement $\overline{G}$ have no odd cycles of length at least 5 as induced subgraphs. However, it seems open if the decision problem "Is G without induced odd cycles of length at least 5?" belongs to ***NP***. Clearly, the problem belongs to the class ***co-NP***. Shmoys [1981] proved that it is ***NP***-complete to decide, for a given graph G and given integer k, if G has an induced odd cycle of length at least k (also, Reed [1990] mentioned this result). A related result was obtained by Bienstock [1991], who proved that the decision problem, with G and a vertex $x \in V(G)$ given as input, "Is there an induced odd cycle of length at least 5 in G passing through x?" is ***NP***-complete. The results of Shmoys and Bienstock seem to indicate that it is a difficult problem to decide if a graph contains an induced odd cycle of length at least 5. Nevertheless, it may still be a polynomially solvable problem to decide if a graph or its complement has such an induced cycle.

For each class $\mathcal{G}$ of graphs in Table 8.1, all graphs in $\mathcal{G}$ are perfect graphs, and there is a polynomial algorithm to decide if any given graph G belongs to $\mathcal{G}$. More results on special classes of perfect graphs can be found in Golumbic [1980] and in a survey by Duchet [1984a].

Lubiw [1991] considered the class of graphs in which every odd cycle $(x_1, x_2, \dots, x_{2k+1})$ $(k \geq 2)$ has a "short diagonal" (i.e., an edge $x_i x_{i+2}$ adding indices modulo $(2k + 1)$) and asked if there exists a polynomial time recognition algorithm for this class of graphs.

For each class of graphs $\mathcal{G}$ in Table 8.2, it is known that the strong perfect graph conjecture holds for $\mathcal{G}$. That is, if $G \in \mathcal{G}$, then G is perfect if and only if G is a Berge graph. For some of these classes $\mathcal{G}$, it is an open problem if there exists a polynomial time algorithm to recognize the graphs G in $\mathcal{G}$ that are perfect.

Table 8.1

$\mathcal{G}$	Proof That Graphs of $\mathcal{G}$ Are Perfect	Polynomial Recognition Algorithm
Triangulated (also called "chordal" or "rigid circuit") graphs and their complements	Hajnal and Surányi [1958] Berge [1961] Dirac [1961]	Leuker [1974] Rose and Tarjan [1975]
Comparability graphs	See, e.g., Berge [1975]	Ghouila-Houri [1962] Gilmore and Hoffman [1964] Gallai [1967]
Graphs in which every odd cycle of length ≥ 5 contains two noncrossing diagonals	Gallai [1962]	Burlet and Fonlupt [1984] and independently Whitesides [1984]
Graphs in which every odd cycle of length ≥ 5 contains two crossing diagonals	E. Olaru and Sachs [1970]	Burlet and Uhry [1982]
Graphs without induced paths on 4 vertices	Seinsche [1974]	Easy
Graphs in which every odd cycle of length ≥ 5 contains at least two diagonals	Meyniel [1976]	Burlet and Fonlupt [1984]
Weakly triangulated graphs (see Problem 8.12)	Hayward [1985]	Hayward [1985]

Table 8.2

$\mathcal{G}$	Proof That $\mathcal{G}$ Satisfies the Strong Perfect Graph Conjecture	Polynomial Recognition Algorithm for Perfect Members of $\mathcal{G}$
Planar graphs	Tucker [1973]	Hsu [1987]
Circular arc graphs (intersection graphs of collections of arcs on a circle)	Tucker [1975]	Seems open
$K_{1,3}$-free graphs	Parthasarathy and Ravindra [1976]	Chvátal and Sbihi [1988]
K_4-free graphs	Tucker [1977a]	Open (Burlet and Uhry [1982])
K_4^--free graphs	Parthasarathy and Ravindra [1979] Tucker [1987]	Seems open
Graphs of maximum degree ≤ 6	Grinstead [1978]	Seems open
Graphs embeddable on the torus	Grinstead [1978, 1981]	Seems open

8.14. *t*-PERFECT GRAPHS

Let G be any graph and let $P_I(G)$ denote the so-called "stable set polytope"—the convex hull in $\mathbf{R}^{V(G)}$ of the set of $(0, 1)$-valued incidence vectors of the independent sets of vertices in G. That is, $P_I(G)$ is the convex hull of all vectors of the form $(x_v = \delta_{v,X} : v \in V(G))$, where $X \subseteq V(G)$ is independent in G, and

$$\delta_{v,X} = \begin{cases} 1 & \text{if } v \in X \\ 0 & \text{if } v \notin X. \end{cases}$$

Let $Q(G)$ denote the set of vectors $(x_v : v \in V(G))$ in $\mathbf{R}^{V(G)}$ satisfying the set of inequalities (a)–(c):

(a) $0 \leq x_v \leq 1$ for all $v \in V(G)$
(b) $x_u + x_v \leq 1$ for all $uv \in E(G)$
(c) $\sum\{x_v : v \in C\} \leq (|C| - 1)/2$ for all vertex sets C of odd cycles in G.

The incidence vector $(x_v = \delta_{v,X} : v \in V(G))$ of any independent set $X \subseteq V(G)$ satisfies (a), (b), and (c), hence $P_I(G) \subseteq Q(G)$. G is called t-perfect if the inclusion holds with equality, $P_I(G) = Q(G)$.

Is it possible to characterize t-perfect graphs in purely graph-theoretic terms?

Is every t-perfect graph 4-colorable? ■

Chvátal [1975] first considered t-perfect graphs, and he conjectured that every series-parallel graph, that is, a graph without a subgraph that can be obtained from K_4 by subdivisions of edges (see Dirac [1952a]), is t-perfect. We note that K_4 is not a t-perfect graph, since $(\frac{1}{3}, \frac{1}{3}, \frac{1}{3}, \frac{1}{3})$ belongs to $Q(K_4)$ but not to $P_I(K_4)$. The conjecture by Chvátal [1975] was proved by Boulala and Uhry [1979], and their result was improved by Gerards and Schrijver [1986], who showed that a graph G is t-perfect if none of its subgraphs is an "odd K_4"—a graph H obtained from K_4 by subdivisions of edges such that every triangle becomes an odd cycle in H.

Catlin [1979] proved that if G has no odd K_4 as a subgraph, then G is 3-colorable. A.M.H. Gerards and F.B. Shepherd [personal communication in 1993] extended the result of Gerards and Schrijver [1986] by proving that any graph G without an odd K_4 as a subgraph is "strongly t-perfect"; that is, every subgraph of G, whether induced or not, is t-perfect. They then extended Catlin's result by proving that any strongly t-perfect graph is 3-colorable.

F.B. Shepherd [personal communication in 1992] conjectured that $P_I(G)$ has the "integer decomposition property" (see Schrijver [1986]) for any t-perfect graph G. This would imply that the constant vector $\mathbf{x} = (\frac{1}{3}, \frac{1}{3}, \ldots, \frac{1}{3})$, which satisfies (a), (b), and (c) and hence belongs to $P_I(G)$, can be written as a sum

$$\mathbf{x} = \frac{1}{3} \cdot (\mathbf{x}_X + \mathbf{x}_Y + \mathbf{x}_Z),$$

where $\mathbf{x}_X, \mathbf{x}_Y, \mathbf{x}_Z \in P_I(G)$ have integer coordinates. Now by (a) and the definition of $P_I(G)$, $\mathbf{x}_X, \mathbf{x}_Y, \mathbf{x}_Z$ are the incidence vectors of independent sets $X, Y, Z \subseteq V(G)$ that partition $V(G)$, hence G would be 3-colored. However, M. Laurent and A. Schrijver disproved Shepherd's conjecture by exhibiting a 4-chromatic t-perfect graph. It remains unknown if every t-perfect graph is 4-colorable [personal communication from Shepherd in 1994].

Grötschel, Lovász, and Schrijver [1988] gave a discussion of t-perfect graphs, and they remarked that the decision problem "Is G a t-perfect graph?" is in ***co-NP***, but not known to be in ***NP*** or in ***P*** (see Garey and Johnson [1979] for definitions).

Combining the results of Lovász [1972b] and Fulkerson [1970, 1973], Chvátal [1975] showed that a graph G is perfect if and only if $P_I(G)$ is determined by the following two sets of inequalities:

$$x_v \geq 0 \qquad \text{for all } v \in V(G), \text{ and}$$

(d) $\sum\{x_v : v \in C\} \leq 1$ for all vertex sets C of complete subgraphs of G.

A graph G is called h-perfect (see Grötschel, Lovász, and Schrijver [1988]) if $P_I(G)$ is determined by the inequalities (a), (c), and (d). Thus both perfect graphs and t-perfect graphs are h-perfect. Grötschel, Lovász, and Schrijver [1986] proved that if G is h-perfect, then it is possible to find a maximum independent set of vertices of G in polynomial time. We do not know if the chromatic number of an h-perfect graph can be determined polynomially.

BIBLIOGRAPHY

[1961] Berge C. Färbung von Graphen, deren sämtliche bzw. deren ungerade Kreise starr sind. *Wiss. Z. Martin-Luther-Univ. Halle–Wittenberg Math.-Natur. Reihe* **10**, 114, 1961.

[1963a] Berge C. Perfect graphs. In: *Six Papers on Graph Theory*, pages 1–21. Indian Statistical Institute, Calcutta, 1963.

[1966] Berge C. Une application de la théorie des graphes à un problème de codage. In: E.R. Caianello, editor, *Automata Theory*, pages 25–34. Academic Press, 1966.

[1975] Berge C. Perfect graphs. In: D.R. Fulkerson, editor, *Studies in Graph Theory, Part 1*, volume 11 of *MAA Studies in Mathematics*, pages 1–22. The Mathematical Association of America, 1975.

[1985] Berge C. *Graphs*. North-Holland, 1985.

[1988a] Berge C. and P. Duchet. Perfect graphs and kernels. *Bull. Inst. Math. Acad. Sinica* **16**, 263–274, 1988.

[1988b] Berge C. and P. Duchet. Recent problems and results about kernels in directed graphs. In: *Applications of Discrete Mathematics (Clemson, SC, 1986)*, pages 200–204. SIAM, Philadelphia, 1988. Reprinted in *Discrete Math.* **86**, 27–31, 1990.

[1991] Bienstock D. On the complexity of testing for odd holes and induced odd paths. *Discrete Math.* **90**, 85–92, 1991. (Corrigendum in *Discrete Math.* **102**, 109, 1992.)

[1979] Bland R.G., H.-C. Huang, and L.E. Trotter. Graphical properties related to minimal imperfection. *Discrete Math.* **27**, 11–22, 1979.

[1986] Blidia M. A parity digraph has a kernel. *Combinatorica* **6**, 23–27, 1986.

[1993] Blidia M., P. Duchet, and F. Maffray. Note on kernels in perfect graphs. *Combinatorica* **13**, 231–233, 1993.

[1992] Blidia M. and K. Engel. Perfectly orderable graphs and almost all perfect graphs are kernel M-solvable. *Graphs Combin.* **8**, 103–108, 1992.

[1979] Boulala M. and J.P. Uhry. Polytope des indépendants d'un graphe série-parallelèle. *Discrete Math.* **27**, 225–243, 1979.

[1984] Burlet M. and J. Fonlupt. Polynomial algorithm to recognize a Meyniel graph. In: W.R. Pulleyblank, editor, *Progress in Combinatorial Optimization*, pages 69–99. Academic Press, 1984.

[1982] Burlet M. and J.P. Uhry. Parity graphs. In: A. Bachem, M. Grötschel, and B. Korte, editors, *Bonn Workshop on Combinatorial Optimization*, volume 16 of *Annals of Discrete Mathematics*, pages 1–26. North-Holland, 1982.

[1986] Cameron K., J. Edmonds, and L. Lovász. A note on perfect graphs. *Period. Math. Hungar.* **17**, 173–175, 1986.

[1979] Catlin P.A. Hajós' graph-coloring conjecture: variations and counterexamples. *J. Combin. Theory Ser. B* **26**, 268–274, 1979.

[1989] Champetier C. Kernels in some orientations of comparability graphs. *J. Combin. Theory Ser. B* **47**, 111–113, 1989.

[1975] Chvátal V. On certain polytopes associated with graphs. *J. Combin. Theory Ser. B* **18**, 138–154, 1975.

[1984a] Chvátal V. An equivalent version of the strong perfect graph conjecture. In: C. Berge and V. Chvátal, editors, *Topics on Perfect Graphs*, volume 21 of *Annals of Discrete Mathematics*, pages 193–195. North-Holland, 1984.

[1985] Chvátal V. Star-cutsets and perfect graphs. *J. Combin. Theory Ser. B* **39**, 189–199, 1985.

[1993] Chvátal V. Problems concerning perfect graphs. Manuscript. Dept. Computer Science, Rutgers University, 1993.

[1979] Chvátal V., R.L. Graham, A.F. Perold, and S.H. Whitesides. Combinatorial designs related to the strong perfect graph conjecture. *Discrete Math.* **26**, 83–92, 1979.

[1990] Chvátal V., W.J. Lenhart, and N. Sbihi. Two-colourings that decompose perfect graphs. *J. Combin. Theory Ser. B* **49**, 1–9, 1990.

[1988] Chvátal V. and N. Sbihi. Recognizing claw-free perfect graphs. *J. Combin. Theory Ser. B* **44**, 154–176, 1988.

[1990] Csiszár I., J. Körner, L. Lovász, K. Marton, and G. Simonyi. Entropy splitting for antiblocking corners and perfect graphs. *Combinatorica* **10**, 27–40, 1990.

[1952a] Dirac G.A. A property of 4-chromatic graphs and some remarks on critical graphs. *J. London Math. Soc.* **27**, 85–92, 1952.

[1961] Dirac G.A. On rigid circuit graphs. *Abh. Math. Sem. Univ. Hamburg* **25**, 71–76, 1961.

[1984a] Duchet P. Classical perfect graphs. In: C. Berge and V. Chvátal, editors, *Topics on Perfect Graphs*, volume 21 of *Annals of Discrete Mathematics*, pages 67–96. North-Holland, 1984.

[1983] Duchet P. and H. Meyniel. Une généralisation du théorème de Richardson sur l'existence de noyaux dans les graphes orientés. *Discrete Math.* **43**, 21–27, 1983.

[1989] Erdős P. and A. Hajnal. Ramsey-type theorems. *Discrete Appl. Math.* **25**, 37–52, 1989.

[1990] Fonlupt J. and A. Sebő. On the clique rank and the coloration of perfect graphs. In: R. Kannan and W.R. Pulleyblank, editors, *Integer Programming and Combinatorial Optimization*, pages 201–216. University of Waterloo Press, 1990.

[1970] Fulkerson D.R. The perfect graph conjecture and pluperfect graph theorem. In: R.C. Bose et al., editors, *Proc. Second Chapel Hill Conference on Combinatorial Mathematics and Its Applications*, pages 171–175, 1970.

[1971] Fulkerson D.R. Blocking and anti-blocking pairs of polyhedra. *Math. Programming* **1**, 168–194, 1971.

[1972] Fulkerson D.R. Anti-blocking polyhedra. *J. Combin. Theory Ser. B* **12**, 50–71, 1972.

[1973] Fulkerson D.R. On the perfect graph theorem. In: T.C. Hu and S. Robinson, editors, *Mathematical Programming*, pages 69–76. Academic Press, 1973.

[1962] Gallai T. Graphen mit triangulierbaren ungeraden Vielecken. *Magyar Tud. Akad. Mat. Kutató Int. Közl.* **7**, 3–36, 1962.

[1967] Gallai T. Transitive orientierbare Graphen. *Acta Math. Acad. Sci. Hungar.* **18**, 25–66, 1967.

[1979] Garey M.R. and D.S. Johnson. *Computers and Intractability: A Guide to the Theory of* **NP**-*Completeness*. W.H. Freeman and Company, 1979.

[1986] Gerards A.M.H. and A. Schrijver. Matrices with the Edmonds–Johnson property. *Combinatorica* **6**, 403–417, 1986.

[1962] Ghouila-Houri A. Caractérisation des graphes non orientés dont on peut orienter les arêtes de manière à obtenir le graphe d'une relation d'ordre. *C.R. Acad. Sci. Paris* **254**, 1370–1371, 1962.

[1964] Gilmore P.C. and A.J. Hoffman. A characterization of comparability graphs and of interval graphs. *Canad. J. Math.* **16**, 539–548, 1964.

[1980] Golumbic M.C. *Algorithmic Graph Theory and Perfect Graphs*. Academic Press, 1980.

[1978] Grinstead C.M. *The strong perfect graph conjecture for a class of graphs*. Ph.D. thesis, UCLA, 1978.

[1981] Grinstead C.M. The strong perfect graph conjecture for toroidal graphs. *J. Combin. Theory Ser. B* **30**, 70–74, 1981.

[1981] Grötschel M., L. Lovász, and A. Schrijver. The ellipsoid method and its consequences in combinatorial optimization. *Combinatorica* **1**, 169–197, 1981.

[1986] Grötschel M., L. Lovász, and A. Schrijver. Relaxations of vertex packing. *J. Combin. Theory Ser. B* **40**, 330–343, 1986.

[1988] Grötschel M., L. Lovász, and A. Schrijver. *Geometric Algorithms and Combinatorial Optimization*, volume 2 of *Algorithms and Combinatorics*. Springer-Verlag, 1988.

[1975] Gyárfás A. On Ramsey covering numbers. In: A. Hajnal, R. Rado, and V.T. Sós, editors, *Infinite and Finite Sets*, volume 10 of *Colloquia Mathematica Societatis János Bolyai*, pages 801–816. North-Holland, 1975.

[1987] Gyárfás A. Problems from the world surrounding perfect graphs. *Zastos. Mat.* XIX, 413–441, 1987.

[1980] Gyárfás A., E. Szemerédi, and Z. Tuza. Induced subtrees in graphs of large chromatic number. *Discrete Math.* **30**, 235–244, 1980.

[1958] Hajnal A. and J. Surányi. Über die Auflösung von Graphen in vollständige Teilgraphen. *Ann. Univ. Sci. Budapest Eötvös Sect. Math.* **1**, 113–121, 1958.

[1985] Hayward R.B. Weakly triangulated graphs. *J. Combin. Theory Ser. B* **39**, 200–208, 1985.

[1987] Hoàng C.T. On a conjecture of Meyniel. *J. Combin. Theory Ser. B* **42**, 302–312, 1987.

[to appear] Hoàng C.T. On the two-edge colourings of perfect graphs. To appear in *J. Graph Theory*.

[1987] Hsu W.-L. Recognizing planar perfect graphs. *J. Assoc. Comput. Mach.* **34**, 255–288, 1987.

[1976] Karapetyan I.A. and S.E. Markosyan. Perfect graphs (in Russian). *Akad. Nauk Armyan. SSR Dokl.* **63**, 292–296, 1976 (Armenian summary).

[1990] Kierstead H.A. and S.G. Penrice. Recent results on a conjecture of Gyárfás. In: *Proc. 21st S–E Conference on Combinatorics, Graph Theory and Computing, Baton Rouge, 1990, Congr. Num., 79*, pages 182–186, 1990.

[1994] Kierstead H.A. and S.G. Penrice. Radius two trees specify χ–bounded classes. *J. Graph Theory* **18**, 119–129, 1994.

[1993] Kierstead H.A. and V. Rödl. Applications of hypergraph coloring to coloring of graphs which do not contain certain trees. Manuscript, 1993. Submitted.

[1936] König D. *Theorie der endlichen und unendlichen Graphen.* Akademische Verlagsgesellschaft M.B.H. Leipzig, 1936. Reprinted by Chelsea 1950 and by B.G. Teubner 1986. English translation published by Birkhäuser 1990.

[1992] Körner J., G. Simonyi, and Z. Tuza. Perfect couples of graphs. *Combinatorica* **12**, 179–192, 1992.

[1979] Lam C.W.H., S. Swiercz, L. Thiel, and E. Regener. A computer search for (α, ω)-graphs. In: *Proc. 9th Manitoba Conference on Numerical Mathematics and Computation, Congr. Num., 27*, pages 285–289, 1979.

[1986] Lehel J. and Z. Tuza. Neighborhood perfect graphs. *Discrete Math.* **61**, 93–101, 1986.

[1974] Leuker G.S. Structured breadth first search and chordal graphs. Technical Report TR–158, Princeton University, 1974.

[1972a] Lovász L. A characterization of perfect graphs. *J. Combin. Theory Ser. B* **13**, 95–98, 1972.

[1972b] Lovász L. Normal hypergraphs and the perfect graph conjecture. *Discrete Math.* **2**, 253–267, 1972.

[1983a] Lovász L. Perfect graphs. In: L.W. Beineke and R.J. Wilson, editors, *Selected Topics in Graph Theory*, volume 2, pages 55–87. Academic Press, 1983.

[1991] Lubiw A. Short-chorded and perfect graphs. *J. Combin. Theory Ser. B* **51**, 24–33, 1991.

[1984] Maffray F. *Sur l'existence de noyaux dans les graphes parfaits*. Ph.D. thesis, Paris 6, 1984.

[1986] Maffray F. On kernels in i–triangulated graphs. *Discrete Math.* **61**, 247–251, 1986.

[1992] Maffray F. Kernels in perfect line-graphs. *J. Combin. Theory Ser. B* **55**, 1–8, 1992.

[1981] Markosyan S.E. Berge's conjecture (in Russian). In: *Applied Mathematics, No. 1*, pages 41–46. Erevan University, 1981.

[1976] Meyniel H. On the perfect graph conjecture. *Discrete Math.* **16**, 339–342, 1976.

[1987] Meyniel H. A new property of critical imperfect graphs and some consequences. *European J. Combin.* **8**, 313–316, 1987.

[1988] Olariu S. All variations on perfectly orderable graphs. *J. Combin. Theory Ser. B* **45**, 150–159, 1988.

[1972] Olaru E. Beiträge zur Theorie der perfekten Graphen. *Elektron. Informationsverarb. Kybernet.* **8**, 147–172, 1972.

[1974] Padberg M.W. Perfect zero–one matrices. *Math. Programming* **6**, 180–196, 1974.

[1976] Parthasarathy K.R. and G. Ravindra. The strong perfect graph conjecture is true for $K_{1,3}$-free graphs. *J. Combin. Theory Ser. B* **21**, 212–223, 1976.

[1979] Parthasarathy K.R. and G. Ravindra. The validity of the strong perfect graph conjecture for $(K_4 - e)$-free graphs. *J. Combin. Theory Ser. B* **26**, 98–100, 1979.

[1992] Prömel H.-J. and A. Steger. Almost all Berge graphs are perfect. *Combin. Probab. Comput.* **1**, 53–79, 1992.

[1990] Reed B.A. Perfection, parity, planarity and packing paths. In: R. Kannan and W.R. Pulleyblank, editors, *Integer Programming and Combinatorial Optimization*, pages 407–419. University of Waterloo Press, 1990.

[1953] Richardson M. Solutions of irreflexive relations. *Ann. of Math.* **58**, 573–580, 1953.

[1975] Rose D.J. and R.E. Tarjan. Algorithmic aspects of vertex elimination. In: *Proc. 7th Annual ACM Symposium on the Theory of Computing*, pages 245–254, 1975.

[1970] Sachs H. On the Berge conjecture concerning perfect graphs. In: R. Guy, H. Hanani, N.W. Sauer, and J. Schönheim, editors, *Combinatorial Structures and Their Applications*, pages 377–384. Gordon and Breach, 1970.

[1993] Sauer N.W. Vertex partition problems. In: D. Miklós, V.T. Sós, and T. Szőnyi, editors, *Combinatorics, Paul Erdős Is Eighty*, volume 1 of *Bolyai Society Mathematical Studies*, pages 361–377. János Bolyai Mathematical Society, 1993.

[1986] Schrijver A. *Theory of Linear and Integer Programming*. Wiley, 1986.

[1992] Sebő A. Forcing colorations and the perfect graph conjecture. In: E. Balas, G. Cornuéjols, and R. Kannan, editors, *Integer Programming and Combinatorial Optimisation*, volume 2. Mathematical Programming Society and Carnegie Mellon University, 1992.

[1974] Seinsche D. On a property of the class of n-colorable graphs. *J. Combin. Theory Ser. B* **16**, 191–193, 1974.

[1981] Shmoys D.B. *Perfect graphs and the strong perfect graph conjecture*. B.S.E. Thesis, Princeton University, 1981.

[1987b] Stiebitz M. Subgraphs of colour-critical graphs. *Combinatorica* **7**, 303–312, 1987.

[1981] Sumner D.P. Subtrees of a graph and chromatic number. In: G. Chartrand, Y. Alavi, D.L. Goldsmith, L. Lesniak-Foster, and D.R. Lick, editors, *The Theory and Applications of Graphs: Proc. 4th International Graph Theory Conference, Kalamazoo, 1980*, pages 557–576. Wiley 1981.

[1991] Sun L. Two classes of perfect graphs. *J. Combin. Theory Ser. B* **53**, 273–292, 1991.

[1985] Toft B. Some problems and results related to subgraphs of colour critical graphs. In: R. Bodendiek, H. Schumacher, and G. Walther, editors, *Graphen in Forschung und Unterricht. Festschrift K. Wagner*, pages 178–186. Barbara Franzbecker Verlag, 1985.

[1977] Trotter L.E. Line perfect graphs. *Math. Programming* **12**, 255–259, 1977.

[1973] Tucker A.C. The strong perfect graph conjecture for planar graphs. *Canad. J. Math.* **25**, 103–114, 1973.

[1975] Tucker A.C. Coloring a family of circular arcs. *SIAM J. Appl. Math.* **3**, 493–502, 1975.

[1977a] Tucker A.C. Critical perfect graphs and perfect 3-chromatic graphs. *J. Combin. Theory Ser. B* **23**, 143–149, 1977.

[1977b] Tucker A.C. Uniquely colorable perfect graphs. *Discrete Math.* **44**, 187–194, 1977.

[1987] Tucker A.C. Coloring perfect $(K_4 - e)$-free graphs. *J. Combin. Theory Ser. B* **42**, 313–318, 1987.

[1990] Tuza Z. Problems and results on graph and hypergraph colorings. In: M. Gionfriddo, editor, *Le Matematiche*, volume XLV, pages 219–238, 1990.

[1991] Tuza Z. Representations of relation algebras and patterns of colored triplets. In: H. Andréka, J.D. Monk, and I. Németi, editors, *Algebraic Logic*, volume 54 of *Colloquia Mathematica Societatis János Bolyai*, pages 671–693. North-Holland, 1991.

[1944] Von Neumann J. and O. Morgenstern. *Theory of Games and Economic Behavior*. Princeton University Press, 1944.

[1980] Wagon S. A bound on the chromatic number of graphs without certain induced subgraphs. *J. Combin. Theory Ser. B* **29**, 345–346, 1980.

[1977] Wessel W. Some color-critical equivalents of the strong perfect graph conjecture. In: *Proc. Internationales Kolloquium Graphentheorie und deren Anwendungen, Oberhof DDR, April 1977*, pages 300–309. Mathematisches Gesellschaft der DDR, 1977.

[1984] Whitesides S.H. A classification of certain graphs with minimal imperfection properties. In: C. Berge and V. Chvátal, editors, *Topics on Perfect Graphs*, volume 21 of *Annals of Discrete Mathematics*, pages 207–218. North-Holland, 1984.

9

Geometric and Combinatorial Graphs

■ 9.1. HADWIGER–NELSON PROBLEM

Let G be the infinite graph with all the points of the plane as vertices and having xy as an edge if and only if the points x and y have distance 1. What is the chromatic number $\chi(G)$ of G? ■

Soifer [1991, to appear] has investigated the history of this problem. It was first thought of in 1950 by E. Nelson, who noted that $\chi(G) \geq 4$. A fellow student, J. Isbell, also in 1950, noted that $\chi(G) \leq 7$. Isbell communicated the problem to several people, among them V. Klee, who passed it on to H. Hadwiger (Hadwiger [1961]). Eventually, it was published for the first time by Gardner [1960]. Erdős [1981] mentioned the problem as one of his favorites. A comprehensive survey was presented recently by Chilakamarri [1993], and another will be given by Soifer [to appear].

Hadwiger [1945] proved, as a special case of a result for higher-dimensional Euclidian space, that if the points of the plane are partitioned into five congruent closed sets, then at least one of the sets contains for every positive real number d a pair of points of distance d. He also proved that the corresponding statement for a partitioning into seven sets is false, and in fact his argument shows that $\chi(G) \leq 7$. This is demonstrated by a tessellation of the plane by regular hexagons of diameter 1 (or slightly less than 1) and a 7-coloring in which the interior points of each hexagon all get the same color (and boundary points get one of the adjacent colors).

To argue the bound $4 \leq \chi(G)$, suppose to the contrary that the plane has a 3-coloring. Then a diamond, consisting of two equilateral triangles ABC and BCD of sidelength 1 and sharing the side BC, shows that any two points A and D of distance $\sqrt{3}$ must have the same color. Then an isosceles triangle ADE with sidelengths $|AD| = |DE| = \sqrt{3}$ and $|AE| = 1$ gives a contradiction, as A, D, and E must all have the same color, which is impossible since A and E have distance 1. Thus there is a system of only seven points in the plane that already requires four colors. This 7-point configuration was first published by Moser and Moser [1961].

G is an infinite graph, but by the theorem of N.G. de Bruijn and P. Erdős (Theorem 1 in Chapter 1) it contains a finite $\chi(G)$-chromatic subgraph. It is therefore sufficient to consider the chromatic numbers of the finite subgraphs of G.

Croft [1967] considered how "dense" a plane set with no two points of unit distance can be. The density δ of such a set is defined as the maximum possible limit of the ratio between the measure of its intersection with a very large disc and the disc's area. Croft [1967] showed $0.2293 \leq \delta \leq 0.2857$. To obtain a four-color theorem for the Hadwiger–Nelson problem thus seems out of reach; not even a set that realizes no unit distance and covers 25% of the plane has been found.

Erdős asked if there is a g such that if G' is a subgraph of G in which all cycles have length at least g, then G' is 3-colorable. Wormald [1979] proved that such a g must be at least 5.

The Hadwiger–Nelson problem may be considered as the overlapping and equal-size case of Ringel's circle problem (Problem 9.2).

Hadwiger and Debrunner [1960] mentioned that A. Heppes studied a very similar problem around 1958, inspired by a suggestion of Erdős. Consider the smallest number of colors in a coloring of the points of the plane such that no color class realizes all distances. This is called the "polychromatic" number of the plane, denoted by χ_p (see Soifer [1992b]). In the Hadwiger–Nelson problem we are looking for a coloring of the plane such that each color class does not realize the distance 1. In the problem of determining the polychromatic number we look for colorings of the plane where each color class i does not realize some distance d_i. Such a coloring of the plane with k colors is said to be of type $(d_1, d_2, \ldots, d_k)$. (The notion of "type" was first introduced by Soifer [1992a, 1992b].) Raĭskiĭ [1970] noted that

$$4 \leq \chi_p \leq 6,$$

where the upper bound is due to S.B. Stechkin (see Soifer [1992b]). This was also proved by Woodall [1973] in an interesting paper on problems related to the chromatic number of the plane. Soifer [1992a, 1992b] obtained a 6-coloring of type $(1, 1, 1, 1, 1, 1/\sqrt{5})$, and Hoffman and Soifer [1993] obtained one of type $(1, 1, 1, 1, 1, \sqrt{2} - 1)$. Recently, Soifer [1994] proved that for any r in the interval $[\sqrt{2} - 1, 1/\sqrt{5}]$ there is a coloring of type $(1, 1, 1, 1, 1, r)$.

By induction it is easy to see that G contains d-regular finite subgraphs G_d for all d: Obtain G_d from two copies of G_{d-1}, such that one is translated a distance of 1 away from the other in some suitably chosen direction. Thus by an application of a theorem of N. Alon (see Section 1.9), it follows that the list-chromatic number of G is infinite (but we do not know if it is $\aleph_0$).

The problem has also been generalized and considered in higher dimensions. Let G_n denote the graph obtained by joining points at unit distance in n-dimensional real space. It is known that

$$5 \leq \chi(G_3) \leq 21,$$

where the lower bound follows from a general result of Raĭskiĭ [1970], who proved that in any decomposition of $\mathbf{R}^n$ into $n + 1$ sets, all distances are realized in one set of the decomposition. The upper bound 21 is an easy generalization of the argument above for $\mathbf{R}^2$ (see also Székely and Wormald [1989]). The best known general bounds on $\chi(G_n)$ are

$$(1 + o(1)) \cdot (6/5)^n \leq \chi(G_n) \leq (3 + o(1))^n.$$

The upper bound was obtained by Larman and Rogers [1972], and the lower bound by Frankl and Wilson [1981], thus solving a problem of Erdős, who asked if $\chi(G_n)$ grows exponentially with n. (See the papers by Erdős and Simonovits [1980], Erdős [1981], and Székely and Wormald [1989] for further results and references.)

The corresponding problem with the vertex set restricted to the set of points in n dimensions with rational coordinates has also been studied (see the paper by Zaks [1992] for references and known results). The smallest n for which this problem remains open is $n = 5$, where at least six colors are needed, as proved by Chilakamarri [1990]. For $n = 2$ the rational points of the plane can be colored with only two colors, as proved by Woodall [1973]. It would also seem natural to consider the graph induced by the set of all points in the plane that are constructible with ruler and compasses; this graph needs four colors, but we do not know if four suffice.

■ 9.2. RINGEL'S CIRCLE PROBLEM

Let $\mathcal{C}$ be a set of circles in the plane. The circles may be of different sizes, and they may overlap; however, no three circles in $\mathcal{C}$ can have a common tangent at a common point. Let G be the graph with vertices $\mathcal{C}$ and edges C_iC_j for circles C_i and C_j having a common tangent at a common point of C_i and C_j. Is there an upper bound on the chromatic number of G, and if so, what is a best possible upper bound? Is it 5? ■

The problem is due to Ringel [1959]. It is known that K_5 is not a graph of this type; however, the graph obtained from two disjoint copies of K_5, say G_1 and G_2, by removing an edge x_iy_i from each G_i, identifying x_1 with x_2 and joining y_1 and y_2 by a new edge (this is the Hajós construction; see Section 1.4) is 5-chromatic and of the desired type. This result is due to Jackson and Ringel [1984a]. No 6-chromatic example seems to be known.

For nonoverlapping circles the maximum chromatic number is 4 by the four-color theorem. Any planar graph can, in fact, be represented as the intersection graph of nonoverlapping circles, as proved by W.P. Thurston. This result has a very interesting history going back to P. Koebe in 1936, as explained by Sachs [1991] (see also the comments to Problem 16 in Bang-Jensen and Toft [1992]).

For nonoverlapping circles of equal size, the maximum is still 4. That it is at most 4 in this case can be proved easily by induction without using the four-color theorem, as pointed out by Jackson and Ringel [1984a], by remarking that two circles at maximum distance correspond to vertices each of degree at most 3 in the corresponding graph. The case of overlapping circles of equal size is unsolved—it is a famous problem due to H. Hadwiger and E. Nelson (see Problem 9.1). For dimension 3 (and higher) all four versions (overlapping or nonoverlapping; equal size or not equal size) of the problem are unsolved (see also Problem 9.3).

■ 9.3. SACHS' UNIT-SPHERE PROBLEM

Let M be a set of unit spheres in $\mathbf{R}^n$ such that no two have interior points in common. Let G be a graph with vertex set M and edges xy whenever spheres x and y touch. What is the maximum chromatic number χ_n for these graphs? ■

This problem is due to Sachs [1968]. It is the nonoverlapping and equal-size higher-dimensional case of Ringel's circle problem (Problem 9.2). It is easy to see that $\chi_2 = 4$, but even χ_3 is unknown. Jackson and Ringel [1984a] reported that $5 \le \chi_3 \le 9$.

■ 9.4. SPHERE COLORINGS

The chromatic number $\chi(S_r^2)$ of a sphere of radius r in $\mathbf{R}^3$ is the minimum number of colors possible in a coloring of the points of the surface of S_r^2 in which any two points at unit (chordal) distance apart are colored differently. Is $\chi(S_r^2) = 4$ for all $r > 1/2$? In particular, is $\chi(S_r^2) = 4$ for $r = 1/\sqrt{3}$? More generally, in dimension n, is $\chi(S_r^{n-1}) = n + 1$ for all $n \ge 3$ and all $r > 1/2$? ■

Answering a question of P. Erdős, Simmons [1974] proved $4 \le \chi(S_{1/\sqrt{3}}^2)$ and reported a coloring due to E. Straus showing $\chi(S_{1/\sqrt{3}}^2) \le 5$. A 4-coloring of the surface of S_1^2 is obtained by projecting the four faces of an inscribed tetrahedron onto the surface of the sphere, and letting the resulting spherical triangles define the four-color classes. The sidelength of the inscribed tetrahedron is $\sqrt{8/3}$, which is also the (chordal) sidelength of each spherical triangle. The diameter can be obtained as the, somewhat larger, (chordal) distance $2/\sqrt{3 - \sqrt{3}}$ from a vertex of the triangle to the midpoint of its opposing side. Therefore, when scaling to radius r, the bound $\chi(S_r^2) \le 4$ holds for $1/2 < r \le \sqrt{3 - \sqrt{3}}/2$. This result is due to Simmons [1975].

Simmons [1975, 1976] conjectured that $\chi(S_r^2) = 4$ for all values of $r > 1/2$ (trivially $\chi(S_r^2) = 1$ for $r < 1/2$ and $\chi(S_{1/2}^2) = 2$). The number 4 as a lower bound for $\chi(S_r^2)$ is an open problem for $1/2 < r < 1/\sqrt{3}$. Simmons [1976] proved that $4 \le \chi(S_r^2)$ for $r \ge 1/\sqrt{3}$ and that equality holds for $r = 1/\sqrt{2}$. A similar problem of Hadwiger [1961] and E. Nelson asks for $\chi(S^2_\infty)$, where the bounds $4 \le \chi(S^2_\infty) \le 7$ are known (see Problem 9.1).

Lovász [1983b] considered the same problem in higher dimensions and proved the deep and interesting result that for the sphere S_r^{n-1} in $\mathbf{R}^n$ of radius r, one has $n \le \chi(S_r^{n-1})$ for all $r > 1/2$, thus answering a question of P. Erdős and R.L. Graham posed at the Hungarian Combinatorial Conference in Eger, Hungary, in 1981. Moreover, Lovász [1983b] observed that by inscribing a regular simplex in the sphere, one gets $\chi(S_r^{n-1}) \le n + 1$ for $1/2 < r \le f(n)$. The value of $f(n)$ given by Lovász is, however, incorrect. For $n = 3$ the correct value obtained from the inscribed simplex is $\sqrt{3 - \sqrt{3}/2}$, as first noticed by Simmons [1975]. The correct values of $f(n)$ for $n \ge 2$ seem to be $\sqrt{(n+2)/(n+1)}/2$ for n even, and $\sqrt{(n + 3 - \sqrt{(n-1)(n+3)})/8}$ for n odd [personal communication from U. Haagerup in 1994]. Lovász [1983b] remarked that it is unknown if the lower bound n is ever attained when $n \ge 3$ (it is attained when $n = 2$). The last question posed above is thus due to Lovász.

■ 9.5. GRAPHS OF LARGE DISTANCES

Let S be a finite set of points in $\mathbf{R}^d$. Let $d_1 > d_2 > \cdots > d_k > \cdots$ be the distances between the points in S, and let $G(S, k)$ be the graph with vertex set S having an edge xy for all $x, y \in S$ such that the distance between x and y is at least d_k. How large can $\chi(G(S, k))$ be as a function of d and k? How large can $\chi(G(S, k))$ be, when S is restricted to the set of vertices of a convex polytope in $\mathbf{R}^d$? ■

These questions were asked by Erdős, Lovász, and Vesztergombi [1988, 1989]. They studied the problem for $d = 2$ and obtained the best possible bounds that if $S \subset \mathbf{R}^2$ has cardinality $|S| \ge c \cdot k^2$, where c is a positive constant, then $\chi(G(S, k)) \le 7$, and if S consists of the vertices of a convex polygon, then $\chi(G(S, k)) \le 3$. They also proved that if S consists of the vertices of a convex polygon, then $\chi(G(S, k)) \le 3k$, independently of the size of S, and conjectured that the correct answer is $2k + 1$, attained by a regular $(2k + 1)$-gon.

For $d > 2$, Erdős, Lovász, and Vesztergombi [1988] showed that $\chi(G(S, k))$ cannot be bounded independently of k, even for large $|S|$. They also remarked that the case when S consists of the vertices of a convex polytope seems to be the most difficult, and noted that even for $k = 1$ the problem is far from solved. A suggested bound of $\chi(G(S, 1)) \le d + 1$ for $k = 1$ would be equivalent to the discrete version of a long-standing open problem of Borsuk [1933], who asked: Can every set in $\mathbf{R}^d$ of diameter 1 be partitioned into $d + 1$ sets of diameter smaller than 1? This question was answered negatively by Kahn and Kalai [1993], who described finite sets S in dimension d with $\chi(G(S, 1)) > (1.2)^{\sqrt{d}}$ when d is sufficiently large, thus giving counterexamples for $d = 1325$ and for all $d \ge 2014$. The best known general upper bound for $k = 1$ seems to be $\chi(G(S, 1)) \le 2^{d-1} + 1$ due to Lassak [1982].

Bounds are known that grow more slowly with d asymptotically, but they all grow exponentially fast (see Problem D14 in Croft, Falconer, and Guy [1991] for further references and results on Borsuk's problem).

For the convex case, and $|S|$ large enough depending on d and k, Erdős, Lovász, and Vesztergombi [1988] made the conjecture that $\chi(G(S,k))$ is bounded by the least number $h(d)$ for which every compact set in $\mathbf{R}^d$ with diameter 1 can be partitioned into $h(d)$ sets with diameter smaller than 1.

■ 9.6. PRIME DISTANCE GRAPHS

For any $\mathcal{D} \subseteq \mathcal{P}$, where $\mathcal{P}$ denotes the set of all primes, let $Z(\mathcal{D})$ denote the graph with the integers as vertex set and with an edge joining i and j if and only if $|i-j| \in \mathcal{D}$. Which sets $\mathcal{D}$ satisfy $\chi(Z(\mathcal{D})) = 4$? ■

Eggleton, Erdős, and Skilton [1985, 1986] posed this problem. They proved that:

(i) $\chi(Z(\mathcal{P})) = 4$,
(ii) $\chi(Z(\mathcal{D})) \leq 2$ if $2 \notin \mathcal{D}$ or $\mathcal{D} = \{2\}$, and
(iii) $\chi(Z(\mathcal{D})) \geq 3$ if $2 \in \mathcal{D}$ and $|\mathcal{D}| \geq 2$; if in addition $3 \notin \mathcal{D}$, then $\chi(Z(\mathcal{D})) = 3$,

and moreover they conjectured that $\chi(Z(\mathcal{D})) = 4$ if and only if $\{2,3\} \subset \mathcal{D}$ and $\mathcal{D}$ contains a pair of twin primes. Eggleton, Erdős, and Skilton [1990] proved that the "if" part of this conjecture is true. However, counterexamples to the "only if" part were discovered by Eggleton [1987], by H. Walther, and by N. Alon and J. Schönheim, as reported in Eggleton, Erdős, and Skilton [1988].

Clearly, $\mathcal{D}' \subseteq \mathcal{D}$ implies that $\chi(Z(\mathcal{D}')) \leq \chi(Z(\mathcal{D}))$. Hence it follows from (i) above that $Z(\mathcal{D})$ can always be 4-colored, and from (ii) and (iii) that $Z(\mathcal{D})$ can be 3-colored if $\{2,3\} \not\subseteq \mathcal{D}$. Eggleton, Erdős, and Skilton [1985] proved that $\mathcal{D} = \{2,3,5\}$ is the only example of $\chi(Z(\mathcal{D})) = 4$ and $|\mathcal{D}| \leq 3$. Voigt and Walther [1991] announced that the minimal sets $\mathcal{D} = \{2,3,p,q\}$ without twin primes for which $\chi(Z(\mathcal{D})) = 4$ can be completely characterized by $(p,q) \in \{(11,19),(11,23),(11,37),(11,41),(17,29),(23,31),(23,41),(29,37)\}$. This may suggest, as an appropriate weakening of the original conjecture by Eggleton, Erdős, and Skilton [1986], to ask the question: Is there, for every fixed cardinality, only a finite collection of minimal sets $\mathcal{D}$ of the given size for which $\mathcal{D}$ is without twin primes and $\chi(Z(\mathcal{D})) = 4$? It is not likely that the condition "without twin primes" can be removed, since by the result of Eggleton, Erdős, and Skilton [1990] this would imply the existence of only finitely many pairs of twin primes, thus solving a long-standing open problem in number theory.

The motivation for this problem arose from a study by Eggleton, Erdős, and Skilton [1985] of natural extensions of the (by itself easy) 1-dimensional version of the problem by Hadwiger [1961] and E. Nelson of coloring the points of the plane so that pairs of points of unit distance are colored differently (Problem 9.1). Eggleton, Erdős, and Skilton [1985] mainly considered coloring the points of the real line subject to various conditions on the distances between pairs of points belonging to the same color class.

Voigt [1992] considered both the problem above and a generalization of it to infinite distance sets not necessarily consisting only of primes.

9.7. CUBE-LIKE GRAPHS

Let S be a set of n elements and $\mathcal{C} \subseteq 2^S$ a family of subsets of S. The "cube-like graph" $Q_S(\mathcal{C})$ is the graph with vertex set 2^S, where two vertices $x, y \subseteq S$ are adjacent if and only if their symmetric difference $x \Delta y$ belongs to $\mathcal{C}$.

Is it possible to give a formula in terms of n and d, where $0 < d \leq n$, for the chromatic number of the graph $Q_n^d = Q_S(\mathcal{C}_d)$, where $\mathcal{C}_d$ denotes the family of all subsets of S of size at most d? Q_n^d can also be described as the graph obtained from the usual n-dimensional cube by joining all pairs of distinct vertices at (graph-theoretic) distance at most d.

What are the possible values of k for which there exist S and $\mathcal{C} \subseteq 2^S$ such that $\chi(Q_S(\mathcal{C})) = k$? ■

Cube-like graphs were introduced by L. Lovász, thereby providing a nontrivial class of graphs having only integer eigenvalues (see Harary [1975]).

The first question asks for the chromatic number of a special class of distance graphs. The distance graph $Q_n(d_1, d_2, \ldots, d_k)$ is the graph whose vertex set consists of the subsets of a set with n elements, with two being adjacent if and only if their symmetric difference has exactly d_i elements for some $i = 1, 2, \ldots, k$. Dvořák, Havel, Laborde, and Liebl [1990] studied such graphs, and they conjectured that the chromatic number of any distance graph is a power of 2. Payan [1992] disproved their conjecture, giving as a counterexample the graph $Q_6(4)$ with $\chi(Q_6(4)) = 7$, and he reported that a computer search has shown that this graph is the smallest such example.

For G any simple graph, define $Q(G) = Q_{V(G)}(E(G))$. F. Jaeger [personal communication in 1992] proved that $\chi(Q(G)) \leq 2^{\lceil \log_2 \chi(G) \rceil}$. In fact, equality holds if $\chi(G) \leq 4$—this follows from a general result of Payan [1992], who proved that a cube-like graph of chromatic number 3 does not exist. Since G is the subgraph of $Q(G)$ induced by the 1-element subsets of $2^{V(G)}$, one has the inequality $\chi(G) \leq \chi(Q(G))$, and thus $\chi(Q(G)) = 2^{\lceil \log_2 \chi(G) \rceil}$ for all graphs G with $\chi(G) \leq 4$. Jaeger [personal communication in 1993] asked if equality holds for all graphs.

However, the answer to Jaeger's question is negative. Let G be the 5-chromatic graph obtained by joining a complete graph K_2 completely to a 5-cycle. G can also be described as the graph obtained from K_7 by deleting the edges of a 5-cycle. Assume that $V(G) = \{1, 2, \ldots, 7\}$ and let the deleted 5-cycle consist of edges 12, 23, 34, 45, and 15. $Q(G)$ has two isomorphic connected components, one corresponding to the even subsets of $V(G)$, the other to the odd subsets. We have found that $Q(G)$ allows a 7-coloring by coloring the component corresponding to the odd subsets as follows:

Color classes

1. 1, 5, 123, 267, 345, 12345, 12567, 13467, 13567, 23467
2. 2, 136, 157, 234, 12467, 13456, 23567, 24567
3. 3, 4, 126, 145, 567, 12347, 12357, 34567
4. 6, 137, 147, 236, 247, 257, 357, 456, 12356, 12456
5. 7, 127, 146, 156, 256, 347, 12346, 12457, 13457, 23456
6. 124, 134, 135, 235, 245, 367, 467, 12367, 14567, 1234567
7. 125, 167, 237, 246, 346, 356, 457, 23457

Thus

$$\chi(Q(G))) < 2^{\lceil \log_2 \chi(G) \rceil} = 8,$$

where the last equality follows from $\chi(G) = 5$.

Linial, Meshulam, and Tarsi [1988] proved that $\chi(Q_{n-1}^2) = 2^{\lceil \log_2 n \rceil}$ if n is of the form $2^t, 2^t - 1, 2^t - 2$, or $2^t - 3$. Hence for the complete graph $G = K_n$, with n of this form, $\chi(Q(G)) = 2^{\lceil \log_2 \chi(G) \rceil}$, since $Q(G)$ has two connected components isomorphic to the graph Q_{n-1}^2, one corresponding to the even-sized subsets of $V(G)$ and the other corresponding to the odd-sized subsets. It follows that $\chi(Q(K_n)) = 2^{\lceil \log_2 n \rceil}$ for all $n \leq 8$. For $n = 9$ the bounds

$$13 \leq \chi(Q(K_9)) \leq 16$$

were noted by Dvořák, Havel, Laborde, and Liebl [1990]. Very recently, G.F. Royle [personal communication in 1993] found a 14-coloring of Q_8^2 (i.e., of $Q(K_9)$) using a randomized computer search. It remains open if a 13-coloring of this graph exists.

It is known that there exist cube-like graphs with chromatic number 2^n for all $n \geq 0$, for example the graphs $Q(K_{2^n})$ or equivalently $Q_{2^n}(2)$, by the above-mentioned result of Linial, Meshulam, and Tarsi [1988]. Payan [1992] showed that cube-like graphs cannot have chromatic number 3, but some have chromatic number 7. Do cube-like graphs of chromatic number 5 or 6 exist? Does there exist a function f such that an m-chromatic cube-like graph exists if and only if $2^n - f(n) \leq m \leq 2^n$ for some n? It follows from the results above that f must satisfy $f(2) = 0, f(3) \geq 1$, and $f(4) \geq 2$.

Graphs that are "cube-like" in a different sense can be obtained as follows. For $q \geq 2$ let $Q(q)_n^d$ be the graph whose vertices are the q^n different n-tuples on a set of q elements, and where two are adjacent if and only if they are distinct and differ in at most d coordinate places. Thus $Q_n^d = Q(2)_n^d$. The chromatic numbers of such graphs seem very little studied for $q > 2$. We note that $\chi(Q(q)_n^2) \geq (q-1)n + 1$, with equality if q is a prime power and if there exists a perfect 1-error-correcting q-ary linear code of length n. A coloring of $Q(q)_n^2$ can be obtained by using such a code together with its $(q-1)n$ distinct co-sets as color classes. A particular case of equality occurs when there exists an integer m so that $n = (q^m - 1)/(q - 1)$, implying the existence of a 1-error-correcting q-ary Hamming code of length n (see MacWilliams and Sloane [1986]).

■ 9.8. ODD GRAPH CONJECTURE

Let $k \geq 2$ and let O_k denote the graph whose vertices are the $(k-1)$-subsets of the set $\{1, 2, \ldots, 2k-1\}$, with an edge joining a pair of vertices if and only if their corresponding subsets are disjoint. Is O_k k-edge-colorable if $k \neq 3$ and k is not a power of 2? ■

The question was asked by Biggs [1972], conjecturing that the answer is affirmative. He noted that O_3 is the Petersen graph, which is not 3-edge-colorable. When k is a power of 2, O_k is a k-regular graph with an odd number of vertices; it can therefore not be k-edge-colored.

O_k is a special case of a "Kneser graph," defined similarly on the set of d-subsets of any set of size n, where $0 < d \leq n/2$. Since O_k is always defined for a set of odd size $2k-1$, it is usual to refer to O_k as an "odd graph." (See Problems 9.13 and 17.10 for questions about vertex colorings of Kneser graphs.)

■ 9.9. CHORD INTERSECTION GRAPHS

Given a geometric circle in the plane, as drawn by compasses, and a set $\mathcal{C}$ of its chords, the corresponding chord intersection graph G has $V(G) = \mathcal{C}$ and two vertices x and y joined by an edge if the two chords x and y cross. Such a graph is often referred to as a "circle graph." Find a best possible upper bound for the chromatic number $\chi(G)$ of a chord intersection graph in terms of the size $\omega(G)$ of a largest complete subgraph of G. ■

As reported by Melnikov [1985] the problem is due to I.A. Karapetyan, who proved that all triangle-free chord intersection graphs are 8-colorable. L.S. Melnikov [personal communication in 1985] informed us that A.V. Kostochka improved "8-colorable" to "5-colorable." It seems unknown if it can be improved to "4-colorable." This problem was mentioned by Gyárfás and Lehel [1985], who also mentioned a 5-coloring algorithm found by Lehel.

Gyárfás [1985, 1986] and Kostochka determined an exponential upper bound in $\omega(G)$ for the chromatic number $\chi(G)$. Kostochka and J. Kratochvíl found an improved such bound [personal communication in 1994]. From the well-known existence of k-chromatic graphs without triangles no such upper bound in $\omega(G)$ for $\chi(G)$ exists in general (see Section 1.5). Does a polynomial upper bound in $\omega(G)$ for $\chi(G)$ exist for chord intersection graphs?

The problem of characterizing chord intersection graphs has been studied by Fournier [1978], de Fraysseix [1984], and Wessel and Pöschel [1985]. Algorithms for recognizing chord intersection graphs in polynomial time were described by

Bouchet [1985, 1987], Gabor, Hsu, and Supowit [1985], Naji [1985], and Spinrad [1994]. For circular arc graphs the corresponding problem by Tucker [1975] has been solved by Karapetyan [1980].

Bielecki [1948] posed a similar problem for 2-dimensional box intersection graphs G; that is, the vertices of G correspond to rectangles with sides parallel to the coordinate axes in the plane, and two vertices are joined by an edge in G whenever the corresponding rectangles overlap. Bielecki's problem was solved by Asplund and Grünbaum [1960], who proved a best possible result: A 2-dimensional box intersection graph G without triangles is 6-colorable. The smallest 6-chromatic example found by Asplund and Grünbaum has about 50,000 vertices, and they remarked that it would be interesting to know whether the size of non-5-colorable families may be substantially reduced. Moreover, Asplund and Grünbaum [1960] obtained the inequality $\chi(G) \leq 4\omega(G)^2$ for all 2-dimensional box intersection graphs G. They remarked that even this bound seems quite crude.

For 3-dimensional box intersection graphs without triangles no upper bound on χ in terms of ω exists. This result follows from a construction due to Burling [1965].

■ 9.10. GYÁRFÁS AND LEHEL'S TRIANGLE-FREE L-GRAPHS

Two line segments AB and AC in the plane with AB vertical (with B above A) and AC horizontal (with C to the right of A) is called an L-shape. Let $\mathcal{L}$ be a set of L-shapes in the plane. The corresponding L-graph G has $V(G) = \mathcal{L}$ and two vertices x and y joined by an edge if and only if the two L-shapes x and y have a common point. Does there exist a constant k such that $\chi(G) \leq k$ for all L-graphs G without triangles? ■

The problem is due to Gyárfás and Lehel [1985], who represented the 4-chromatic triangle-free Mycielski graph on 11 vertices as an L-graph, thus proving $k \geq 4$. The problem for L-graphs with infinite vertical segments has been studied by S. McGuinness [personal communication in 1993].

Gyárfás and Lehel [1985] considered several similar problems for other classes of graphs, among them chord intersection graphs, where $\mathcal{L}$ is replaced by a set $\mathcal{C}$ of chords of a geometric circle and two vertices are joined by an edge if and only if the two corresponding chords cross. For triangle-free such graphs G the upper bound $\chi(G) \leq 5$ has been obtained by Lehel and A.V. Kostochka; however, no 5-chromatic example is known (see Problem 9.9).

For "boxes" instead of L-shapes, that is, rectangular regions with sides parallel to the coordinate axes, the corresponding problem due to Bielecki [1948] was solved by Asplund and Grünbaum [1960], who proved that the best possible upper bound for the chromatic number of a triangle-free 2-dimensional box intersection graph is 6.

For triangle-free graphs, in general, no upper bound on χ exists, and this holds true even for triangle-free 3-dimensional box intersection graphs, as proved by

Burling [1965]. The opposite extreme holds for interval graphs (i.e., 1-dimensional box intersection graphs). These are perfect and hence satisfy $\chi = \omega$ (see the monograph by Golumbic [1980]).

■ 9.11. ERDŐS–FABER–LOVÁSZ PROBLEM

Let $K^1, K^2, \ldots, K^k$ be complete graphs, all of size at most k, and assume that for any two of them the intersection is either empty or exactly one vertex. Let G be the union of $K^1, K^2, \ldots, K^k$. Is G k-colorable? ■

The conjecture that the answer is affirmative was made at a party in Boulder, Colorado, in September 1972. Erdős [1981] has offered \$500 for a proof or a counterexample.

The class described above contains perhaps the simplest graphs for which the chromatic numbers are not known. Using the hypergraph H with vertex set $\{K^1, K^2, \ldots, K^k\}$ and edges E_x consisting of all those K^j sharing a common vertex x of G, the Erdős–Faber–Lovász conjecture can be formulated as follows: Let H be a hypergraph with k vertices and with edges each pair of which has at most one vertex in common. Then H is edge-colorable in k colors (where incident edges as usual get different colors). Kahn and Seymour [1992] proved that one can assign nonnegative real weights to sets of mutually disjoint edges of H, such that the weights add up to at most k, and for every edge the sum of the weights of the sets containing it is at least 1. This is the fractional version of the Erdős–Faber–Lovász conjecture, which in effect asserts that this can be done with (0,1)-valued weights. In particular, for a graph H the theorem of V.G. Vizing (Theorem 29 in Chapter 1) implies that $\chi'(H) \leq \Delta(H) + 1 \leq k$, where χ' denotes the edge-chromatic number and Δ the maximum degree.

Mitchem [1978], and independently Chang and Lawler [1988], obtained the bound $\chi(G) \leq \lceil 3k/2 - 2 \rceil$ for $k \geq 3$. Kahn [1992] proved that asymptotically the conjecture is almost true; more precisely he proved that $\chi(G) < k + o(k)$. Kahn [1994b] mentioned a suggestion by Erdős to remove the restriction on the intersection of the complete graphs: If G is the union of k complete graphs of size k, is it then possible to bound $\chi(G)$ as a function of $\omega(G)$, the size of a largest complete subgraph in G? Kahn [1994b] added that it is conceivable that $\chi(G) = \omega(G)$ might always hold, which would imply the Erdős–Faber–Lovász conjecture. Horák and Tuza [1990] obtained a best possible bound on $\chi(G)$ in terms of k: If G is the union of k complete k-graphs, then $\chi(G) \leq k^{3/2}$.

For the union of critical graphs in general, without restrictions on the intersections, Erdős recently asked for the maximum chromatic number $f(k, \ell)$ of the union of a k-critical and ℓ-critical graph [personal communication in 1994]. It is easy to see, by removing a common vertex of the two graphs, that $f(k, \ell) \leq (k - 1)(\ell - 1) + 1$. Is this the best possible? Since K_5 is the union of two 5-cycles, one gets $f(3, 3) = 5$. Is $f(4, 4) = 10$?

The following problem was suggested by Klotz [1989]: If G is the union of complete subgraphs of size at most k so that each vertex of G is contained in at most r of them, does

$$\chi(G) \leq \left(r - 1 + \frac{1}{r}\right) \cdot k$$

hold? Klotz proved that the right-hand side is an upper bound for $\omega(G)$ and characterized the graphs for which it is equal to $\omega(G)$. He remarked that for $r = 2$ the answer is affirmative by the theorem of Shannon [1949], that $\chi'(H) \leq 3\Delta(H)/2$ for a multigraph H.

■ 9.12. ALON–SAKS–SEYMOUR PROBLEM

If the edge set of G is the disjoint union of the edge sets of m complete bipartite graphs, is it true that $\chi(G) \leq m + 1$? ■

Kahn [1994b] reported that N. Alon, M. Saks, and P.D. Seymour raised this question and conjectured that the answer is affirmative.

The question generalizes a theorem of Graham and Pollak [1972] stating that the edge set of the complete graph K_n cannot be partitioned into the edge sets of fewer than $n - 1$ complete bipartite graphs. A very elegant short algebraic proof of the theorem of Graham and Pollak is due to Tverberg [1982], but no combinatorial proof seems known. Kahn [1994b] wrote that the conjecture of Alon, Saks, and Seymour "*...presents the nice challenge of either extending the algebraic proofs (which would be very appealing, but does not seem easy), or at least finding a combinatorial proof of the Graham–Pollak Theorem, a goal which has also proved fairly elusive.*"

Replacing "complete bipartite graphs" with "complete graphs on m vertices" in the Alon–Saks–Seymour problem, and replacing "$\chi(G) \leq m + 1$" with "$\chi(G) \leq m$," leads to a formulation of the Erdős–Faber–Lovász problem (Problem 9.11).

■ 9.13. GENERAL KNESER GRAPHS

For integers $n > k > t > 0$ the general Kneser graph $K(n, k, t)$ is defined as the graph with the set of all k-subsets of the set $\{1, 2, \ldots, n\}$ as vertex set and two such sets X and Y joined by an edge if and only if $|X \cap Y| < t$. What is the chromatic number $\chi(K(n, k, t))$ of $K(n, k, t)$? ■

Frankl [1985] asked this question and conjectured that $\chi(K(n, k, t)) = T(n, k, t)$, for n sufficiently large and $t \geq 2$, where $T(n, k, t)$ is the smallest number of t-subsets of an n-set X such that any k-subset of X contains at least one of the t-subsets. Thus $T(n, k, t)$ is an obvious upper bound for $\chi(K(n, k, t))$. Frankl [1985] proved the conjecture for $t = 2$, and he showed that $\chi(K(n,k,t)) = (1 + o(1))T(n,k,t)$ holds for all $t \geq 3$. The problem of determining $T(n,k,t)$ is a famous extremal problem due to Turán [1941, 1954, 1961]. It was solved for $t = 2$ by Turán [1941, 1954] (this is the famous Turán's Theorem, to be found in most graph theory textbooks, e.g., Bondy and Murty [1976]), and remains open for $t \geq 3$. For some of the known estimates for $T(n,k,t)$, see Brouwer and Voorhoeve [1979] and Füredi [1991]. Füredi [1991] gave a comprehensive survey of Turán-type problems, drawing attention to the special case $k = 4$ and $t = 3$. He mentioned that, in memory of Turán, P. Erdős has offered \$1000 for the determination of $T(n,4,3)$.

Kneser [1955] conjectured $\chi(K(n,k,1)) = n - 2k + 2$ for all $n > k > 1$. This was proved by Lovász [1978], and with a shorter proof by Bárány [1978]. A more general version of Kneser's conjecture was stated and proved by Sarkaria [1990].

Tort [1983] proved that $\chi(K(n,3,2)) = T(n,3,2) \left(= \left\lfloor \left(\frac{n-1}{2}\right)^2 \right\rfloor \right)$ for $n \geq 6$.

■ 9.14. QUESTION OF GALLAI RELATED TO SPERNER'S LEMMA

Let $n \geq 1$ and let $\mathcal{S} = \{S_i : i = 1,2,\ldots,m\}$ $(m \geq 1)$ be a set of n-dimensional simplexes together forming a subdivision of an n-dimensional simplex S, that is, $S = \bigcup_{1 \leq i \leq m} S_i$, where each pair S_i and S_j have disjoint interiors $(1 \leq i < j \leq m)$, and if some point $p \in \mathbf{R}^n$ is a vertex (a zero-dimensional face) of S_i for some $i = 1,2,\ldots,m$, then p is a vertex of every S_j such that $p \in S_j$.

Let G be the graph obtained from $\mathcal{S}$ as follows. The vertices of G are of three kinds:

(a) a vertex v_p for each $p \in \mathbf{R}^n$ that is a vertex of some $S_i \in \mathcal{S}$,
(b) vertices v_i, $i = 1,2,\ldots,m$, each corresponding to a simplex of $\mathcal{S}$, and
(c) $n + 1$ vertices $v^1, v^2, \ldots, v^{n+1}$, corresponding to the $n + 1$ $(n - 1)$-dimensional faces $F_1, F_2, \ldots, F_{n+1}$ of S.

The edges of G join v_p to v_i if $p \in S_i$, and v_p to v^j whenever $p \in F_j$. Finally, the vertices $v^1, v^2, \ldots, v^{n+1}$ form a complete $(n + 1)$-subgraph of G.

It follows from Sperner's simplex lemma in n dimensions that G is an $(n + 2)$-chromatic graph. Is G an $(n + 2)$-critical graph? Is G naturally a member of a larger class of $(n + 2)$-chromatic graphs defined in purely graph-theoretical terms? Is G a member of such a class of $(n + 2)$-critical graphs? ■

The statement $\chi(G) = n + 2$ is, in fact, equivalent to the n-dimensional version of Sperner's lemma proved by Sperner [1928]. This fact was first observed by T. Gallai, who also raised the question of criticality and proved that the answer is affirmative in the 2-dimensional case [personal communication in 1969] (see the paper by Nielsen and Toft [1975]). The answer to the question of criticality may possibly be known to topologists to be affirmative for $n \geq 3$, but we have not found any references to support this.

In one dimension G is an odd cycle of length at least 5, hence G is 3-critical. In two dimensions G can be represented as a plane graph Γ consisting of a 6-cycle C with vertices $s_1, t_1, s_2, t_2, s_3, t_3$ in cyclic order, together with edges (s_1, s_2), (s_2, s_3), and (s_3, s_1) outside C together with vertices and edges inside C such that all faces inside C are bounded by 4-cycles. Gallai [personal communication in 1969] proved that such a plane graph Γ is critical 4-chromatic if and only if (i) Γ has no vertex of degree 2, and (ii) every 4-cycle of the graph $\Gamma - (s_1, s_2) - (s_2, s_3) - (s_3, s_1)$ is the boundary of a face in Γ. It is not hard to see that conditions (i) and (ii) are satisfied by the graph Γ representing G. The critical members of the class of 4-chromatic planar graphs obtained by replacing the 6-cycle C by any cycle of length 2 modulo 4 were characterized by Nielsen and Toft [1975].

Tompkins [1964] gave an extremely well-written introduction to Sperner's lemma and some of its applications, including a proof using a counting argument. As far as we know, the first proof using graph theory was presented by Wagner [1970], and such a proof was also included, for the 2-dimensional case, in the introductory text on graph theory by Bondy and Murty [1976]. These proofs rely on the very elementary graph-theoretic proposition that the number of odd-degree vertices in a graph is always even. Thus Sperner's lemma has become a classical example of applying graph theory to proofs of nontrivial results in other mathematical fields.

BIBLIOGRAPHY

[1960] Asplund E. and B. Grünbaum. On a coloring problem. *Math. Scand.* **8**, 181–188, 1960.

[1992] Bang-Jensen J. and B. Toft. Unsolved problems presented at the Julius Petersen Graph Theory Conference. *Discrete Math.* **101**, 351–360, 1992.

[1978] Bárány I. A short proof of Kneser's conjecture. *J. Combin. Theory Ser. A* **25**, 325–326, 1978.

[1948] Bielecki A. Problems 56. *Colloq. Math.* **1**, 333, 1948.

[1972] Biggs N.L. An edge-colouring problem. *Amer. Math. Monthly* **79**, 1018–1020, 1972.

[1976] Bondy J.A. and U.S.R. Murty. *Graph Theory with Applications*. Macmillan, 1976.

[1933] Borsuk K. Drei Sätze über die n-dimensionale euklidische Sphäre. *Fund. Math.* **20**, 177–190, 1933.

[1985] Bouchet A. Un algorithme polynomial pour reconnaitre les graphes d'alternance. *C.R. Acad. Sci. Paris Sér. I Math.* **300**, 569–572, 1985.

[1987] Bouchet A. Reducing prime graphs and recognizing circle graphs. *Combinatorica* **7**, 243–254, 1987.

[1979] Brouwer A.E. and M. Voorhoeve. Turán theory and the lotto problem. In: A. Schrijver, editor, *Packing and Covering in Combinatorics*, volume 106 of *Mathematical Centre Tracts*, pages 99–105. Mathematisch Centrum, 1979.

[1965] Burling J.P. *On coloring problems of families of prototypes*. Ph.D. thesis, University of Colorado, 1965.

[1988] Chang W.I. and E.L. Lawler. Edge coloring of hypergraphs and a conjecture of Erdős, Faber, Lovász. *Combinatorica* **8**, 293–295, 1988.

[1990] Chilakamarri K.B. On the chromatic number of rational five-space. *Aequationes Math.* **39**, 146–148, 1990.

[1993] Chilakamarri K.B. The unit distance graph problem: a brief survey and some new results. *Bull. Inst. Combin. Appl.* **8**, 39–60, 1993.

[1967] Croft H.T. Incidence incidents. *Eureka* **30**, 22–26, 1967.

[1991] Croft H.T., K.J. Falconer, and R.K. Guy. *Unsolved Problems in Geometry*. Springer-Verlag, 1991.

[1990] Dvořák T., I. Havel, J.M. Laborde, and P. Liebl. Generalized hypercubes and graph embedding with dilation. *Rostock. Math. Kolloq.* **39**, 13–20, 1990.

[1987] Eggleton R.B. Three unsolved problems in graph theory. *Ars Combin.* **23–A**, 105–121, 1987.

[1985] Eggleton R.B., P. Erdős, and D.K. Skilton. Coloring the real line. *J. Combin. Theory Ser. B* **39**, 86–100, 1985.

[1986] Eggleton R.B., P. Erdős, and D.K. Skilton. Research Problem 77. *Discrete Math.* **58**, 323, 1986.

[1988] Eggleton R.B., P. Erdős, and D.K. Skilton. Update information on Research Problem 77. *Discrete Math.* **69**, 105–106, 1988.

[1990] Eggleton R.B., P. Erdős, and D.K. Skilton. Colouring prime distance graphs. *Graphs Combin.* **6**, 17–32, 1990.

[1981] Erdős P. On the combinatorial problems which I would most like to see solved. *Combinatorica* **1**, 25–42, 1981.

[1988] Erdős P., L. Lovász, and K. Vesztergombi. The chromatic number of the graph of large distances. In: A. Hajnal, L. Lovász, and V.T. Sós, editors, *Combinatorics*, volume 52 of *Colloquia Mathematica Societatis János Bolyai*, pages 547–551. North-Holland, 1988.

[1989] Erdős P., L. Lovász, and K. Vesztergombi. On the graph of large distances. *Discrete Comput. Geom.* **4**, 541–549, 1989.

[1980] Erdős P. and M. Simonovits. On the chromatic number of geometric graphs. *Ars Combin.* **9**, 229–246, 1980.

[1978] Fournier J.C. Une caracterisation des graphes de cordes. *C.R. Acad. Sci. Paris Sér. A–B* **286**, A811–A813, 1978.

[1985] Frankl P. On the chromatic number of the general Kneser-graph. *J. Graph Theory* **9**, 217–220, 1985.

[1981] Frankl P. and R.M. Wilson. Intersection theorems with geometric consequences. *Combinatorica* **1**, 357–368, 1981.

[1984] de Fraysseix H. A characterization of circle graphs. *European J. Combin.* **5**, 223–238, 1984.

[1991] Füredi Z. Turán type problems. In: A.D. Keedwell, editor, *Surveys in Combinatorics: Proc. 13th British Combinatorial Conference*, pages 253–300. Cambridge University Press, 1991.

[1985] Gabor C.P., W.-L. Hsu, and K.J. Supowit. Recognizing circle graphs in polynomial time. In: *Proc. 26th IEEE Annual Symposium*, 1985.

[1960] Gardner M. Mathematical games. *Sci. Amer.* **206**, 172–180, October 1960.

[1980] Golumbic M.C. *Algorithmic Graph Theory and Perfect Graphs*. Academic Press, 1980.

[1972] Graham R.L. and H.O. Pollak. On embedding graphs in squashed cubes. In: Y. Alavi, D.R. Lick, and A.T. White, editors, *Graph Theory and Applications*, volume 303 of *Lecture Notes in Mathematics*, pages 99–110. Springer-Verlag, 1972.

[1985] Gyárfás A. On the chromatic number of multiple interval graphs and overlap graphs. *Discrete Math.* **55**, 161–166, 1985.

[1986] Gyárfás A. Corrigendum. *Discrete Math.* **62**, 333, 1986.

[1985] Gyárfás A. and J. Lehel. Covering and coloring problems for relatives of intervals. *Discrete Math.* **55**, 167–180, 1985.

[1945] Hadwiger H. Überdeckung des euklidischen Raumes durch kongruente Mengen. *Portugal. Math.* **4**, 238–242, 1945.

[1961] Hadwiger H. Ungelöste Probleme No. 40. *Elem. Math.* **16**, 103–104, 1961.

[1960] Hadwiger H. and H. Debrunner. *Kombinatorische Geometrie in der Ebene*. L'Enseignement Mathématique, Genève, 1960.

[1975] Harary F. Four difficult unsolved problems in graph theory. In: M. Fiedler, editor, *Recent Advances in Graph Theory, Proc. Symposium, Prague, June 1974*, pages 249–256. Academia Praha, 1975.

[1993] Hoffman I. and A. Soifer. Almost chromatic number of the plane. *Geombinatorics* **III**, 38–40, 1993.

[1990] Horák P. and Z. Tuza. A coloring problem related to the Erdős–Faber–Lovász conjecture. *J. Combin. Theory Ser. B* **50**, 321–322, 1990. (Erratum in *J. Combin. Theory Ser. B* **51**, 329, 1991.)

[1984a] Jackson B. and G. Ringel. Colorings of circles. *Amer. Math. Monthly* **91**, 42–49, 1984.

[1992] Kahn J. Coloring nearly-disjoint hypergraphs with $n + o(n)$ colors. *J. Combin. Theory Ser. A* **59**, 31–39, 1992.

[1994b] Kahn J. Recent results on some not–so–recent hypergraph matching and covering problems. In: P. Frankl, Z. Füredi, G.O.H. Katona, and D. Miklós, editors, *Extremal Problems for Finite Sets*, volume 3 of *Bolyai Society Mathematical Studies*. János Bolyai Mathematical Society, 1994.

[1993] Kahn J. and G. Kalai. A counterexample to Borsuk's conjecture. *Bull. Amer. Math. Soc.* **29**, 60–62, 1993.

[1992] Kahn J. and P.D. Seymour. A fractional version of the Erdős–Faber–Lovász conjecture. *Combinatorica* **12**, 155–160, 1992.

[1980] Karapetyan I.A. Coloring of arc graphs (in Russian). *Akad. Nauk Armyan. SSR Dokl.* **70**, 306–311, 1980. (Armenian summary).

[1989] Klotz W. Clique covers and coloring problems of graphs. *J. Combin. Theory Ser. B* **46**, 338–345, 1989.

[1955] Kneser M. Aufgabe 360. *Jahresber. Deutsch. Math.-Verein.* **58**, 2. Abteilung, 27, 1955.

[1972] Larman D.G. and C.A. Rogers. The realization of distances within sets in Euclidian space. *Mathematika* **19**, 1–24, 1972.

[1982] Lassak M. An estimate concerning Borsuk's partition problem. *Bull. Acad. Polon. Sci. Sér. Math.* **30**, 449–451, 1982.

[1988] Linial N., R. Meshulam, and M. Tarsi. Matroidal bijections between graphs. *J. Combin. Theory Ser. B* **45**, 31–44, 1988.

[1978] Lovász L. Kneser's conjecture, chromatic number, and homotopy. *J. Combin. Theory Ser. A* **25**, 319–324, 1978.

[1983b] Lovász L. Self-dual polytopes and the chromatic number of distance graphs on the sphere. *Acta Sci. Math. (Szeged)* **45**, 317–323, 1983.

[1986] MacWilliams F.J. and N.J.A. Sloane. *The Theory of Error-Correcting Codes*, 5th ed. North-Holland, 1986.

[1985] Melnikov L.S. Problems 11 and 12. In: H. Sachs, editor, *Graphs, Hypergraphs and Applications: Proc. Conference on Graph Theory, Eyba, DDR, October 1984*, page 214, 1985. *Teubner Texte zur Mathematik*, 73.

[1978] Mitchem J. On n-coloring certain finite set systems. *Ars Combin.* **5**, 207–212, 1978.

[1961] Moser L. and W. Moser. Problem and solution P10. *Canad. Math. Bull.* **4**, 187–189, 1961.

[1985] Naji W. Reconnaissance des graphes de cordes. *Discrete Math.* **54**, 329–337, 1985.

[1975] Nielsen F. and B. Toft. On a class of planar 4-chromatic graphs due to T. Gallai. In: M. Fiedler, editor, *Recent Advances in Graph Theory: Proc. Symposium, Prague, June 1974*, pages 425–430. Academia Praha, 1975.

[1992] Payan C. On the chromatic number of cube-like graphs. *Discrete Math.* **103**, 271–277, 1992.

[1970] Raĭskiĭ D.E. The realisation of all distances in a decomposition of the space $\mathbf{R}^n$ into $n + 1$ parts (in Russian). *Math. Z.*, **7**, 319–323, 1970. English translation in *Math. Notes* **7**, 194–196, 1970.

[1959] Ringel G. *Färbungsprobleme auf Flächen und Graphen.* VEB Deutscher Verlag der Wissenschaften, 1959.

[1968] Sachs H. Problem 6. In: H. Sachs, H.-J. Voss, and H. Walther, editors, *Beiträge zur Graphentheorie vorgetragen auf dem internationalen Kolloquium, Manebach, DDR, Mai 1967*, page 225. B.G. Teubner, 1968.

[1991] Sachs H. Coin graphs, polyhedra and conformal mapping. Manuscript, 1991. Submitted to *Proc. Workshop on Algebraic and Topological Methods in Graph Theory, Yugoslavia, June 1991*. To appear in *Discrete Math.* **134**.

[1990] Sarkaria K.S. A generalized Kneser conjecture. *J. Combin. Theory Ser. B* **49**, 236–240, 1990.

[1949] Shannon C.E. A theorem on coloring the lines of a network. *J. Math. Phys.* **28**, 148–151, 1949.

[1974] Simmons G.J. On a problem of Erdős concerning a 3-coloring of the unit sphere. *Discrete Math.* **8**, 81–84, 1974.

[1975] Simmons G.J. Bounds on the chromatic number of the sphere. In: *Proc. 6th S–E Conference on Combinatorics, Graph Theory and Computing, Boca Raton, 1975, Congr. Num., 14*, pages 541–548, 1975.

[1976] Simmons G.J. The chromatic number of the sphere. *J. Austral. Math. Soc. Ser. A* **21**, 473–480, 1976.

[1991] Soifer A. Chromatic number of the plane: a historical essay. *Geombinatorics* **I**(3), 13–15, 1991.

[1992a] Soifer A. A six-coloring of the plane. *J. Combin. Theory Ser. A* **61**, 292–294, 1992.

[1992b] Soifer A. Relatives of the chromatic number of the plane I. *Geombinatorics* **I**(4), 13–17, 1992.

[1994] Soifer A. Six-realizable set X_6. *Geombinatorics* **III**, 140–145, 1994.

[to appear] Soifer A. *Mathematical Coloring Book.* Center for Excellence in Mathematical Education. To appear.

[1928] Sperner E. Neuer Beweis für die Invarianz der Dimensionszahl und des Gebietes. *Hamburger Abh.* **6**, 265–272, 1928.

[1994] Spinrad J. Recognition of circle graphs *J. Algorithms* **16**, 264–282, 1994.

[1989] Székely L.A. and N.C. Wormald. Bounds on the measurable chromatic number of $\mathbf{R}^n$. *Discrete Math.* **75**, 343–372, 1989.

[1964] Tompkins C.B. Sperner's lemma and some extensions. In: E.F. Beckenbach, editor, *Applied Combinatorial Mathematics*, Chap. 15, pages 416–455. Wiley, 1964.

[1983] Tort J.R. Un problème de partition de l'ensemble des parties à trois éléments d'un ensemble fini. *Discrete Math.* **44**, 181–185, 1983.

[1975] Tucker A.C. Coloring a family of circular arcs. *SIAM J. Appl. Math.* **3**, 493–502, 1975.

[1941] Turán P. On an extremal problem in graph theory (in Hungarian). *Mat. Fiz. Lapok.* **48**, 436–452, 1941. English translation in P. Erdős, editor, *Collected Papers of Paul Turan*, Volume 1, pages 231–240, Akadémiai Kiadó Budapest, 1990.

[1954] Turán P. On the theory of graphs. *Colloq. Math.* **3**, 19–30, 1954.

[1961] Turán P. Research problems. *Magyar Tud. Akad. Mat. Kutató Int. Közl.* **6**, 417–423, 1961.

[1982] Tverberg H. On the decomposition of K_n into complete bipartite graphs. *J. Graph Theory* **6**, 493–494, 1982.

[1992] Voigt M. On the chromatic number of distance graphs. *J. Inform. Process. Cybernet.* EIK **28**, 21–28, 1992.

[1991] Voigt M. and H. Walther. On the chromatic number of special distance graphs. *Discrete Math.* **97**, 395–397, 1991.

[1970] Wagner K. *Graphentheorie.* Number 248/248a* in B.I. Hochschultaschenbücher. Bibliographisches Institut, 1970.

[1985] Wessel W. and R. Pöschel. On circle graphs. In: H. Sachs, editor, *Graphs, Hypergraphs and Applications, Proc. Conference on Graph Theory, Eyba, DDR, October 1984*, pages 207–210, 1985. *Teubner Texte zur Mathematik*, 73.

[1973] Woodall D.R. Distances realized by sets covering the plane. *J. Combin. Theory Ser. A* **14**, 187–200, 1973.

[1979] Wormald N.C. A 4-chromatic graph with a special drawing. *J. Austral. Math. Soc. Ser. A* **28**, 1–8, 1979.

[1992] Zaks J. On the chromatic number of some rational spaces. *Ars Combin.* **33**, 253–256, 1992.

10

Algorithms

10.1. POLYNOMIAL GRAPH COLORING

Does there exist a polynomial graph coloring algorithm producing for every graph G a coloring of G using at most $c \cdot \chi(G) \cdot |V(G)|^{1-\varepsilon}$ colors, where c and ε are constants satisfying $c > 0$ and $0 < \varepsilon \leq 1$? ∎

The question is due to Johnson [1978]. Based on an argument of P. Erdős, Johnson [1974] showed the existence of a polynomial algorithm A using $A(G)$ colors to color any arbitrary graph G of chromatic number $\chi(G)$ at least 2, where

$$A(G) \leq c \cdot \log \chi(G) \cdot |V(G)| / \log |V(G)|,$$

implying that the performance guarantee $A(G)/\chi(G)$ is at most $c \cdot |V(G)| / \log |V(G)|$. Halldórsson [1990] improved the bound to

$$A(G)/\chi(G) \leq c \cdot |V(G)| \cdot (\log \log |V(G)|)^2 / (\log |V(G)|)^3,$$

and this seems to be the currently best known order of magnitude for a polynomial graph coloring algorithm.

Based on a general method by Arora, Lund, Motwani, Sudan, and Szegedy [1992] for proving the intractability of a range of combinatorial approximation problems, Lund and Yannakakis [1993] proved that for some constant $\varepsilon_0 > 0$, no polynomial graph coloring algorithm A has a performance guarantee $A(G)/\chi(G) \leq |V(G)|^{\varepsilon_0}$, unless $\boldsymbol{NP} = \boldsymbol{P}$ (Theorem 20 in Chapter 1). A corollary of this result is an affirmative answer to a question of Johnson [1978]:

Theorem. *If there is a polynomial graph coloring algorithm that uses at most $c \cdot \chi(G)$ colors (where c is a positive constant) to color any given graph G, then there exists a polynomial algorithm to determine $\chi(G)$.*

If there is a polynomial algorithm to decide for any graph G if G is 3-colorable, then $\boldsymbol{NP} = \boldsymbol{P}$ and there is a polynomial algorithm to determine $\chi(G)$ (see Theorem 19 in Chapter 1). Using this, Garey and Johnson [1976] proved the theorem for $c < 2$. For $c < 4/3$ there is a very simple proof (see Garey and Johnson [1979]).

Restricting the discussion to 3-colorable graphs, Wigderson [1983] proved that there exists a polynomial algorithm for coloring any such graph on n vertices with $O(\sqrt{n})$ colors. This was improved by Blum [1990] to $O(n^{3/8} \log^{5/8} n)$ colors, the so far best known bound. We do not know if there exists a constant k and a polynomial algorithm for coloring any 3-colorable graph with k colors. Khanna, Linial, and Safra [1993] have shown that such a k must be at least 5, unless $\boldsymbol{NP} = \boldsymbol{P}$.

Even if there is a polynomial algorithm to decide for any planar graph G of maximum degree 4 if G is 3-colorable, then $\boldsymbol{NP}=\boldsymbol{P}$ would follow (see Theorem 19 in Chapter 1). But for graphs in general of maximum degree 3 Brooks' theorem (Theorem 11 in Chapter 1) gives a very simple such polynomial algorithm. If, in contrast to asking for just the existence of a 3-coloring, the number of different 3-colorings is wanted, then the situation changes drastically. Edwards [1992] proved that asking for the precise number of different 3-colorings of a 3-regular bipartite graph is already a **#*P***-complete problem. Enumeration problems in this class are considered even more unlikely to be polynomially solvable than their counterparts among the decision problems, the ***NP***-complete problems (see Garey and Johnson [1979]).

■ 10.2. POLYNOMIAL APPROXIMATION

Does there exist a function g and a polynomial algorithm that for any given input graph G will find a number s, such that the chromatic number of G satisfies $s \leq \chi(G) \leq g(s)$? ■

This question was asked by Alon [1993], who proved that if the chromatic number $\chi(G)$ is replaced by the list-chromatic number $\chi_\ell(G)$, then the corresponding question has a positive answer (see Problem 17.2). The proof uses the existence of a polynomial algorithm for the coloring number (see Sections 1.3 and 1.9).

■ 10.3. EVEN CHROMATIC GRAPHS

If $\chi(G)$ is known to be even, does there exist a polynomial algorithm to decide if $\chi(G) \leq 4$? ■

This problem is due to Kratochvíl and Nešetřil [1989], and it was posed by Nešetřil at the graph theory meeting in memory of G.A. Dirac, held at Sandbjerg, Denmark, in June 1985.

The corresponding problem with $\chi(G) \leq 2$ can be solved by a simple polynomial algorithm. The corresponding problem with $\chi(G) \leq 6$ is ***NP***-complete, since a polynomial algorithm to decide the answer to this could be used to decide for any graph G whether $\chi(G) \leq 3$ (take two disjoint copies of G joined completely and decide if this even chromatic graph is 6-colorable), and this 3-coloring problem is known to be ***NP***-complete (see Theorem 19 in Chapter 1).

Khanna, Linial, and Safra [1993] proved that if there is a polynomial algorithm with output YES for all 3-colorable graphs and output NO for all non-4-colorable graphs (and either of the two answers for 4-chromatic graphs), then ***NP*** = ***P***. We express this by saying that it is ***NP***-hard to distinguish between "$\chi \leq 3$" and "$\chi \geq 5$." It follows that it is ***NP***-hard to 4-color 3-colorable graphs (and more generally, that it is ***NP***-hard to $(5k - 1)$-color $3k$-colorable graphs), and also that it is ***NP***-hard to distinguish between "$\chi \leq 4$" and "$\chi \geq 6$." However, this does not exclude the possibility that there is a polynomial algorithm that can distinguish between "$\chi \leq 4$" and "$\chi \geq 6$" for the special class of even-chromatic graphs (although it makes it seem unlikely).

A related question was asked by Thostrup [1992]. If it is known for some $g > 3$ that

(i) $\chi(G) \neq 7$, and
(ii) every cycle in G has length at least g,

does there exist a polynomial algorithm to decide if $\chi(G) \leq 6$? If condition (ii) is removed, this problem becomes ***NP***-complete by the construction mentioned above. If condition (i) is removed, the problem becomes ***NP***-complete, since then the ***NP***-complete problem of deciding whether a given graph G is 6-colorable can be reduced to it. This follows from the existence of 7-critical graphs H with cycles of length at least g (Theorem 22 in Chapter 1): The reduction consists in applying Hajós' construction (see Theorem 18 in Chapter 1) with such an H for each edge of G—the obtained graph is 7-chromatic if and only if $\chi(G) > 6$. However, neither reduction shows that the combined problem given above is ***NP***-complete. Thostrup [1992] showed that if this problem is ***NP***-complete (with $g = 43$), then the problem of deciding whether a given graph has an orientation as a Hasse diagram of some partially ordered set is ***NP***-complete. (A directed graph is the Hasse diagram of a partially ordered set if and only if none of its cycles has all, or all but one, of the edges of the cycle of the same direction on the cycle.) A proof that the problem of deciding whether a given graph has an orientation as a Hasse diagram is indeed ***NP***-complete was published by Nešetřil and Rödl [1987]. Thostrup [1992] pointed out a mistake in the proof, and it has since been corrected by Nešetřil and Rödl [1993] using a deep result of Lund and Yannakakis [1993] (Theorem 20 in Chapter 1). A more elementary proof was given by Brightwell [1993].

■ 10.4. GRUNDY NUMBER

A Grundy coloring of order k of a graph G is a k-coloring of G with colors $1, 2, \ldots, k$ such that for each vertex x the color of x is the smallest positive integer not used as a color on any neighbor of x in G. The Grundy number $\Gamma(G)$ is the largest integer k for which G has a Grundy coloring of order k. Is it an ***NP***-complete problem to decide for given G and k if $\Gamma(G) \geq k$? ■

This problem was first raised by Hedetniemi, Hedetniemi, and Beyer [1982], who proved that there is a linear algorithm to determine $\Gamma(T)$ for any tree T. The Grundy number was perhaps first defined by Christen and Selkow [1979]. Grundy colorings go back to Grundy [1939], who studied such colorings for directed graphs corresponding to games, and they were used by Berge [1958] to study kernels of directed graphs.

The usual sequential polynomial graph coloring algorithms, as described by Matula, Marble, and Isaacson [1972] and Johnson [1974], give Grundy colorings. The smallest integer for which G has a Grundy coloring is the chromatic number $\chi(G)$. Moreover,

$$\chi(G) \leq \Gamma(G) \leq \psi(G)$$

where $\psi(G)$ is the achromatic number (defined in Problem 10.5). Both $\chi(G)$ and $\psi(G)$ are difficult to determine, as their determination has been shown ***NP***-complete by Karp [1972] (see Theorem 19 in Chapter 1) and Yannakakis and Gavril [1980], respectively.

The so-called ordered chromatic number of Simmons [1982, 1983, 1985] is equal to the Grundy number. This was noted by P. Erdős in collaboration with Hare, Hedetniemi, Laskar, and Pfaff [1986] and by Erdős, Hare, Hedetniemi, and Laskar [1987]. They reported that this was also observed by E. Brickell.

■ 10.5. ACHROMATIC NUMBER OF A TREE

The achromatic number $\psi(G)$ of a graph G is the maximum number of colors possible in a coloring of G with any two different colors adjacent somewhere in G. Is it an ***NP***-hard problem to determine for a given tree T and a given number k if $\psi(T) \geq k$? ■

The achromatic number was introduced by Harary, Hedetniemi, and Prins [1967]. That the answer to the question above is affirmative was suspected by Hedetniemi, Hedetniemi, and Beyer [1982]. For general graphs G it is an ***NP***-hard problem to determine $\psi(G)$ as proved by Yannakakis and Gavril [1980].

The most important reference to achromatic numbers is the paper by Farber, Hahn, Hell, and Miller [1986]. They proved among many other things that for any fixed k there is a polynomial algorithm to decide if $\psi(G) = k$. Moreover, the problem for trees has a polynomial algorithm for certain special classes of trees.

10.6. ON-LINE COLORING

Given a graph G with an ordering $x_1 < x_2 < \cdots < x_n$ of the vertices, an on-line algorithm A with input $(G, <)$ produces a coloring of G by assigning a color to each vertex $x_1, x_2, \ldots, x_n$ one at a time, where the color assigned to x_i depends only on the subgraph $G[x_1, x_2, \ldots, x_i]$ induced by the vertices up to and including x_i together with their induced ordering. Over all possible orderings $<$ let $A(G)$ denote the maximum number of colors used by A to color G.

The performance ratio $r_A(G)$ of A on G is the ratio $A(G)/\chi(G)$, and the performance function ρ_A of A is defined by

$$\rho_A(n) = \max r_A(G),$$

where the maximum is over all graphs G with n vertices.

What is a best possible bound for the performance function ρ_A of an on-line algorithm A? ■

The question has been considered by Lovász, Saks, and Trotter [1989]. It is clear that $\rho_A(n) \leq n$ for any on-line algorithm A. Lovász, Saks, and Trotter explicitly described an on-line algorithm A with a better than linear performance function

$$\rho_A(n) \leq (1 + o(1)) \cdot \frac{2n}{\log^* n},$$

where $\log^* n$ denotes the smallest k for which the k times iterated logarithm

$$\log^{(k)} n = \log\log \ldots \log n \ (k \text{ times})$$

gives $\log^{(k)} n \leq 1$. They remarked that the algorithm can be modified to color any 3-colorable graph on n vertices with at most

$$c \cdot \frac{n \log\log\log n}{\log\log n}$$

colors, where $c > 0$ is a constant.

A first lower bound for the performance function of an on-line coloring algorithm A was obtained by Bean [1976], who proved the result (originally stated in terms of recursive functions) that for any integer $t > 0$ there exists a tree T on 2^t vertices such that $A(T) > t$. It follows that

$$\rho_A(n) > \frac{1}{2} \log_2 n.$$

Lovász, Saks, and Trotter [1989] pointed out that there exists an on-line algorithm A such that $A(G) \leq 2\log_2 n$ for any bipartite graph G on n vertices. Thus when restricting to bipartite graphs on n vertices the best possible performance ratio is close to $\log_2 n$. Improved lower bounds for graphs in general have been obtained by Vishwanathan [1992] and by M. Szegedy, who, as reported by Lovász, Saks, and Trotter [1989], proved that

$$\rho_A(n) \geq \frac{n}{(\log_2 n)^2}$$

for any on-line coloring algorithm A. Vishwanathan [1992] in particular studied the lower and upper bounds that can be obtained when restricting to graphs of bounded chromatic numbers. The informative survey by Kierstead and Trotter [1992] contains a discussion of the bounds given above and additional results and open problems.

For any class of graphs $\mathcal{G}$ the on-line chromatic number $\chi^*(\mathcal{G})$ is defined as the smallest number k such that there exists an on-line algorithm A for which $\chi_A(G) \leq k$ for every member G of $\mathcal{G}$ (if no such number k exists, define $\chi^*(\mathcal{G}) = \infty$). A graph G has on-line chromatic number $\chi^*(G) = \chi^*(\{G\})$. Gyárfás, Király, and Lehel [1993] proved that deciding if a graph has on-line chromatic number at most 3 is polynomial for bipartite graphs, for triangle-free graphs, and for connected graphs, and they conjectured that it is polynomially decidable also in general. However, they made another conjecture that it is an ***NP***-complete problem to decide for a given graph if it has on-line chromatic number at most 4.

For the class $OL(3)$ of graphs with on-line chromatic number at most 3, Gyárfás, Király, and Lehel [1993] announced the result $\chi^*(OL(3)) = 4$. The inequality $\chi^*(OL(3)) \leq 5$ was proved independently by K. Kolossa. Gyárfás, Király, and Lehel conjectured that $\chi^*(OL(k))$ is finite for all $k \geq 4$.

Because of applications to dynamic storage allocation, many authors have studied the problem of on-line coloring restricted to interval graphs, the intersection graphs of finite collections of intervals on the real line. In particular, the on-line algorithm "First Fit" FF (also sometimes called "the greedy algorithm") has been studied by many authors. Given $(G, <)$ as input, FF assigns to every vertex x_i $(1 \leq i \leq n)$ of G the smallest possible color from $\mathbf{Z}_+$, that is, the smallest color not yet assigned to any vertex adjacent to x_i among the previously colored vertices. Woodall [1974] and independently Chrobak and Ślusarek [1984] proved that if G is an interval graph, then $FF(G)$ is bounded by a quadratic function of $\chi(G)$ (note that $\chi(G) = \omega(G)$, since interval graphs are perfect; see Section 1.6), and they asked if a linear upper bound exists. W. Just improved the upper bound to

$$FF(G) \leq c \cdot \chi(G) \cdot \log \chi(G),$$

where $c > 0$ is constant (see Gyárfás and Lehel [1988]). The question by Woodall, Chrobak, and Ślusarek was answered in the affirmative by Kierstead [1988], who proved that

$$r_{FF}(G) \leq 40$$

for any interval graph G. The bound of 40 was further improved by Kierstead and J. Qin to $129/5$ as reported by Kierstead and Trotter [1992]. The best known lower bound

$$FF(G) \geq \frac{22}{5} \cdot \chi(G) + c'$$

for any interval graph G, and c' a constant, is due to M. Ślusarek (see Gyárfás and Lehel [1988]).

Kierstead and Trotter [1981] obtained an exact answer for the on-line coloring problem restricted to interval graphs: There exists an on-line coloring algorithm A satisfying

$$A(G) \leq 3\chi(G) - 2$$

for any interval graph G. Moreover, this bound is best possible.

A probabilistic result obtained by McDiarmid [1979] states that for any $\varepsilon > 0$ almost all graphs G satisfy

$$r_{FF}(G) \leq 2 + \varepsilon.$$

Results on lower bounds for the algorithm FF were obtained by Kučera [1991]. Anthony and Biggs [1993] studied the expected number of colors used by the algorithm FF to color G when all possible orderings of the vertex set are considered equally likely to appear in the input. They called this number the "mean chromatic number" $\bar{\chi}(G)$, and proved that for paths and even cycles it approaches the value 3 as the length tends to infinity.

■ 10.7. EDGE-COLORING MULTIGRAPHS

An edge-coloring of a graph or a multigraph G is an assignment of colors to the edges of G such that no two adjacent edges are assigned the same color. The edge-chromatic number $\chi'(G)$ is the minimum number of colors needed in an edge-coloring of G. Does there exist a polynomial edge-coloring algorithm A such that either $A(G) = \chi'(G)$ or $A(G) \leq \Delta(G) + 1$, where $A(G)$ is the number of colors A uses on G and $\Delta(G)$ is the maximum degree of G? ■

The conjecture that the answer is affirmative is due to Goldberg [1984b], who proved that the problem of finding $\chi'(G)$ is polynomial on the class of multigraphs with $\chi'(G) > (9\Delta(G) + 6)/8$. This result was also obtained by Hochbaum, Nishizeki, and Shmoys [1986]. In addition, Goldberg [1984a, 1984b] and Hochbaum, Nishizeki, and Shmoys [1986] obtained polynomial edge-coloring algorithms A with $A(G) \leq (9\chi'(G) + 6)/8$.

Hochbaum, Nishizeki, and Shmoys [1986] conjectured that there is a polynomial edge-coloring algorithm A with $A(G) \leq \chi'(G) + 1$. The truth of this would follow from the truth of Goldberg's conjecture stated above.

For a simple graph G it was proved by V.G. Vizing that $\chi'(G)$ is either $\Delta(G)$ or $\Delta(G) + 1$ (Theorem 29 in Chapter 1). It is an ***NP***-complete problem to decide for a given 3-regular simple graph G whether $\chi'(G) = 3$ or $\chi'(G) = 4$, as proved by Holyer [1981].

For the usual chromatic number χ no polynomial vertex-coloring algorithm A for which $A(G) \leq c \cdot \chi(G) + d$, where c and d are constants and $c < 2$, can exist (unless ***NP*** = ***P***). This is a result of Garey and Johnson [1976]. (For further information, see the discussion and the Theorem in Problem 10.1.)

■ 10.8. COMPLEXITY OF DIRECTED-GRAPH COLORING

Let $\vec{H}$ be a fixed directed graph. An $\vec{H}$-coloring of a directed graph $\vec{G}$ is a mapping $f : V(\vec{G}) \rightarrow V(\vec{H})$ such that $\overrightarrow{f(x)f(y)}$ is a directed edge of $\vec{H}$ whenever $\vec{xy}$ is a directed edge of $\vec{G}$. Such an f is also called a homomorphism from $\vec{G}$ to $\vec{H}$. For which $\vec{H}$ is there a polynomial algorithm to decide for any $\vec{G}$ whether $\vec{G}$ has an $\vec{H}$-coloring? For which $\vec{H}$ is this decision problem ***NP***-complete? ■

The corresponding H-coloring problem for undirected graphs G and H with $H = K_k$, the complete graph with k vertices, asks whether there exists a usual k-coloring of G. This problem is polynomially solvable for $k \leq 2$ and ***NP***-complete for $k \geq 3$ (see Theorem 19 in Chapter 1).

For undirected graphs and H bipartite, it is clear that G has an H-coloring if and only if G is bipartite. Hence the H-coloring problem is polynomially solvable if H is bipartite. Maurer, Sudborough, and Welzl [1981] and independently Nešetřil [1981] proved that if H is an odd cycle, then the problem of deciding H-colorability is ***NP***-complete, and they conjectured that it remains an ***NP***-complete problem for any nonbipartite graph H. This conjecture was proved by Hell and Nešetřil [1989, 1990]. Thus the complexity of undirected H-coloring is well understood.

Nešetřil [1981] proved that it is an ***NP***-complete problem to decide, given any C_3-colorable graph G (i.e., G is 3-colorable), whether G is C_5-colorable, where C_k denotes the cycle of length k. This special problem on undirected H-coloring was also considered independently by Füredi, Griggs, and Kleitman [1989]. More general problems in this direction were treated by Brewster and MacGillivray [1991].

Gutjahr, Welzl, and Woeginger [1992] proved that the $\vec{H}$-colorability problem can be solved by a polynomial algorithm when the underlying undirected graph of $\vec{H}$ is a path. However, they also proved that there exists an $\vec{H}$ for which the underlying graph is a tree, and the problem is ***NP***-complete.

Bang-Jensen, Hell, and MacGillivray [1988] considered the $\vec{H}$-colorability problem for semicomplete digraphs $\vec{H}$. These are directed graphs without any loops and with each pair of vertices joined by a directed edge in at least one direction. They proved that $\vec{H}$-colorability is an ***NP***-complete problem if $\vec{H}$ is a semicomplete digraph containing at least two directed cycles, and that the problem can be solved by a polynomial algorithm otherwise.

Bang-Jensen and Hell [1990] formulated the following conjecture:

Conjecture. *Let $\vec{H}$ be a directed graph without vertices of in-degree or out-degree zero. If $\vec{H}$ does not allow a homomorphism to its shortest directed cycle, then $\vec{H}$-colorability is **NP**-complete. Otherwise, $\vec{H}$-colorability is polynomially decidable.*

10.9. PRECEDENCE CONSTRAINED 3-PROCESSOR SCHEDULING

Is there a polynomial algorithm to solve the decision problem: Given an acyclic directed graph $\vec{G}$ and an integer k, is it possible to give a k-coloring of the vertices of $\vec{G}$, $\varphi : V(\vec{G}) \to \{1, 2, \ldots, k\}$, so that

(i) $\varphi(x) < \varphi(y)$ for all $\vec{xy} \in E(\vec{G})$, and
(ii) the number of vertices colored i is at most 3 for every $i = 1, 2, \ldots, k$?

■

The question seems first asked by Garey and Johnson [1977, 1979 (listed as open problem 8)]. They formulated an equivalent problem in terms of multiprocessor scheduling: When given a set T of tasks with a partial order $\rangle$ on T describing the order in which pairs of tasks may be carried out, and a number m of processors that each can carry out one task per unit of time, can all tasks in T be carried out by the processors within a deadline D? With $m = 3$, $D = k$, and letting $\vec{G}$ be the directed graph corresponding to $(T, \rangle)$, one gets the formulation described above.

Fujii, Kasami, and Ninomiya [1969] described a polynomial algorithm to solve the corresponding 2-processor problem, and thus answered the question for $m = 2$. Other algorithms for the 2-processor problem were presented by Coffman and Graham [1972] and Garey and Johnson [1977] (see also the informal essay on combinatorial scheduling theory by Graham [1978] containing outlines of all three algorithms).

Concerning the m-processor problem for $m > 2$, Garey and Johnson [1979] remarked that it is not known for any fixed such m whether the m-processor problem can be solved in polynomial time or is ***NP***-complete. If m itself is given as part of the input, the problem becomes ***NP***-complete, as shown by Ullman [1975].

With condition (ii) left out, the problem becomes easy. In fact, it becomes equivalent to determining the length of a longest directed path in the acyclic $\vec{G}$.

With condition (i) left out, the problem can be restated in terms of undirected graph G: Is it possible to k-color G so that at most 3 (in general m) vertices receive the same color? For $m = 2$ this problem is solvable in polynomial time, since it is equivalent to finding a maximum matching in $\overline{G}$, the complement of G, which can be done by the polynomial algorithm of Edmonds [1965b]. For $m = 3$ the problem of partitioning the vertices of $\overline{G}$ into triangles reduces to this problem, and this shows that it is ***NP***-complete by a result of Garey and Johnson [1979].

BIBLIOGRAPHY

[1993] Alon N. Restricted colorings of graphs. In: K. Walker, editor, *Surveys in Combinatorics: Proc. 14th British Combinatorial Conference*, pages 1–33. Cambridge University Press, 1993.

[1993] Anthony M. and N. Biggs. The mean chromatic number of paths and cycles. *Discrete Math.* **120**, 227–231, 1993.

[1992] Arora S., C. Lund, R. Motwani, M. Sudan, and M. Szegedy. Proof verification and hardness of approximation problems. In: *Proc. 33rd Annual Symposium on the Foundations of Computer Science*, pages 14–23, 1992.

[1990] Bang-Jensen J. and P. Hell. The effect of two cycles on the complexity of colourings by directed graphs. *Discrete Appl. Math.* **26**, 1–23, 1990.

[1988] Bang-Jensen J., P. Hell, and G. MacGillivray. The complexity of colourings by semicomplete digraphs. *SIAM J. Discrete Math.* **1**, 281–298, 1988.

[1976] Bean D.R. Effective coloration. *J. Symbolic Logic* **41**, 469–480, 1976.

[1958] Berge C. *Théorie des Graphes et ses Applications.* Dunod, 1958.

[1990] Blum A. Some tools for approximate 3-coloring. In: *Proc. 31st Annual Symposium on the Foundations of Computer Science*, pages 554–562, 1990.

[1991] Brewster R. and G. MacGillivray. A note on restricted H-colouring. *Inform. Process. Lett.* **40**, 149–151, 1991.

[1993] Brightwell G.R. On the complexity of diagram testing. *Order* **10**, 297–303, 1993.

[1979] Christen C.A. and S.M. Selkow. Some perfect colouring properties of graphs. *J. Combin. Theory Ser. B* **27**, 49–59, 1979.

[1984] Chrobak M. and M. Ślusarek. Problem 84–23. *J. Algorithms* **5**, 588, 1984.

[1972] Coffman E.G. Jr and R.L. Graham. Optimal scheduling for two-processor systems. *Acta Inform.* **1**, 200–213, 1972.

[1965b] Edmonds J. Paths, trees and flowers. *Canad. J. Math.* **17**, 449–467, 1965.

[1992] Edwards K. The complexity of some graph colouring problems. *Discrete Appl. Math.* **36**, 131–140, 1992.

[1987] Erdős P., W.R. Hare, S.T. Hedetniemi, and R. Laskar. On the equality of the Grundy and ochromatic numbers of a graph. *J. Graph Theory* **11**, 157–159, 1987.

[1986] Farber M., G. Hahn, P. Hell, and D.J. Miller. Concerning the achromatic number of graphs. *J. Combin. Theory Ser. B* **40**, 21–39, 1986.

[1969] Fujii M., T. Kasami, and K. Ninomiya. Optimal sequencing of two equivalent processors. *SIAM J. Appl. Math.* **17**, 784–789, 1969. (Erratum in *SIAM J. Appl. Math.* **20**, 141, 1971.)

[1989] Füredi Z., J.R. Griggs, and D.J. Kleitman. Pair labellings with given distance. *SIAM J. Discrete Math.* **2**, 491–499, 1989.

[1976] Garey M.R. and D.S. Johnson. The complexity of near-optimal graph coloring. *J. Assoc. Comput. Mach.* **23**, 43–49, 1976.

[1977] Garey M.R. and D.S. Johnson. Two-processor scheduling with start-times and deadlines. *SIAM J. Comput.* **6**, 416–426, 1977.

[1979] Garey M.R. and D.S. Johnson. *Computers and Intractability. A Guide to the Theory of* ***NP****-Completeness.* W.H. Freeman and Company, 1979.

[1984a] Goldberg M.K. An approximate algorithm for the edge-coloring problem. In: *Proc. 15th S–E Conference on Combinatorics, Graph Theory and Computing, Baton Rouge, 1984, Congr. Num., 43*, pages 317–319, 1984.

[1984b] Goldberg M.K. Edge-coloring of multigraphs: recoloring technique. *J. Graph Theory* **8**, 123–137, 1984.

[1978] Graham R.L. Combinatorial scheduling theory. In: L.A. Steen, editor, *Mathematics Today: Twelve Informal Essays*, pages 183–211. Springer-Verlag, 1978.

[1939] Grundy P.M. Mathematics and games. *Eureka* **2**, 6–8, 1939.

[1992] Gutjahr W., E. Welzl, and G. Woeginger. Polynomial graph-colorings. *Discrete Appl. Math.* **35**, 29–45, 1992.

[1993] Gyárfás A., Z. Király, and J. Lehel. On-line graph coloring and finite basis problems. In: D. Miklós, V.T. Sós, and T. Szőnyi, editors, *Combinatorics, Paul Erdős Is Eighty*, volume 1 of *Bolyai Society Mathematical Studies*, pages 207–214. János Bolyai Mathematical Society, 1993.

[1988] Gyárfás A. and J. Lehel. On-line and first-fit coloring of graphs. *J. Graph Theory* **12**, 217–227, 1988.

[1990] Halldórsson M.M. A still better performance guarantee for approximate graph coloring. Technical Report 91–35, DIMACS, New Brunswick, N.J., 1990.

[1967] Harary F., S.T. Hedetniemi, and G. Prins. An interpolation theorem for graphical homomorphisms. *Portugal. Math.* **26**, 453–462, 1967.

[1986] Hare W.R., S.T. Hedetniemi, R. Laskar, and J. Pfaff. Complete coloring parameters of graphs. In: *Proc. 16th S–E Conference on Combinatorics, Graph Theory and Computing, Boca Raton, 1986, Congr. Num., 48*, pages 171–178, 1986.

[1982] Hedetniemi S.M., S.T. Hedetniemi, and T. Beyer. A linear algorithm for the Grundy (coloring) number of a tree. In: *Proc. 13th S–E Conference on Combinatorics, Graph Theory and Computing, Boca Raton 1982, Congr. Num., 36*, pages 351–363, 1982.

[1989] Hell P. and J. Nešetřil. The existence problem for graph homomorphisms. In: L.D. Andersen, I.T. Jakobsen, C. Thomassen, B. Toft, and P.D. Vestergaard, editors, *Graph Theory in Memory of G.A. Dirac*, volume 41 of *Annals of Discrete Mathematics*, pages 255–266. North-Holland, 1989.

[1990] Hell P. and J. Nešetřil. On the complexity of H-coloring. *J. Combin. Theory Ser. B* **48**, 92–110, 1990.

[1986] Hochbaum D.S., T. Nishizeki, and D.B. Shmoys. A better than "best possible" algorithm to edge color multigraphs. *J. Algorithms* **7**, 79–104, 1986.

[1981] Holyer I. The ***NP***-completeness of edge-coloring. *SIAM J. Comput.* **10**, 718–720, 1981.

[1974] Johnson D.S. Worst case behavior of graph coloring algorithms. In: *Proc. 5th S–E Conference on Combinatorics, Graph Theory and Computing, Boca Raton, Congr. Num., 10*, pages 513–527, 1974.

[1978] Johnson D.S. Research problem no. 13. In: B. Alspach, P. Hell, and D.J. Miller, editors, *Algorithmic Aspects of Combinatorics*, volume 2 of *Annals of Discrete Mathematics*, page 243. North-Holland, 1978.

[1972] Karp R. Reducibility among combinatorial problems. In: R.E. Miller and J.W. Thatcher, editors, *Complexity of Computer Computations*, pages 85–104. Plenum Press, 1972.

[1993] Khanna S., N. Linial, and S. Safra. On the hardness of approximating the chromatic number. In: *2nd Israel Symposium on the Theory of Computing Systems*, pages 250–260, 1993.

[1988] Kierstead H.A. The linearity of first-fit for coloring interval graphs. *SIAM J. Discrete Math.* **1**, 526–530, 1988.

[1981] Kierstead H.A. and W.T. Trotter. An extremal problem in recursive combinatorics. In: *Proc. 12th S–E Conference on Combinatorics, Graph Theory and Computing, Baton Rouge, 1981, Congr. Num., 33*, pages 143–153, 1981.

[1992] Kierstead H.A. and W.T. Trotter. On-line graph coloring. In: L.A. McGeoch and D.D. Sleator, editors, *On-Line Algorithms: Proc. DIMACS Workshop, February 11–13, 1991*, pages 85–92. American Mathematical Society, 1992.

[1989] Kratochvíl J. and J. Nešetřil. Problem 13. In: L.D. Andersen, I.T. Jakobsen, C. Thomassen, B. Toft, and P.D. Vestergaard, editors, *Graph Theory in Memory of G.A. Dirac*, volume 41 of *Annals of Discrete Mathematics*, page 517. North-Holland, 1989.

[1991] Kučera L. The greedy coloring is a bad probabilistic algorithm. *J. Algorithms* **12**, 674–684, 1991.

[1989] Lovász L., M. Saks, and W.T. Trotter. An on-line graph coloring algorithm with sublinear performance ratio. *Discrete Math.* **75**, 319–325, 1989.

[1993] Lund C. and M. Yannakakis. On the hardness of approximating minimization problems. In: *Proc. 25th ACM Symposium on Theory of Computing*, pages 286–293. ACM, New York, 1993.

[1972] Matula D.W., G. Marble, and J.D. Isaacson. Graph coloring algorithms. In: R.C. Read, editor, *Graph Theory and Computing*, pages 109–122. Academic Press, 1972.

[1981] Maurer H.A., J.H. Sudborough, and E. Welzl. On the complexity of the general coloring problem. *Inform. and Control* **51**, 123–145, 1981.

[1979] McDiarmid C.J.H. Coloring random graphs badly. In: R.J. Wilson, editor, *Graph Theory and Combinatorics*, pages 76–86. Pitman, 1979.

[1981] Nešetřil J. Representations of graphs by means of products and their complexity. In: J. Gruska and M. Chytil, editors, *Mathematical Foundations of Computer Science*, volume 118 of *Lecture Notes in Computer Science*, pages 94–102. Springer-Verlag, 1981.

[1987] Nešetřil J. and V. Rödl. Complexity of diagrams. *Order* **3**, 321–330, 1987.

[1993] Nešetřil J. and V. Rödl. More on complexity of diagrams. Manuscript, 1993.

[1982] Simmons G.J. The ordered chromatic number of planar maps. In: *Proc. 13th S–E Conference on Combinatorics, Graph Theory and Computing, Boca Raton, 1982, Congr. Num., 36*, pages 59–67, 1982.

[1983] Simmons G.J. On the ochromatic number of a graph. In: *Proc. 14th S–E Conference on Combinatorics, Graph Theory and Computing, Boca Raton, 1983, Congr. Num., 40*, pages 339–366, 1983.

[1985] Simmons G.J. The ochromatic number of planar graphs. In: F. Harary and J.S. Maybee, editors, *Graphs and Applications: Proc. First Colorado Symposium on Graph Theory*, pages 295–316. Wiley, 1985.

[1992] Thostrup J. *Partially ordered sets and graphs* (in Danish). Master's thesis, Odense University, 1992.

[1975] Ullman J.D. ***NP***-complete scheduling problems. *J. Comput. System Sci.* **10**, 384–393, 1975.

[1992] Vishwanathan S. Randomized online graph coloring. *J. Algorithms* **13**, 657–669, 1992.

[1983] Wigderson A. Improving the performance guarantee for approximate graph coloring. *J. Assoc. Comput. Mach.* **30**, 729–735, 1983.

[1974] Woodall D.R. Problem no. 4. In: T.P. McDonough and V.C. Mavron, editors, *Combinatorics: Proc. British Combinatorial Conference, 1973*, page 202. Cambridge University Press, 1974.

[1980] Yannakakis M. and F. Gavril. Edge dominating sets in graphs. *SIAM J. Appl. Math.* **38**, 364–372, 1980.

11

Constructions

■ 11.1. DIRECT PRODUCT

Let $G \times G'$ denote the direct product of G and G', that is, the graph with vertex set $V(G) \times V(G')$ and edges $(a, a')(b, b')$ for all $ab \in E(G)$ and $a'b' \in E(G')$. It is easy to see that

$$\chi(G \times G') \leq \min\{\chi(G), \chi(G')\}.$$

Does equality hold? ■

This question was asked by Hedetniemi [1966]. It first went relatively unnoticed, but was later raised independently by others; in particular, L. Lovász suggested the problem and attracted greater attention to it around 1972 [personal communication]. Also, Greenwell and Lovász [1974] mentioned the question.

Hedetniemi [1966] noted that the answer is affirmative for $\chi(G) = \chi(G') = 3$, and conjectured that the answer is affirmative in general. El-Zahar and Sauer [1985] proved this in the case $\chi(G) = \chi(G') = 4$. Duffus, Sands, and Woodrow [1985] gave a well-written and comprehensive survey of the problem.

A number of partial results have been obtained by Duffus, Sands, and Woodrow [1985], and more recently, Sauer and Zhu [1992] attacked the conjecture with an original approach using Hajós' construction (see Theorem 18 in Chapter 1).

The conjecture cannot be extended to infinite graphs. Miller [1968] proved that the direct product of all odd cycles C_{2n+1}, for $n \geq 1$, is a bipartite infinite graph. Hajnal [1985] proved that the product of two $\aleph_1$-chromatic graphs may have countable chromatic number (see also Problem 16.13).

The direct product was defined by Berge [1958] who called it simply the "product" of graphs; independently Čulik [1958] defined it as the "cardinal product." It was referred to by Ore [1962] as the "Cartesian product." Weichsel [1962] called it the "Kronecker product," and other authors (Harary and Trauth [1966], Harary and Wilcox [1967]) named it the "tensor product." Miller [1968] used the term "categorical product," Borowiecki [1972] the term "conjunction," Lovász [1979] the term "(weak) direct product," and Nešetřil and Rödl [1983] the term "direct product." The latter seems to have become the most preferred.

Nešetřil and Rödl [1985] described 2^8 (= 256) different products of two graphs G and G', all defined on the vertex set $V(G) \times V(G')$, by considering for each pair of vertices x, y of G and each pair x', y' of G' the 2^2 combinations of xy and $x'y'$ being or not being edges of G and G', respectively, and the 2^6 possible graphs on four labeled vertices. Puš [1988] studied the chromatic properties of these special products, and he proved that for every such product $*$, if there exists a function $f(i, j)$ such that

$$\chi(G * G') = f(\chi(G), \chi(G')),$$

then either $f(i, j) = 1$, $f(i, j) = i$, $f(i, j) = j$, or $f(i, j) = \max(i, j)$, or possibly $*$ is the direct product and $f(i, j) = \min(i, j)$. We note, however, that even with the vertex set restricted to $V(G) \times V(G')$, products have been defined that do not belong to this particular class of products: for example, the "wreath product" by Anderson and Lipman [1985] (see Problem 11.2).

■ 11.2. WREATH PRODUCT

Let $G\rho G'$ denote the graph with vertex set $V(G) \times V(G')$ and an edge $(x, x')(y, y')$ if and only if either $x = y$ and $x'y' \in E(G')$, or $xy \in E(G)$ and $\varphi(x') = y'$ for some automorphism $\varphi : V(G') \to V(G')$ of G'.

Is it true that if $\chi'(G) = \Delta(G)$, then $\chi'(G\rho G') = \Delta(G\rho G')$, where χ' denotes the edge-chromatic number and Δ the maximum degree? ■

This was conjectured by Anderson and Lipman [1985]. The theorem of V.G. Vizing (Theorem 29 in Chapter 1) implies that $\chi'(G\rho G')$ is equal to either $\Delta(G\rho G')$ or $\Delta(G\rho G') + 1$.

Sabidussi [1957, 1960] defined the "Cartesian product" $G\Box G'$ with vertex set $V(G) \times V(G')$, and an edge $(x, x')(y, y')$ if and only if either $x = y$ and $x'y' \in E(G')$, or $xy \in E(G)$ and $x' = y'$ (this product was studied earlier by Shapiro [1953]). If $\chi'(G) = \Delta(G)$, then $\chi'(G\Box G') = \Delta(G\Box G')$ (see exercise 6.2.6 in Bondy and Murty [1976]). It is clear that $G\Box G'$ is a subgraph of $G\rho G'$, and that they are identical if and only if G' has trivial automorphism group.

The "lexicographic product" (sometimes referred to as the "composition") $G[G']$ of G with G', first defined by Harary [1959], has vertex set $V(G) \times V(G')$, and an edge $(x, x')(y, y')$ if and only if $x = y$ and $x'y' \in E(G')$, or $xy \in E(G)$. Anderson and Lipman [1985] proved for this (noncommutative) graph product that if $\chi'(G) = \Delta(G)$, then $\chi'(G[G']) = \Delta(G[G'])$. The wreath product and the lexicographic product are related by $G\rho G' \subseteq G[G']$ with equality if and only if G' has transitive automorphism group. Vertex colorings of lexicographic products have also been investigated (see Čižek and Klavžar [1993] and Kaschek and Klavžar [1993]).

11.3. A VERY STRONG PRODUCT

A product $*$ for graphs such that

(i) $G * G'$ can be obtained in a polynomial number of steps in the sizes $|V(G)|$ and $|V(G')|$,
(ii) $\chi(G) \cdot \chi(G') \geq \chi(G * G') \geq c \cdot \chi(G) \cdot \chi(G')$ for some fixed constant $c > 0$,

appears unknown. Does such a product exist? With $c = 1$? With $c > 3/4$? ■

If the answer is affirmative with $c > 3/4$, then this would give a new simple proof of the consequence of the theorem of C. Lund and M. Yannakakis (Theorem 20 in Chapter 1), conjectured by Johnson [1978], that the existence of a polynomial graph coloring algorithm A using at most $r\chi(G) + d$ colors to color any graph G (r and d being fixed constants) implies the existence of a polynomial algorithm A' to determine the exact value of $\chi(G)$: Suppose that such an A exists and let N be any integer satisfying $c^{N-1}4^N > r3^N + d$. Apply A to the graph $G * G * \cdots * G$ (N factors) and determine if $\chi(G) \leq 3$, thus solving an ***NP***-complete problem polynomially; this implies the existence of A' (see Theorem 19 in Chapter 1).

The theorem of C. Lund and M. Yannakakis (Theorem 20 in Chapter 1) is in fact a much more general theorem on approximative algorithms. However, the more general theorem does not seem to follow directly from the existence of a strong product. On the other hand, the existence of a strong product as described might perhaps be so strong as to imply the existence of a polynomial algorithm to determine the chromatic number of a graph, which would imply that ***NP*** = ***P*** (see Theorem 19 in Chapter 1).

The special product $\oplus$ ("Cartesian sum") defined by Ore [1962]:

(a) $V(G \oplus G') = V(G) \times V(G')$, and
(b) $(x, x')(y, y') \in E(G \oplus G')$ if and only if either $xy \in E(G)$ or $x'y' \in E(G')$ or both,

might be a possible first guess. The chromatic properties of this product were investigated by Yang [1968], Borowiecki [1972], and Puš [1988]. It is not difficult to see that $\chi(G \oplus G') \leq \chi(G) \cdot \chi(G')$, and that equality does not always hold. For example, $\chi(C_5 \oplus C_5) \leq 8$, as pointed out by Borowiecki [1972]. Čižek and Klavžar [1993] proved that if G and G' are vertex-critical and not complete, then $\chi(G \oplus G') < \chi(G) \cdot \chi(G')$. There does not seem to be any known lower bound better than $\chi(G \oplus G') \geq c \cdot \max\{\chi(G), \chi(G')\}$ for a constant $c > 0$ (see the survey by Čižek and Klavžar [1993]).

A weaker product, usually called "the strong product," has been investigated in detail by Vesztergombi [1981] and recently by Klavžar [1993]. This product of G and G' is defined as the graph with vertex set $V(G) \times V(G')$ and having an edge

$(x, x')(y, y')$ whenever either [$x = y$ and $x'y' \in E(G')$] or [$xy \in E(G)$ and $x' = y'$] or [$xy \in E(G)$ and $x'y' \in E(G')$]. The strong product of G and G' is a subgraph of the Cartesian sum $G \oplus G'$.

■ 11.4. GALLAI'S PROBLEM ON DIRAC'S CONSTRUCTION

Let G be a graph obtained from two disjoint graphs G_1 and G_2 by joining some of the vertices of G_1 to some of the vertices of G_2. Consider the statements

(a) G_1 and G_2 are both critical,
(b) G_1 and G_2 are completely joined in G,
(c) G is critical.

Under what circumstances do (a) and (c) together imply (b)? ■

This question is due to T. Gallai [personal communication in 1969]. That (a) and (b) together imply (c) is the construction by Dirac [1952a] (see also the important paper of Gallai [1963a]). Moreover, (b) and (c) together imply (a).

As noted by Gallai, the question has an affirmative answer when G_1 and G_2 are both complete. This is in fact the statement that the complement of a bipartite graph is perfect, first proved by D. König and Gallai [1959]. This result was used as a lemma by Brown and Jung [1969] and Toft [1972a, 1974a]. Stiebitz and Wessel [1993] answered Gallai's question completely in the case when G_1 is a complete graph and G_2 an odd cycle.

As shown by Youngs [1992], one can obtain a 4-critical graph from two 7-cycles by joining some of the vertices of one cycle to some of the vertices of the other. Thus (a) and (c) do not imply (b) when G_1 and G_2 are odd cycles. It remains open to characterize the pairs of cycle lengths for which it is possible to obtain 4-critical or 5-critical graphs in this fashion.

■ 11.5. HAJÓS VERSUS ORE

A graph is Hajós–k-constructible if it can be obtained from complete k-graphs K_k by repeated applications of the following two operations:

(a) (Hajós' construction) Let G_1 and G_2 be already obtained disjoint graphs with edges x_1y_1 and x_2y_2. Remove x_1y_1 and x_2y_2, identify x_1 and x_2, and join y_1 and y_2 by an edge.
(b) Identify independent vertices (replace multiple by single edges).

Such a Hajós–k-constructible graph is said to be Ore–k-constructible if operation (b) is used following operation (a) only on pairs $(z_1^1, z_1^2), \ldots, (z_t^1, z_t^2)$ of vertices, where $z_1^1, z_2^1, \ldots, z_t^1$ are distinct vertices of $G_1 - x_1$ and $z_1^2, z_2^2, \ldots, z_t^2$ are distinct vertices of $G_2 - x_2$.

Is every Hajós–k-constructible graph Ore–k-constructible? ■

Hajós [1961] proved that a graph has chromatic number at least k if and only if it contains a Hajós–k-constructible subgraph (Theorem 18 in Chapter 1). Ore [1967] noted that the proof shows that it must in fact contain an Ore–k-constructible subgraph, and remarked that for some applications it is essential to use the more specific form, but he did not give any examples to show this.

From the definition it seems that the class of Ore–k-constructible graphs is much more restricted than the class of Hajós–k-constructible graphs. But no example of a Hajós–k-constructible graph that is not Ore–k-constructible seems to be known. This problem was mentioned by Hanson, Robinson, and Toft [1986]. A striking difference between the two definitions is that there seems to be no immediate finite algorithm to decide whether a given graph is Hajós–k-constructible or not (for Ore–k-constructibility it is a finite process to check all possibilities). One might speculate that perhaps there exist graphs for which this is an undecidable problem. Undecidable graph coloring problems were studied by Burr [1984].

In his talk given at the 25th anniversary celebration in 1992 at the Department of Combinatorics and Optimization, University of Waterloo, Canada, W.T. Tutte addressed the peculiar feature of Hajós' theorem that it seems to promise so much and yet has achieved so little. Tutte [1992] proved Hajós' theorem in a slightly stronger form than the original; sufficiently strong to obtain Brooks' theorem (Theorem 11 in Chapter 1) as a consequence.

■ 11.6. LENGTH OF HAJÓS PROOFS

If a graph G has chromatic number at least k, then this can be proved by a Hajós proof, which consists in constructing G from complete k-graphs by the following operations:

(a) (Hajós' construction) If G_1 and G_2 are already obtained disjoint graphs, then remove an edge x_1y_1 from G_1 and an edge x_2y_2 from G_2, identify x_1 and x_2 and add the edge y_1y_2.

(b) Identify two vertices not joined by an edge in an already obtained graph (replace multiple by single edges).

(c) Add new vertices and edges to an already obtained graph.

How long is a Hajós proof for a graph on n vertices? Are there interesting classes of graphs for which Hajós proofs are short? ■

Lovász [1982] wrote that the algorithmic complexity aspect of the theorem of Hajós [1961] (Theorem 18 in Chapter 1) is an unexplored territory. He also remarked that the perfect graphs are those obtainable in just one step. Mansfield and Welsh [1982] initiated a complexity study. Among other things they noted that a polynomial upper bound on the length of a Hajós proof in terms of $|V(G)|$ would imply that $\boldsymbol{NP} = \boldsymbol{co\text{-}NP}$, which is considered unlikely.

A more restricted type of Hajós proof was introduced by Ore [1967], who proved that one may restrict the use of operation (b) to identifying vertices of G_1 with vertices of G_2 following a use of operation (a). Using the result that there are $c^{(n^2)}$ nonisomorphic k-critical graphs on n vertices for $k \geq 4$, due to V. Rödl and published by Toft [1985], Hanson, Robinson, and Toft [1986] observed that there are graphs on n vertices for which a shortest Hajós proof of Ore type requires at least $cn/\log n$ uses of operation (a).

■ 11.7. HAJÓS CONSTRUCTIONS OF CRITICAL GRAPHS

A k-critical graph can be obtained from complete k-graphs by the following two operations:

(a) (Hajós' construction) If G_1 and G_2 are already obtained disjoint graphs, then remove an edge x_1y_1 from G_1 and an edge x_2y_2 from G_2, identify x_1 and x_2 and add the edge y_1y_2.

(b) Identify two vertices not joined by an edge in an already obtained graph (replace multiple by single edges).

Can every k-critical graph G be obtained in this way such that the graphs in each step of the construction of G are all k-critical, when $4 \leq k \leq 7$? ■

The result given above follows from the theorem of Hajós [1961] (Theorem 18 in Chapter 1), and the problem is due to Hajós [personal communication from T. Gallai in 1969].

For general k the answer is NO, as shown by the 8-chromatic counterexample to Hajós' conjecture due to Catlin [1979] (see Problem 6.2). But perhaps it is true for general k that operation (a) can always be restricted to be used on k-critical graphs exclusively? For the restriction of Hajós' construction due to Ore [1967] (see Problem 11.5), this last question has a negative answer for $k \geq 6$, as noted by Hanson, Robinson, and Toft [1986].

■ 11.8. CONSTRUCTION OF HAJÓS GENERALIZED BY DIRAC

Let G_1 and G_2 be disjoint k-critical graphs ($k \geq 4$) and for $i = 1, 2$ let H_i be a complete h-graph in G_i where $1 \leq h \leq k - 2$. For $i = 1, 2$ let y_i be a vertex of $G_i - H_i$ joined to a vertex x_i of H_i. Obtain G from G_1 and G_2 by identifying H_1 and H_2 to a new complete h-graph H so that x_1 and x_2 are identified, delete the edges $x_i y_i$ and join y_1 and y_2 by a new edge. Is G k-critical? ■

This question was asked in the excellent survey of critical graphs with vertices of degree $k - 1$ by Sachs and Stiebitz [1989]. The construction is due to G.A. Dirac, who may have known the answer to the problem.

For $h = 1$ the construction is the well-known construction of Hajós [1961] (see Theorem 18 in Chapter 1). For $h \leq 2$ the answer is affirmative, and this is also the case in general when each vertex x of H is joined to either y_1 or y_2 in G. Thus the question asks if criticality is preserved, also when this condition is not fulfilled.

■ 11.9. FOUR-CHROMATICITY IN TERMS OF 3-COLORABILITY

Let $\chi(G) \geq 4$ and $G \not\supseteq K_4$. Does G contain two vertices x and y and two 3-colorable subgraphs G_1 and G_2, both containing x and y, but not containing the edge xy such that

(a) in every 3-coloring of G_1 the independent vertices x and y get different colors, and

(b) in every 3-coloring of G_2 the independent vertices x and y get the same color? ■

This question was raised by B. Toft in lectures on graph coloring theory at the University of Regina in 1985.

With 4 replaced by 3 (and K_4 by K_3) the answer is affirmative. With 4 replaced by any $k \geq 6$ (and K_4 by K_k) the answer is negative. For example, for $k = 8$ this is shown by the 8-chromatic counterexample to Hajós' conjecture, due to Catlin [1979], by an argument of Hanson, Robinson, and Toft [1986] (it does not contain a subgraph of type G_2).

Tutte [1992] observed that for every pair of vertices x and y of distance 2 in a k-critical graph, there is always a $(k - 1)$-colorable subgraph containing x and y, in which x and y must receive different colors in any $(k - 1)$-coloring. Hence subgraphs of type G_1 always exist. Tutte's observation implies that if $\chi(G) \geq k$, then G contains k vertices $x_1, x_2, \ldots, x_k$ and a collection of $(k - 1)$-colorable subgraphs G_{ij}, $1 \leq i < j \leq k$, such that G_{ij} contains x_i and x_j, and x_i and x_j receive different colors in any $(k - 1)$-coloring of G_{ij}. (For a proof, consider a k-critical subgraph, any one of its vertices x_1, and any $k - 1$ neighbors $x_2, x_3, \ldots x_k$, and use Tutte's observation for those pairs i and j such that x_i and x_j are nonadjacent. Note that $G_{1j} = K_2$ for all j.)

This gives a characterization of graphs of chromatic number at least k in terms of $(k - 1)$-colorability. This characterization does not give a polynomial explanation of $\chi(G) \geq 4$, that is, the characterization is not "good" in the sense of Edmonds [1965b]. A polynomial explanation of $\chi(G) \geq 4$ cannot be expected, since then the ***NP***-complete problem "$\chi(G) \leq 3$?" would also be in ***co-NP*** and hence ***NP*** = ***co-NP***. This is considered unlikely (see Garey and Johnson [1979]).

BIBLIOGRAPHY

[1985] Anderson M. and M. Lipman. The wreath product of graphs. In: F. Harary and J.S. Maybee, editors, *Graphs and Applications: Proc. First Colorado Symposium on Graph Theory*, pages 23–39. Wiley, 1985.

[1958] Berge C. *Théorie des Graphes et ses Applications*. Dunod, Paris, 1958.

[1976] Bondy J.A. and U.S.R. Murty. *Graph Theory with Applications*. Macmillan, 1976.

[1972] Borowiecki M. On chromatic number of products of two graphs. *Colloq. Math.* **25**, 49–52, 1972.

[1969] Brown W.G. and H.A. Jung. On odd circuits in chromatic graphs. *Acta Math. Acad. Sci. Hungar.* **20**, 129–134, 1969.

[1984] Burr S.A. Some undecidable problems involving the edge-coloring and vertex-coloring of graphs. *Discrete Math.* **50**, 171–177, 1984.

[1979] Catlin P.A. Hajós' graph-coloring conjecture: variations and counterexamples. *J. Combin. Theory Ser. B* **26**, 268–274, 1979.

[1993] Čižek N. and S. Klavžar. On the chromatic number of the lexicographic product and the Cartesian sum of graphs. Manuscript, 1993. To appear in *Discrete Math.*

[1958] Čulik K. Zur Theorie der Graphen. *Časopis Pěst. Mat.* **83**, 133–155, 1958.

[1952a] Dirac G.A. A property of 4-chromatic graphs and some remarks on critical graphs. *J. London Math. Soc.* **27**, 85–92, 1952.

[1985] Duffus D., B. Sands, and R.E. Woodrow. On the chromatic number of the product of graphs. *J. Graph Theory* **9**, 487–495, 1985.

[1965b] Edmonds J. Paths, trees and flowers. *Canad. J. Math.* **17**, 449–467, 1965.

[1985] El-Zahar M. and N.W. Sauer. The chromatic number of the product of two 4-chromatic graphs is 4. *Combinatorica* **5**, 121–126, 1985.

[1959] Gallai T. Über extreme Punkt- und Kantenmengen. *Ann. Univ. Sci. Budapest Eötvös Sect. Math.* **2**, 133–138, 1959.

[1963a] Gallai T. Kritische Graphen I. *Publ. Math. Inst. Hungar. Acad. Sci.* **8**, 165–192, 1963.

[1979] Garey M.R. and D.S. Johnson. *Computers and Intractability: A Guide to the Theory of* ***NP****-Completeness*. W.H. Freeman and Company, 1979.

[1974] Greenwell D. and L. Lovász. Applications of product coloring. *Acta Math. Acad. Sci. Hungar.* **25**, 335–340, 1974.

[1985] Hajnal A. The chromatic number of the product of two $\aleph_1$-chromatic graphs can be countable. *Combinatorica* **5**, 137–139, 1985.

[1961] Hajós G. Über eine Konstruktion nicht n-färbbarer Graphen. *Wiss. Z. Martin-Luther-Univ. Halle–Wittenberg Math.-Natur. Reihe* **10**, 116–117, 1961.

[1986] Hanson D., G.C. Robinson, and B. Toft. Remarks on the graph colour theorem of Hajós. In: *Proc. 17th S–E Conference on Combinatorics, Graph Theory and Computing, Boca Raton, 1986, Congr. Num., 55*, pages 69–76, 1986.

[1959] Harary F. On the group of the composition of two graphs. *Duke Math. J.* **26**, 29–34, 1959.

[1966] Harary F. and C.A. Trauth Jr. Connectedness of products of two directed graphs. *SIAM J. Appl. Math.* **14**, 250–254, 1966.

[1967] Harary F. and G. Wilcox. Boolean operations on graphs. *Math. Scand.* **20**, 41–54, 1967.

[1966] Hedetniemi S.T. Homomorphisms of graphs and automata. Technical Report 03105–44–T, University of Michigan, 1966.

[1978] Johnson D.S. Research problem no. 13. In: B. Alspach, P. Hell, and D.J. Miller, editors, *Algorithmic Aspects of Combinatorics*, volume 2 of *Annals of Discrete Mathematics*, page 243. North-Holland, 1978.

[1993] Kaschek R. and S. Klavžar. Improved bounds for the chromatic number of the lexicographic product of graphs. Technical report, University of Ljubljana, Slovenia, January 1993. Submitted to *Graphs Combin.*

[1993] Klavžar S. Strong products of χ-critical graphs. *Aequationes Math.* **45**, 153–162, 1993.

[1979] Lovász L. *Combinatorial Problems and Exercises*. North-Holland, 1979.

[1982] Lovász L. Bounding the independence number of a graph. In: A. Bachem, M. Grötschel, and B. Korte, editors, *Bonn Workshop on Combinatorial Optimization*, volume 16 of *Annals of Discrete Mathematics*, pages 213–223. North-Holland, 1982.

[1982] Mansfield A.J. and D.J.A. Welsh. Some colouring problems and their complexity. In: B. Bollobás, editor, *Graph Theory*, volume 13 of *Annals of Discrete Mathematics*, pages 159–170. North-Holland, 1982.

[1968] Miller D.J. The categorical product of graphs. *Canad. J. Math.* **20**, 1511–1521, 1968.

[1983] Nešetřil J. and V. Rödl. Products of graphs and their applications. In: M. Borowiecki, J.W. Kennedy, and M.M. Sysło, editors, *Graph Theory, Łagów 1981*, volume 1018 of *Lecture Notes in Mathematics*, pages 151–160. Springer-Verlag, 1983.

[1985] Nešetřil J. and V. Rödl. Three remarks on dimensions of graphs. In: M. Karoński and A. Ruciński, editors, *Random Graphs '83*, volume 28 of *Annals of Discrete Mathematics*, pages 199–207. North-Holland, 1985.

[1962] Ore O. *Theory of Graphs*. American Mathematical Society, 1962.

[1967] Ore O. *The Four-Color Problem*. Academic Press, 1967.

[1988] Puš V. Chromatic number of products of graphs. Technical Report 88–85, Charles University, Prague, 1988.

[1957] Sabidussi G. Graphs with given group and given graph-theoretical properties. *Canad. J. Math.* **9**, 515–525, 1957.

[1960] Sabidussi G. Graph multiplication. *Math. Z.* **72**, 446–457, 1960.

[1989] Sachs H. and M. Stiebitz. Colour-critical graphs with vertices of low valency. In: L.D. Andersen, I.T. Jakobsen, C. Thomassen, B. Toft, and P.D. Vestergaard, editors, *Graph Theory in Memory of G.A. Dirac*, volume 41 of *Annals of Discrete Mathematics*, pages 371–396. North-Holland, 1989.

[1992] Sauer N.W. and X. Zhu. An approach to Hedetniemi's conjecture. *J. Graph Theory* **16**, 423–436, 1992.

[1953] Shapiro H.S. The embedding of graphs in cubes and the design of sequential relay circuits. Technical report, Bell Telephone Laboratories, July 1953.

[1993] Stiebitz M. and W. Wessel. On coloring partial joins of a complete graph and a cycle. *Math. Nachr.* **163**, 109–116, 1993.

[1972a] Toft B. On separating sets of edges in contraction-critical graphs. *Math. Ann.* **196**, 129–147, 1972.

[1974a] Toft B. Color-critical graphs and hypergraphs. *J. Combin. Theory Ser. B* **16**, 145–161, 1974.

[1985] Toft B. Some problems and results related to subgraphs of colour critical graphs. In: R. Bodendiek, H. Schumacher, and G. Walther, editors, *Graphen in Forschung und Unterricht: Festschrift K. Wagner*, pages 178–186. Barbara Franzbecker Verlag, 1985.

[1992] Tutte W.T. A lecture on graph-colourings. Dept. Combinatorics and Optimization, University of Waterloo, Canada, 1992.

[1981] Vesztergombi K. Chromatic number of strong product of graphs. In: L. Lovász and V. Sós, editors, *Algebraic Methods in Graph Theory*, volume 25 of *Colloquia Mathematica Societatis János Bolyai*, pages 819–825. North-Holland, 1981.

[1962] Weichsel P.M. The Kronecker product of graphs. *Proc. Amer. Math. Soc.* **13**, 47–52, 1962.

[1968] Yang K.W. Chromatic number of Cartesian sum of two graphs. *Proc. Amer. Math. Soc.* **19**, 617–618, 1968.

[1992] Youngs D.A. Gallai's problem on Dirac's construction. *Discrete Math.* **101**, 343–350, 1992.

12

Edge Colorings

■ 12.1. GOLDBERG'S CONJECTURE

If $\chi'(G) \geq \Delta(G) + 2$, is it then true that $\chi'(G) = w(G)$ with

$$w(G) = \max_{H \subseteq G} \left\lceil \frac{e(H)}{\lfloor \frac{1}{2} v(H) \rfloor} \right\rceil,$$

where $e(H)$ and $v(H)$ denote the number of edges and vertices of H, respectively? ■

If $\chi'(G) \geq \Delta(G) + 2$, then by the theorem of Vizing [1964] (Theorem 29 in Chapter 1) G has multiple edges. All graphs G satisfy $w(G) \leq \chi'(G)$.

The question seems to have been thought of first by Goldberg [1973, 1984b] around 1970. An early related result is due to Vizing, who proved in his doctoral dissertation in 1968 that if G is a multigraph for which $\chi'(G)$ equals the general upper bound $3\Delta(G)/2$ of Shannon [1949], then G contains a subgraph on three vertices and $3\Delta(G)/2$ edges, implying that $\chi'(G) = w(G)$ [personal communication from Goldberg in 1986].

Equivalent conjectures have been stated by others, in particular by Andersen [1977] and Seymour [1979a].

Conjecture (Andersen [1977]). *Let G be critical w.r.t. χ' and let $\chi'(G) \geq \Delta(G) + 2$. Then for any edge $e \in E(G)$ there is a $(\chi'(G) - 1)$-edge coloring of $G - e$ such that no color is missing at two different vertices.*

The fact that Andersen's conjecture is equivalent to Goldberg's conjecture was proved by Goldberg [1984b].

Giving another equivalent formulation, Seymour [1979a] asked, independently of Goldberg and Andersen: If G is any graph or multigraph, is

$$\chi'(G) \leq \max\{\Delta(G) + 1, w(G)\}?$$

Seymour also suggested a weaker conjecture (Problem 12.3) equivalent to the statement that any graph or multigraph G satisfies $\chi'(G) \leq \max\{\Delta(G), w(G)\} + 1$.

An affirmative answer to Goldberg's conjecture has been proved for $\Delta(G) \leq 7$ by Andersen [1977], for $\Delta(G) \leq 9$ by Goldberg [1977, 1984b], and for $\Delta(G) \leq 11$ by Nishizeki and Kashiwagi [1985, 1990]. Marcotte [1990] has proved that the conjecture is true for any graph that does not contain the graph K_5^- (the complete 5-graph minus one edge) as a minor (i.e., as a subcontraction, a graph obtained by successive deletions of edges and vertices and contractions of edges).

Plantholt and Tipnis [1991] verified the conjecture for all multigraphs G with sufficiently high maximum degree $\Delta(G)$ relative to the number of vertices and the multiplicity of G.

Goldberg [1984b] also stated his problem with the roles of Δ and w interchanged: If $\chi'(G) \geq w(G) + 2$, is it then true that $\chi'(G) = \Delta(G)$? Equivalently, is

$$\chi'(G) \leq \max\{\Delta(G), w(G) + 1\}?$$

■ 12.2. JAKOBSEN'S CONJECTURE

Let G be critical w.r.t. χ'. If m is odd, $m \geq 3$, and

$$\chi'(G) > \frac{m \cdot \Delta(G) + (m - 3)}{m - 1},$$

is it then true that G has at most $m - 2$ vertices? ■

The conjecture that the answer is affirmative was formulated by Jakobsen [1975], who noted from the so-called "ring multigraphs" (e.g., a 5-cycle with two more copies of each edge added) that it would be best possible. Andersen [1977] stated a conjecture (equivalent to a conjecture of Goldberg [1973, 1984b]; see Problem 12.1 for both conjectures), and he proved that an affirmative answer to that conjecture implies an affirmative answer to Jakobsen's question. Hence Jakobsen's question is a weaker version of Goldberg's conjecture.

For $m = 3$ the affirmative answer is the theorem of Shannon [1949]. For $m = 5$ it was obtained by B.Aa. Sørensen [unpublished], Andersen [1977], and Goldberg [1973], for $m = 7$ by B.Aa. Sørensen [unpublished] and Andersen [1977], for $m = 9$ by Goldberg [1977, 1984b], and for $m = 11$ by Nishizeki and Kashiwagi [1985, 1990].

An affirmative answer to the problem in general implies that if G is critical w.r.t. χ' and $\chi'(G) \geq \Delta(G) + 2$, then

$$|V(G)| \leq \Delta(G) - 1 \text{ if } \Delta(G) \geq 2 \text{ and } \Delta(G) \text{ is even, and}$$

$$|V(G)| \leq \Delta(G) - 2 \text{ if } \Delta(G) \geq 3 \text{ and } \Delta(G) \text{ is odd.}$$

Hence for a fixed $\Delta(G)$ (or a fixed $\chi'(G)$) there are only finitely many critical graphs with $\chi'(G) \geq \Delta(G) + 2$, assuming an affirmative answer to Jakobsen's conjecture. Also this weak version of Jakobsen's conjecture (and of Goldberg's conjecture) seems unsolved.

■ 12.3. SEYMOUR'S r-MULTIGRAPH CONJECTURE

Let $r \geq 1$. A multigraph G is said to be an r-multigraph if G is r-regular and for all $X \subseteq V(G)$ with $|X|$ odd, the number of edges between X and $V(G) - X$ is at least r (in particular, $|V(G)|$ is even). If G is an r-multigraph, is it true that $\chi'(G) \leq r + 1$? ■

This problem is due to Seymour [1979a], who mentioned that the answer is affirmative for $r \leq 6$, and conjectured that the answer is affirmative in general. Moreover, he related the conjecture to other problems; in particular, a positive answer to the conjecture of Goldberg [1973, 1984b] implies a positive answer to Seymour's question (see Problem 12.1): Let for any graph or multigraph G

$$w(G) = \max_{H \subseteq G} \left\lceil \frac{e(H)}{\lfloor \frac{1}{2} v(H) \rfloor} \right\rceil ,$$

where $e(H)$ and $v(H)$ denote the number of edges and vertices of H, respectively. Then, as proved by Seymour [1979a], the conjecture on r-multigraphs is equivalent to

$$\chi'(G) \leq \max\{\Delta(G), w(G)\} + 1.$$

The conjecture by Goldberg is equivalent to

$$\chi'(G) \leq \max\{\Delta(G) + 1, w(G)\}.$$

From the partial results for Goldberg's conjecture, the answer to Seymour's problem is affirmative for $r \leq 11$.

■ 12.4. WEAK CRITICAL GRAPH CONJECTURE

Does there exist a constant $c > 2$ such that if G is a multigraph of maximum degree Δ that is critical w.r.t. χ' and satisfies $\chi'(G) \geq \Delta(G) + 1$ and $|V(G)| \leq c \cdot \Delta$, then $|V(G)|$ must be odd? ■

Jakobsen [1974] formulated what came to be known as "the critical graph conjecture," that any graph or multigraph critical w.r.t. χ' and satisfying $\chi' > \Delta$ must have an odd number of vertices. Beineke and Wilson [1973] independently formulated a similar conjecture. The critical graph conjecture was eventually disproved: Goldberg [1979, 1981] found infinitely many counterexamples for $\Delta = 3$. Counterexamples for $\Delta = 4$ were found by Fiol [1980], who obtained such an example with only 18 vertices. However, counterexamples with $\Delta \geq 5$ have not been found. For a detailed discussion of the critical graph conjecture, see the expository paper by Chetwynd and Wilson [1983].

According to Yap [1980], Goldberg formulated the question above and conjectured that the answer is affirmative. From the known counterexamples to the critical graph conjecture, c would have to be strictly smaller than $9/2$. We do not know if the conjecture can be proved for $c = 2$ or any value of c between 1 and 2.

■ 12.5. CRITICAL MULTIGRAPH CONJECTURE

If G is critical w.r.t. χ' and satisfies $\chi'(G) \geq \Delta(G) + 2$, is it then true that $|V(G)|$ is odd? ■

G has multiple edges by the theorem of Vizing [1964] (Theorem 29 in Chapter 1), which states that if G is simple, then $\chi'(G) \leq \Delta(G) + 1$. The corresponding question with the assumption $\chi'(G) \geq \Delta(G) + 2$ replaced by $\chi'(G) \geq \Delta(G) + 1$ was asked by Jakobsen [1974]. Beineke and Wilson [1973] independently asked a very similar question. But counterexamples with $\chi'(G) = \Delta(G) + 1 = 4$ and 5 were found by Goldberg [1979, 1981] and Fiol [1980] (see also Yap [1980]).

The question has an affirmative answer if Goldberg's conjecture (Problem 12.1) has an affirmative answer. This was pointed out by Andersen [1977].

■ 12.6. VIZING'S 2-FACTOR CONJECTURE

Let G be a simple graph. If G is critical w.r.t. χ' and $\chi'(G) = \Delta(G) + 1$, does G have a 2-factor, that is, a spanning 2-regular subgraph? ■

Vizing [1965b] conjectured that the answer is affirmative.

A related conjecture by Fiorini and Wilson [1978], that every critical graph with an even number of vertices has a 1-factor, was disproved by Choudum [1993].

■ 12.7. VIZING'S PLANAR GRAPH CONJECTURE

Let G be a simple planar graph of maximum degree $\Delta(G)$ equal to either 6 or 7. Is $\chi'(G) = \Delta(G)$? ■

This question is due to Vizing [1965a, 1968], who proved that the answer is positive for $\Delta(G) \geq 8$ and negative for $\Delta(G) \leq 5$. Since G is simple, G satisfies

$\Delta(G) \leq \chi'(G) \leq \Delta(G) + 1$ by the theorem of Vizing [1964] (Theorem 29 in Chapter 1).

O.V. Borodin [personal communication in 1993] attracted our attention to a much more general result proved by Vizing [1965a]: For $k > 0$ and G any simple graph, not necessarily planar,

$$\left.\begin{array}{r}\mathrm{col}(G) \leq k+1 \\ \Delta(G) \geq 2k\end{array}\right\} \Longrightarrow \chi'(G) = \Delta(G).$$

Since any planar graph G satisfies $\mathrm{col}(G) \leq 6$ by Euler's formula, the positive answer follows for $\Delta(G) \geq 10$. Vizing provided additional arguments showing that the bound $\Delta(G) \geq 10$ can be weakened to $\Delta(G) \geq 8$.

■ 12.8. MINIMAL NUMBER OF EDGES IN χ'-CRITICAL GRAPHS

Let G be a simple graph on n vertices. If G is critical w.r.t. χ' and $\chi'(G) = \Delta(G) + 1$, is it true that

$$2|E(G)| \geq n \cdot (\Delta(G) - 1) + 3?$$ ■

This question is due to Vizing [1965b, 1968]. Vizing [1968] remarked that its truth would imply a positive answer to the case $\Delta(G) = 7$ of the planar graph conjecture of Vizing [1965a] (Problem 12.7). The cases $\Delta(G) \leq 5$ have affirmative solutions, see K. Kayathri's paper in *Graphs Combin.* **10**, 139–144, 1994.

The truth of the statement implies that $|E(G)| > (\Delta(G))^2/2$. Vizing [1968] remarked: *"It is not clear how to prove this even in the case when $\Delta(G)$ is even and the critical graph has $\Delta(G) + 1$ vertices."* This case was, however, solved by Plantholt [1981]. Further results in this direction, when n is small compared to Δ, were obtained by Plantholt [1983] and Chetwynd and Hilton [1984a, 1984b, 1989b]. For $\Delta > n/3$ an affirmative answer to Vizing's question would follow from an affirmative answer to Hilton's overfull subgraph conjecture (Problem 12.10).

The same question for multigraphs has a negative answer. Chetwynd and Hilton [1988a] found a critical multigraph of maximum degree 7 with nine vertices and twenty-eight edges. This example is obtained from the Petersen graph by replacing every edge xy of a 1-factor by five parallel edges joining x to y, followed by the deletion of one vertex from the resulting graph.

■ 12.9. INDEPENDENT SETS IN χ'-CRITICAL GRAPHS

Let G be a simple graph on n vertices, with $\chi'(G) = \Delta(G) + 1$, and $\chi'(G') \leq \Delta(G)$ for every proper subgraph G' of G. Is it true that G cannot have more than $n/2$ pairwise nonadjacent vertices? ■

Vizing [1968] remarked: *"It appears perfectly natural (...) How can this be proved?"*

■ 12.10. HILTON'S OVERFULL SUBGRAPH CONJECTURE

A simple graph H is called overfull if $|E(H)| > \Delta(H) \cdot \lfloor |V(H)|/2 \rfloor$ (implying that $\chi'(H) > \Delta(H)$). Let G be any simple graph. If G contains an overfull subgraph H with $\Delta(H) = \Delta(G)$, then this implies $\chi'(G) = \Delta(G) + 1$. Is the converse implication true, assuming that $\Delta(G) > |V(G)|/3$? ■

The question was raised by A.J.W. Hilton at the meeting in Sandbjerg, Denmark, June 1985. It appeared in print in Chetwynd and Hilton [1986, 1989b].

The theorem of Vizing [1964] (Theorem 29 in Chapter 1) states that if G is simple, then $\chi'(G) \leq \Delta(G) + 1$. Thus equality holds in $\chi'(G) \geq \chi'(H) \geq \Delta(H) + 1 = \Delta(G) + 1$ when H is a subgraph of G as above.

If $\Delta(G) \leq |V(G)|/3$, the converse implication is not always true, as can be seen from the Petersen graph with one vertex removed.

The conjecture is true for all G such that $\Delta(G) \geq |V(G)| - 3$, as proved in papers by Plantholt [1981, 1983] and by Chetwynd and Hilton [1984a, 1984b, 1989b]. Hilton [1989] has announced that the conjecture has been proved by Chetwynd and himself in the case when $\Delta(G) \geq (\sqrt{21} - 1) \cdot (|V(G)| + 1)/4 + 1$ and $|E(G)| = \lfloor |V(G)|/2 \rfloor$.

Many interesting consequences of Hilton's overfull subgraph conjecture were described by Hilton and Johnson [1987]. In particular, the following two conjectures on graphs G that are vertex-critical w.r.t. χ'; that is, G is simple, $\chi'(G) = \Delta(G) + 1$, and $\chi'(G - v) < \chi'(G)$ for all $v \in V(G)$:

Conjecture 1. *Let G be a simple graph with an even number of vertices and satisfying $\Delta(G) > |V(G)|/3$. Then G is not vertex-critical w.r.t. χ'.*

Conjecture 2. *Let G be a simple graph with $\Delta(G) \geq |V(G)|/2$. Then G is vertex-critical w.r.t. χ' if and only if G is overfull.*

It seems unknown both how to prove the "if" part and how to prove the "only if" part of the latter conjecture without assuming Hilton's overfull subgraph conjecture. The condition $\Delta(G) \geq |V(G)|/2$ cannot be sharpened, as can be seen from infinitely many examples, constructed by Hilton and Johnson [1987], of overfull graphs G with $\Delta(G) = (|V(G)| - 1)/2$ that are not vertex-critical w.r.t. χ'.

■ 12.11. THE Δ-SUBGRAPH CONJECTURE

Let G be a simple graph of maximum degree Δ, and let G_Δ denote the subgraph induced by the vertices of degree Δ in G. If

(i) $\Delta > \frac{2}{3}(|V(G)| - 3)$, and
(ii) G_Δ has minimum degree at most 1,

is $\chi'(G) = \Delta$? ■

This was conjectured by Chetwynd and Hilton [1989a], who proved that the answer is affirmative if (i) is replaced by

$$\Delta \geq \left\lfloor \frac{1}{2}|V(G)| \right\rfloor + \frac{7}{2}|V(G_\Delta)| - 3,$$

and that (i) would be best possible.

If G is a simple graph such that G_Δ is a forest (i.e., without cycles), then $\chi'(G) = \Delta(G)$, as proved by Fournier [1973]. This also follows from Vizing's adjacency lemma (Vizing [1965a]; see also Chapter 11 in the book by Fiorini and Wilson [1977]): If H is a critical graph with maximum degree Δ, that is, $\chi'(H) = \Delta + 1$ and $\chi'(H - e) = \Delta$ for every edge $e \in E(H)$, and if u and v are adjacent vertices of H, where the degree of v is d, then

(i) if $d < \Delta$, then u is adjacent to at least $\Delta - d + 1$ vertices of degree Δ, and
(ii) if $d = \Delta$, then u is adjacent to at least two vertices of degree Δ.

The truth of the conjecture would follow from the truth of Hilton's overfull subgraph conjecture (Problem 12.10), as proved by Chetwynd and Hilton [1989a].

Generalizations of Fournier's result were obtained by Hoffman and Rodger [1988].

■ 12.12. REGULAR GRAPHS OF HIGH DEGREE

Let G be a k-regular simple graph with an even number of vertices and such that $k \geq |V(G)|/2$. Is G 1-factorizable (i.e., k-edge-colorable)? ■

The conjecture that the answer is affirmative appeared in Chetwynd and Hilton [1985]. The question may go back to G.A. Dirac in the early 1950s (see Chetwynd and Hilton [1989c]). An interesting consequence is that for any regular graph G with an even number of vertices, either G or its complement $\overline{G}$ has a 1-factorization. One might possibly consider a similar statement for nonregular graphs: Is it true for every simple graph G with an even number of vertices that either $\chi'(G) = \Delta(G)$ or $\chi'(\overline{G}) = \Delta(\overline{G})$? Hilton [1989] proved that Hilton's overfull subgraph conjecture (Problem 12.10) would imply the conjecture above on k-regular simple graphs.

Chetwynd and Hilton [1985] proved the conjecture for $k \geq |V(G)| - 5$. They also proved that if G is k-regular with $|V(G)|$ even and $k \geq 0.849 \cdot |V(G)|$, then G is 1-factorizable; later Chetwynd and Hilton [1989c] improved the bound to $k \geq \frac{1}{2}(\sqrt{7} - 1) \cdot |V(G)|$, and this appears to be the best known result of this type. Chetwynd and Hilton [1985] remarked that R. Häggkvist has announced that for any $\varepsilon > 0$ there exists $N > 0$ so that every k-regular graph G with an even number of vertices and satisfying $|V(G)| \geq N$ and $k \geq (\frac{1}{2} + \varepsilon) \cdot |V(G)|$ is 1-factorizable.

A slightly stronger version of the conjecture was mentioned by Chetwynd and Hilton [1985]: If G is k-regular on $2n$ vertices, where $k \geq 2\lfloor(n+1)/2\rfloor - 1$, is G 1-factorizable?

Chetwynd and Hilton [1985] made an even stronger conjecture: Let $n \geq k \geq 2$. If G is k-regular and $|V(G)| = 2n$, is it true that the complement $\overline{G}$ has a 1-factor F such that both $G \cup F$ and $\overline{G} - E(F)$ have 1-factorizations, unless k is odd and G is the graph $K_{k,k}$? Faudree and Sheehan [1984] asked the same question without the condition on $\overline{G} - E(F)$.

■ 12.13. BERGE AND FULKERSON'S CONJECTURE

If G is a 3-regular simple graph without bridges and H is obtained from G by duplicating each edge, is it then true that H is 6-edge-colorable? ■

The conjecture that the answer is affirmative is due to C. Berge (see Seymour [1979a]) and Fulkerson [1971], who first formulated it in print. A weaker version of the conjecture suggests that the edges of G can be covered with five 1-factors. This would immediately follow from a 6-edge coloring of H after deleting one color class. The weaker conjecture is also due to C. Berge [personal communication from F. Jaeger in 1994].

An attractive equivalent reformulation of the conjecture is the following, where a graph is called "even" if each of its vertices has an even degree. Is it true that the edge set of any graph without bridges can be covered with a family of six even subgraphs, with repetitions allowed, such that every edge is contained in exactly four members of the family? The equivalence of the two questions was explained by Jaeger [1988]. The 8-flow theorem (see Section 1.8) is equivalent to the weaker statement obtained by replacing "six even subgraphs" by "seven even subgraphs." See also Problems 12.14 and 13.2 for related questions and discussion.

Support of the conjecture was found by Seymour [1979a], who proved that it is possible to find two families $\mathcal{F}_1$ and $\mathcal{F}_2$ of 1-factors of H such that every edge is contained exactly once more in a member of $\mathcal{F}_1$ than in a member of $\mathcal{F}_2$. Berge and Fulkerson's conjecture is equivalent to the same statement with $\mathcal{F}_2 = \emptyset$.

■ 12.14. PETERSEN COLORING

If G is a 3-regular bridge-less graph, is it possible to color the edges of G, using the edges of the Petersen graph as colors, in such a way that any three mutually adjacent edges of G are also colored with three edges that are mutually adjacent in the Petersen graph? ■

The question was asked by Jaeger [1988], who conjectured that the answer is affirmative. This would imply the conjecture of C. Berge and Fulkerson [1971] on multicolorings of cubic graphs (Problem 12.13). It would also imply the cycle double cover conjecture by Szekeres [1973] and Seymour [1979b], who conjectured independently that every bridge-less graph has a family of cycles, possibly with repetitions, such that every edge of the graph is contained in precisely two of the cycles.

■ 12.15. TUTTE'S CONJECTURE ON 3-EDGE COLORINGS

Let G be a 3-regular simple graph without bridges and without a subdivision of the Petersen graph as a subgraph. Is $\chi'(G) = 3$? ■

The conjecture that the answer is affirmative is due to Tutte [1966, 1969] (see also early roots of the question in Descartes [1948a], as well as the delightful paper by Tutte [1978]).

It is a consequence of the conjecture that if G is a planar 3-regular simple graph without bridges, then $\chi'(G) = 3$. This statement is equivalent to the four-color theorem by the theorem of P.G. Tait (Theorem 26 in Chapter 1; see also Fiorini and Wilson [1977]).

Tutte's conjecture implies the conjecture of H. Grötzsch, that if G is a planar graph of maximum degree three, then G is 3-edge-colorable if and only if G does not contain a subgraph all of whose vertices have degree three, except for one vertex of degree two (see Seymour [1981b]). Seymour [1981b] observed that the conjecture of Grötzsch implies that if G is a 3-regular bridge-less graph with a vertex v so that $G - v$ is planar, then G can be 3-edge-colored. This is directly implied by Tutte's conjecture, since the graph obtained from the Petersen graph by deleting a vertex is nonplanar. See also Problem 13.2.

Examples of 3-regular simple graphs G without bridges having $\chi'(G) = 4$ (called "snarks" by Gardner [1976], and usually taken to contain no edge cut of size 3 or less that separates two cycles, and to be without cycles of length less than five) are difficult to find (Descartes [1948a]), but since the paper of Isaacs [1975], several families of snarks have been found (see the excellent survey by Watkins [1989]). However, it is unlikely that a simple characterization of such graphs can be found; it is an ***NP***-complete problem to recognize a snark by a result of Holyer [1981].

Jaeger and Swart [1980] conjectured that every snark has a cycle of length at most six. This conjecture was disproved by M. Kochol, who produced examples of snarks with arbitrarily long shortest cycles, that is, of arbitrarily high girth. This was done using a general construction of graphs without k-flows from existing ones, for $k \leq 4$, due to M. Kochol and M. Škoviera [personal communication from M. Škoviera in 1994]. See Problem 13.2 for the extension, in terms of k-flows, of Tutte's conjecture to bridge-less graphs in general.

Supporting results for Tutte's conjecture were given by Seymour [1979a]. An interesting partial result was obtained by Ellingham [1984], who noted that the 2-

factor theorem of Petersen [1891] can be used to classify all 3-regular bridge-less graphs according to the number of odd and even cycles in 2-factors. In terms of 2-factors, Tutte's conjecture is equivalent to the statement that every bridge-less 3-regular graph without a subdivision of the Petersen graph as a subgraph has a 2-factor consisting of a disjoint union of even cycles. Ellingham considered 3-regular graphs with 2-factors consisting of precisely two odd cycles without diagonals. He proved Tutte's conjecture for these graphs by proving that each either contains a subdivision of the Petersen graph as a subgraph or has a Hamilton cycle (i.e., a 2-factor consisting of only one even cycle).

■ 12.16. GRÖTZSCH AND SEYMOUR'S CONJECTURE

A simple graph H is called overfull if $|E(H)| > \Delta(H) \cdot \lfloor |V(H)|/2 \rfloor$ (implying that $\chi'(H) > \Delta(H)$). If G is any simple planar graph, is it true that either $\chi'(G) = \Delta(G)$ or G has an overfull subgraph H with $\Delta(H) = \Delta(G)$? ■

The affirmative answer was conjectured by Seymour [1979c]. The special case of the conjecture with $\Delta(G) = 3$ is due to H. Grötzsch (see Seymour [1981b]): Is it true that if G is any planar graph of maximum degree three, then G is 3-edge-colorable if and only if no subgraph of G can be obtained from a 3-regular graph by deleting an edge and replacing it by a path of length two?

Grötzsch and Seymour's conjecture was formulated independently by Hoffman, Mitchem, and Schmeichel [1992], who also listed a number of its implications, among them the following:

(i) The four-color theorem.
(ii) The planar graph conjecture of Vizing [1965a] (Problem 12.7).
(iii) The critical graph conjecture (Jakobsen [1974], Beineke and Wilson [1973]) for planar graphs (see Problems 12.4 and 12.5).
(iv) A polynomial algorithm for determining $\chi'(G)$ for any planar graph G.
(v) The statement that for any $(\Delta(G) - 1)$-edge-connected planar graph G, either $\chi'(G) = \Delta(G)$, or G is overfull.

Each of (ii)–(v) seems open. A further implication (also open) is

(vi) Every planar graph G satisfies $\chi'(G) = \max\{\Delta(G), w(G)\}$, where $w(G)$ is the parameter of G defined in Problem 12.1.

Kilakos and Shepherd [1993] have suggested the strengthening of (vi) that the conclusion holds not only for planar graphs, but for all graphs that do not contain a subdivision of the graph obtained from the Petersen graph by deleting any vertex.

■ 12.17. CYCLE-DECOMPOSABLE 4-REGULAR PLANE MULTIGRAPHS

Let G be a 4-regular plane multigraph, and suppose that G has a cycle decomposition S; that is, each edge is contained in exactly one cycle of S, such that every pair of adjacent edges in the same face are always in different cycles of the decomposition. Is $\chi'(G) = 4$? ■

The question was asked by Jaeger [1976b]. It was proved by Jaeger and Shank [1981], and with a short proof by Jaeger and Koester [1990], that if S can be partitioned into four classes of pairwise disjoint cycles, then $\chi'(G) \leq 4$. Jaeger and Shank [1981] proved that the corresponding statement with five classes implies the four-color theorem.

■ 12.18. SEYMOUR'S PLANAR 4-MULTIGRAPH CONJECTURE

Let G be a planar 4-multigraph; that is, G is 4-regular, and for every $X \subseteq V(G)$ with $|X|$ odd, the number of edges between X and $V(G) - X$ is at least 4 (in particular $|V(G)|$ is even). Is $\chi'(G) \leq 4$? ■

This problem is due to Seymour [1979a], who noted that an affirmative answer implies the four-color theorem by a result of Kotzig [1957]. Also, an affirmative answer implies an affirmative answer to the problem of Jaeger [1976b] on cycle-decomposable 4-regular plane multigraphs (Problem 12.17).

■ 12.19. UNIQUELY 3-EDGE-COLORABLE PLANAR GRAPHS

Does every uniquely 3-edge-colorable planar graph different from $K_{1,3}$ contain a triangle? ■

The conjecture that the answer is affirmative can be found in Fiorini and Wilson [1978] and it originated from a paper on edge-colorings by Fiorini [1975], who omitted a detail when formulating the conjecture "a uniquely 3-colorable planar graph contains a 3-circuit." This has led to an alternative question that is seemingly weaker than the one stated above and also remains open: Is it true that every uniquely 3-edge-colorable 3-regular planar graph contains a triangle?

Restricting the discussion to 3-regular graphs, every uniquely 3-edge-colorable graph has exactly three Hamilton cycles; hence the conjecture above would follow from a conjecture by R. Cantoni (see Tutte [1976]), also posed by Greenwell and Kronk [1973], that every 3-regular planar graph with exactly three Hamilton cycles contains a triangle. Thomason [1982] found examples of 3-regular nonplanar graphs with exactly three Hamilton cycles, but with more than one 3-edge-coloring; it follows from these examples of certain generalized Petersen graphs that Cantoni's conjecture cannot be extended to nonplanar graphs.

If the answer is affirmative, then all uniquely 3-edge-colorable 3-regular planar graphs can be obtained from K_4 by repeatedly replacing vertices by triangles. The word "planar" cannot be removed as shown by the generalized Petersen graph $P(9, 2)$. This was first observed by Tutte [1976]. It seems unknown if there are other nonplanar counterexamples (see Fiorini and Wilson [1978]). Zhang [1993] has conjectured that the planarity condition may be weakened, and that the answer to the following question is affirmative: Does every uniquely 3-edge-colorable 3-regular graph without any subdivision of the Petersen graph as a subgraph contain a triangle?

12.20. LIST-EDGE-CHROMATIC NUMBERS

Let each edge xy of G be assigned a list $\Lambda(xy)$ of λ permissible colors. A Λ-coloring is an edge-coloring in which each edge is assigned a color from its list. The list-edge-chromatic number $\chi'_\ell(G)$ is the minimum number λ for which such a coloring always exists no matter what the lists look like (but they are all of size λ). It is clear that $\chi'_\ell(G) \geq \chi'(G)$.
Is $\chi'_\ell(G) = \chi'(G)$?

This conjecture first appeared in print in a paper by Bollobás and Harris [1985]. Chetwynd and Häggkvist [1989] mentioned that the problem may have been studied earlier by R.P. Gupta and referred to implicitly by Erdős [1979]. The conjecture has also been attributed to M.O. Albertson and K.L. Coliins, and it was independently formulated by V.G. Vizing as early as in 1975 (see the excellent survey of this and related problems by Häggkvist and Chetwynd [1992]).

Hind [1988] proved $\chi'_\ell(G) \leq (9/5) \cdot \Delta(G)$ for any multigraph G and $\chi'_\ell(G) \leq (5/3) \cdot \Delta(G)$ for any triangle-free multigraph G. The conjecture above and the bound $\chi'(G) \leq (3/2) \cdot \Delta(G)$ by Shannon [1949] would imply that the constant 9/5 can be improved to 3/2. For simple graphs G with sufficiently large maximum degree Δ, Bollobás and Hind [1989] proved that

$$\chi'_\ell(G) \leq (7/4) \cdot \Delta + \lceil 25 \log \Delta \rceil .$$

R. Häggkvist [personal communication in 1993] and J.C.M. Janssen were the first to obtain an upper bound behaving asymptotically as $\Delta + o(1)$ for a simple graph, namely

$$\chi'_\ell \leq \Delta + c \cdot \Delta^{2/3} \cdot \log \Delta$$

where $c > 0$ is a constant.

Kostochka [1992] proved that if all the cycles in G are sufficiently long relative to $\Delta(G)$, then $\chi'_\ell(G) \leq \Delta(G) + 1$. If the conjecture above is true, the same bound would hold for any simple graph, since by the theorem of Vizing [1964] (Theorem 29 in Chapter 1), a simple graph G satisfies $\chi'(G) \leq \Delta(G) + 1$.

A famous special case of the conjecture is due to J. Dinitz (see Chetwynd and Häggkvist [1989]). Dinitz conjectured in 1978 that given any collection of sets S_{ij} $(1 \leq i, j \leq n)$ all of size n, there exists a Latin square $(\ell_{ij} : 1 \leq i, j \leq n)$, so that $\ell_{ij} \in S_{ij}$ for all i, j. In terms of list coloring, the Dinitz conjecture translates to $\chi'_\ell(K_{n,n}) = n$. Janssen [1993] obtained partial results using the theorem of N. Alon and M. Tarsi (Theorem 34 in Chapter 1). Recently, F. Galvin proved the conjecture with a very short proof using similar ideas, but based on less powerful results [personal communications in 1994 from N. Alon and L. Goddyn]. In fact, J.A. Bondy [personal communication in 1994] has pointed out that the proof can be obtained by combining a theorem of M. Richardson dating back to 1953 (see Problem 13.9) with a result of Gale and Shapley [1962]. Galvin's proof of Dinitz' conjecture can be extended to show $\chi'_\ell(G) = \chi'(G)$ for every bipartite graph G.

■ 12.21. STRONG CHROMATIC INDEX

A strong k-edge coloring of a graph or multigraph G is an assignment of k colors to the edges of G in such a way that any two adjacent vertices in G are not incident to two distinct edges of the same color. Equivalently, the vertices incident to any color class induce a 1-regular subgraph of G. The strong chromatic index $s\chi'(G)$ is the smallest number k for which G has a strong k-edge coloring.

If the maximum degree of G is Δ, does

$$s\chi'(G) \leq \begin{cases} 5\Delta^2/4 & \text{if } \Delta \text{ is even,} \\ 5\Delta^2/4 - \Delta/2 + 1/4 & \text{if } \Delta \text{ is odd} \end{cases}$$

hold?

If G is bipartite, is $s\chi'(G) \leq \Delta^2$?

If G is 3-regular and planar, is $s\chi'(G) \leq 9$? ■

The first question seems first considered at a conference in Prague in 1985 by P. Erdős and J. Nešetřil, who conjectured that the answer is affirmative (see Faudree, Gyárfás, Schelp, and Tuza [1989]). The last two questions were asked by Faudree, Gyárfás, Schelp, and Tuza [1990].

It is easy to find a strong edge coloring of G with at most $2\Delta^2 - 2\Delta + 1$ colors. This shows that the conjecture of Erdős and Nešetřil is true for $\Delta \leq 2$. However, as remarked by Horák, Qing, and Trotter [1993], it appears very hard to prove $s\chi'(G) \leq (2 - \varepsilon)\Delta^2$ for some $\varepsilon > 0$. A result in support of the conjecture of Erdős and Nešetřil was obtained by Chung, Gyárfás, Trotter, and Tuza [1990], who proved that if G contains no induced 1-regular subgraph with more than one edge, then G has at most $5\Delta^2/4$ edges.

For 3-regular graphs, the best possible inequality $s\chi'(G) \leq 10$ was obtained independently by Andersen [1992] and by Horák, Qing, and Trotter [1993]. Horák [1990] proved $s\chi'(G) \leq 23$ for $\Delta = 4$. The conjecture of Erdős and Nešetřil is unsolved for $\Delta \geq 4$.

For bipartite G the second question above remains open even if $\Delta = 3$. Faudree, Gyárfás, Schelp, and Tuza [1990] also asked if $s\chi'(G) \leq 7$ if G is bipartite without a 4-cycle, and if $s\chi'(G) \leq 5$ if G is bipartite and without cycles of sufficiently large lengths.

The inequality $s\chi'(G) \leq 9$ for G 3-regular and planar would be best possible, since equality holds for the prism graph on six vertices.

■ 12.22. VIZING'S INTERCHANGE PROBLEM

Let φ be any given k-edge coloring of a multigraph G, where G satisfies $\chi'(G) \geq \Delta(G) + 2$. An "interchange" with respect to colors α and β, $1 \leq \alpha < \beta \leq k$, consists in swapping the two colors on the edges of a connected component of the subgraph of G induced by all edges colored α or β, thus obtaining a new edge coloring of G using at most k colors. Is it possible to obtain an edge coloring of G with $\chi'(G)$ colors from φ by a sequence of interchanges? ■

The question was asked by Vizing [1965b, 1968].

If $\chi'(G) \geq \Delta(G) + 2$ then, by the theorem of Vizing [1964] (Theorem 29 in Chapter 1), G has multiple edges. Vizing [1965b] proved that $\chi'(G) \leq \Delta(G) + \mu(G)$, where $\mu(G)$ is the multiplicity of G (i.e., the maximum number of edges joining the same pair of vertices) and he noted that a $(\Delta(G) + \mu(G))$-edge coloring and a $3\Delta(G)/2$-edge coloring (corresponding to the bound proved by Shannon [1949]) can be obtained from an arbitrary edge coloring by interchanges.

It is not clear to us if it is possible in general to obtain a $\chi'(G)$-edge coloring from an arbitrary edge coloring of a graph or multigraph G by a sequence of interchanges. For vertex coloring this is not the case, as can be seen from a 4-coloring of the graph obtained from the complete tripartite graph $K_{4,4,4}$ by removing the edges of four disjoint triangles.

■ 12.23. SCHEDULING WITHOUT WAITING PERIODS

Call a coloring of the edges of a graph with colors $1, 2, 3, \ldots$ a "consecutive coloring" if the colors received by the edges incident to each vertex are distinct and form an interval of integers. Let G be a bipartite graph with vertex partition (X, Y). If all vertices of X have the same degree d_X and all vertices of Y have degree d_Y, does G allow a consecutive coloring? ■

This question arose from a practical scheduling problem: At the Saint Canute High School in Odense, Denmark, parent consultations are arranged by letting each parent, or couple of parents, decide beforehand on a list of teachers that he/she/they would like to consult. Each meeting between parent(s) and a teacher lasts for the same fixed amount of time. The problem is to create a schedule without waiting periods for neither parents nor teachers. Letting X represent the set of teachers and Y the set of parents/couples of parents, and letting an edge xy represent a meeting of teacher x with parent(s) y, we obtain an instance of this coloring problem (without the degree constraints), where each color corresponds to an assigned time-slot.

This particular variation of edge coloring for multigraphs in general seems first studied under the name of "interval coloring" by Asratyan and Kamalyan [1987] and by Sevast'yanov [1990].

Let us first consider the problem for bipartite graphs without degree constraints. Then it is not always possible to color the edges consecutively. The first example showing this seems due to Sevast'yanov [1990]. In the fall of 1991 we received examples found independently by P. Erdős and by A. Hertz and D. de Werra. To obtain Erdős' example take a finite projective plane P of order p and let X represent the set of points of P and Y the set of lines, with xy an edge if and only if the point x belongs to the line y. Finally, join one new vertex z to all the vertices of Y. For $p \geq 3$, this bipartite graph G cannot be consecutively colored. The vertices of the part $X \cup \{z\}$ of the bipartition contains vertices of two different degrees $p + 1$ and $p^2 + p + 1$, whereas the vertices in Y all have degree $p + 2$. For $p = 3$, G has 27 vertices and $\Delta = 13$. The example by Sevast'yanov has 28 vertices and $\Delta = 21$. A smaller example, due to Hertz and de Werra, has 21 vertices and $\Delta = 14$. We do not know any examples with $\Delta \leq 12$. For $\Delta \leq 3$ such examples do not exist, as proved by Hansen [1992].

If G is Δ-regular, then by the 1-factor theorem of D. König (Theorem 27 in Chapter 1) it is always possible to Δ-edge-color G, and of course such a coloring is also a consecutive coloring.

For bipartite graphs where all vertices in X have degree d_X and all vertices in Y have degree d_Y, Hansen [1992] noted that the answer to the question above is affirmative in the case $d_X = 2$ and d_Y is any even number, and that this statement is in fact equivalent to the 2-factor theorem of Petersen [1891]: Every $2r$-regular graph has a decomposition into 2-regular edge-disjoint subgraphs. The simplest unsolved case of the problem stated above is $(d_X, d_Y) = (3, 4)$. The $(2, 5)$ case has recently been settled in the affirmative by D. Hanson, C. Loten, and B. Toft. The solution uses six colors and is based on a simple application of the theorem of Bäbler [1938] and Gallai [1950] that a bridge-less 5-regular graph has a 2-factor. This is in fact the key to a complete affirmative solution for the case $(2, d_Y)$ with d_Y odd.

Hansen [1992] proved that the complete bipartite graph $K_{m,n}$ has a consecutive coloring using $m + n - \gcd(m, n)$ colors, and claimed that in general $d_X + d_Y - \gcd(d_X, d_Y)$ colors are necessary. However, he noted that this number of colors is not always sufficient.

Asratyan and Kamalyan [1987] proved that for a given bipartite graph with bipartition (X, Y) and a given number k of colors, it is an ***NP***-complete problem to decide if there exists a k-coloring that is consecutive on X. Sevast'yanov [1990] proved that in general it is an ***NP***-complete problem to decide for a given bipartite graph if it has a consecutive coloring (allowing any number of colors).[1]

For nonbipartite graphs, consecutive colorings were studied by Asratyan and Kamalyan [1987]. We note that a k-regular graph is consecutively colorable if and only if it can be k-edge-colored.

BIBLIOGRAPHY

[1977] Andersen L.D. On edge-colourings of graphs. *Math. Scand.* **40**, 161–175, 1977.

[1992] Andersen L.D. The strong chromatic index of a cubic graph is at most 10. *Discrete Math.* **108**, 231–252, 1992.

[1987] Asratyan A.S. and R.R. Kamalyan. Interval colorings of the edges of a multigraph (in Russian). In: *Applied Mathematics, No. 5*, pages 25–34 and 130–131. Erevan University, 1987 (Armenian summary).

[1938] Bäbler F. Über die Zerlegung regulärer Streckenkomplexe ungerader Ordnung. *Math. Helv.* **10**, 275–287, 1938.

[1973] Beineke L.W. and R.J. Wilson. On the edge-chromatic number of a graph. *Discrete Math.* **5**, 15–20, 1973.

[1985] Bollobás B. and A.J. Harris. List-colourings of graphs. *Graphs Combin.* **1**, 115–127, 1985.

[1989] Bollobás B. and H.R. Hind. A new upper bound for the list chromatic number. *Discrete Math.* **74**, 65–75, 1989.

[1989] Chetwynd A.G. and R. Häggkvist. A note on list-colourings. *J. Graph Theory* **13**, 87–95, 1989.

[1984a] Chetwynd A.G. and A.J.W. Hilton. Partial edge-colourings of complete graphs or of graphs which are nearly complete. In: B. Bollobás, editor, *Graph Theory and Combinatorics*, pages 81–98. Academic Press, 1984.

[1984b] Chetwynd A.G. and A.J.W. Hilton. The chromatic index of graphs of even order with many edges. *J. Graph Theory* **8**, 463–470, 1984.

[1985] Chetwynd A.G. and A.J.W. Hilton. Regular graphs of high degree are 1-factorizable. *Proc. London Math. Soc. (3)* **50**, 193–206, 1985.

[1986] Chetwynd A.G. and A.J.W. Hilton. Star multigraphs with three vertices of maximum degree. *Math. Proc. Cambridge Philos. Soc.* **100**, 303–317, 1986.

[1988a] Chetwynd A.G. and A.J.W. Hilton. Snarks and k-snarks. *Ars Combin.* **25-C**, 39–54, 1988.

[1989a] Chetwynd A.G. and A.J.W. Hilton. A Δ-subgraph condition for a graph to be class 1. *J. Combin. Theory Ser. B* **46**, 37–45, 1989.

[1989b] Chetwynd A.G. and A.J.W. Hilton. The edge-chromatic class of graphs with maximum degree at least $|V| - 3$. In: L.D. Andersen, I.T. Jakobsen, C. Thomassen, B. Toft, and P.D. Vestergaard, editors, *Graph Theory in Memory of G.A. Dirac*, volume 41 of *Annals of Discrete Mathematics*, pages 91–110. North-Holland, 1989.

[1989c] Chetwynd A.G. and A.J.W. Hilton. 1-factorizing regular graphs of high degree: an improved bound. *Discrete Math.* **75**, 103–112, 1989.

[1983] Chetwynd A.G. and R.J. Wilson. The rise and fall of the critical graph conjecture. *J. Graph Theory* **7**, 153–157, 1983.

[1]We are grateful to G. Gutin for rendering assistance in reading the paper of Sevast'yanov written in Russian.

[1993] Choudum S.A. Edge-chromatic critical graphs and the existence of 1-factors. *J. Graph Theory* **17**, 23–29, 1993.

[1990] Chung F.R.K., A. Gyárfás, W.T. Trotter, and Z. Tuza. The maximum number of edges in $2K_2$-free graphs of bounded degree. *Discrete Math.* **81**, 129–135, 1990.

[1948a] Descartes B. Network-colouring. *Math. Gaz.* **32**, 67–69, 1948.

[1984] Ellingham M.N. Petersen subdivisions in some regular graphs. In: *Proc. 15th S–E Conference on Combinatorics, Graph Theory and Computing, Baton Rouge, 1984, Congr. Num., 44*, pages 33–40, 1984.

[1979] Erdős P. Some old and new problems in various branches of combinatorics. In: *Proc. 10th S–E Conference on Combinatorics, Graph Theory and Computing, Boca Raton, 1979, Congr. Num., 23*, pages 19–37, 1979.

[1989] Faudree R.J., A. Gyárfás, R.H. Schelp, and Z. Tuza. Induced matchings in bipartite graphs. *Discrete Math.* **78**, 83–87, 1989.

[1990] Faudree R.J., A. Gyárfás, R.H. Schelp, and Z. Tuza. The strong chromatic index of graphs. *Ars Combin.* **29-B**, 205–211, 1990.

[1984] Faudree R.J. and J. Sheehan. Regular graphs and edge-chromatic number. *Discrete Math.* **48**, 197–204, 1984.

[1980] Fiol M.A. *3-grafos criticos*. Ph.D. thesis, Barcelona University, 1980.

[1975] Fiorini S. On the chromatic index of a graph. III. Uniquely edge-colourable graphs. *Quart. J. Math. Oxford Ser. (2)* **26**, 129–140, 1975.

[1977] Fiorini S. and R.J. Wilson. *Edge-Colourings of Graphs*. Pitman, 1977.

[1978] Fiorini S. and R.J. Wilson. Edge colourings of graphs. In: L.W. Beineke and R.J. Wilson, editors, *Selected Topics in Graph Theory*, pages 103–126. Academic Press, 1978.

[1973] Fournier J.C. Colorations des arêtes d'un graphe. *Cahiers Centre Études Rech. Opér.* **15**, 311–314, 1973.

[1971] Fulkerson D.R. Blocking and anti-blocking pairs of polyhedra. *Math. Programming* **1**, 168–194, 1971.

[1962] Gale D. and L.S. Shapley. College admissions and the stability of marriage. *Amer. Math. Monthly* **69**, 9–15, 1962.

[1950] Gallai T. On the factorization of graphs. *Acta Math. Acad. Sci. Hungar.* **1**, 133–153, 1950.

[1976] Gardner M. Mathematical games. *Sci. Amer.* **234** and **235**, 1976. April issue, pp. 126–130; September issue, pp. 210–211.

[1973] Goldberg M.K. On multigraphs of almost maximal chromatic class (in Russian). *Metody Diskret. Analiz.* **23**, 3–7, 1973.

[1977] Goldberg M.K. Structure of multigraphs with restrictions on the chromatic class (in Russian). *Metody Diskret. Analiz.* **30**, 3–12, 1977.

[1979] Goldberg M.K. Critical graphs with an even number of vertices (in Russian). *Soobšč. Akad. Nauk Gruzin. SSR* **94**, 25–27, 1979.

[1981] Goldberg M.K. Construction of class 2 graphs with maximum vertex degree 3. *J. Combin. Theory Ser. B* **31**, 282–291, 1981.

[1984b] Goldberg M.K. Edge-coloring of multigraphs: recoloring technique. *J. Graph Theory* **8**, 123–137, 1984.

[1973] Greenwell D. and H.V. Kronk. Uniquely line-colorable graphs. *Canad. Math. Bull.* **16**, 525–529, 1973.

[1992] Häggkvist R. and A.G. Chetwynd. Some upper bounds on the total and list chromatic numbers of multigraphs. *J. Graph Theory* **16**, 503–516, 1992.

[1992] Hansen H.M. *Scheduling with minimization of waiting periods* (in Danish). Master's thesis, Odense University, 1992.

[1989] Hilton A.J.W. Two conjectures on edge-colouring. *Discrete Math.* **74**, 61–64, 1989.

[1987] Hilton A.J.W. and P.D. Johnson. Graphs which are vertex-critical with respect to the edge-chromatic number. *Math. Proc. Cambridge Philos. Soc.* **102**, 211–221, 1987.

[1988] Hind H.R. *Restricted edge-colourings.* Ph.D. thesis, Peterhouse College, Cambridge, 1988.

[1992] Hoffman T., J. Mitchem, and E.F. Schmeichel. On edge-coloring graphs. *Ars Combin.* **33**, 119–128, 1992.

[1988] Hoffman D.G. and C.A. Rodger. Class one graphs. *J. Combin. Theory Ser. B* **44**, 372–376, 1988.

[1981] Holyer I. The ***NP***-completeness of edge-coloring. *SIAM J. Comput.* **10**, 718–720, 1981.

[1990] Horák P. The strong chromatic index of graphs with maximum degree four. In: R. Bodendiek, editor, *Contemporary Methods in Graph Theory*, pages 399–403. B.I. Wissenschaftsverlag, 1990.

[1993] Horák P., H. Qing, and W.T. Trotter. Induced matchings in cubic graphs. *J. Graph Theory* **17**, 151–160, 1993.

[1975] Isaacs R. Infinite families of non-trivial trivalent graphs which are not Tait colorable. *Amer. Math. Monthly* **82**, 221–239, 1975.

[1976b] Jaeger F. Problem. In: C. St. J.A. Nash-Williams and J. Sheehan, editors, *Proc. Fifth British Combinatorial Conference in Aberdeen, 1975*, pages 682–683. Utilitas Mathematics Publications Inc., 1976.

[1988] Jaeger F. Nowhere-zero flow problems. In: L.W. Beineke and R.J. Wilson, editors, *Selected Topics in Graph Theory*, volume 3, pages 71–95. Academic Press, 1988.

[1990] Jaeger F. and G. Koester. Vertex signatures and edge-4-colorings of 4-regular plane graphs. *J. Graph Theory* **14**, 399–403, 1990.

[1981] Jaeger F. and H. Shank. On the edge-coloring problem for a class of 4-regular maps. *J. Graph Theory* **5**, 269–275, 1981.

[1980] Jaeger F. and T. Swart. Conjecture 1. In: M. Deza and I.G. Rosenberg, editors, *Combinatorics 79, part II*, volume 8–9 of *Annals of Discrete Mathematics*, page 305. North-Holland, 1980.

[1974] Jakobsen I.T. On critical graphs with chromatic index 4. *Discrete Math.* **9**, 265–276, 1974.

[1975] Jakobsen I.T. On graphs critical with respect to edge-colouring. In: A. Hajnal, R. Rado, and V.T. Sós, editors, *Infinite and Finite Sets*, volume 10 of *Colloquia Mathematica Societatis János Bolyai*, pages 927–934. North-Holland, 1975.

[1993] Janssen J.C.M. The Dinitz problem solved for rectangles. *Bull. Amer. Math. Soc.* **29**, 243–249, 1993.

[1993] Kilakos K. and F.B. Shepherd. Minors and the chromatic index of r-graphs. Manuscript, 1993.

[1992] Kostochka A.V. List edge chromatic number of graphs with large girth. *Discrete Math.* **101**, 189–201, 1992.

[1957] Kotzig A. From the theory of finite regular graphs of degree three and four. *Časopis Pěst. Mat.* **82**, 76–92, 1957.

[1990] Marcotte O. On the chromatic index of multigraphs and a conjecture of Seymour, II. In: W. Cook and P.D. Seymour, editors, *Polyhedral Combinatorics, Proc. DIMACS Workshop, Morristown, N.J., June 12–16, 1989*, pages 245–279. American Mathematical Society, 1990.

[1985] Nishizeki T. and K. Kashiwagi. An upper bound on the chromatic index of multigraphs. In: Y. Alavi, G. Chartrand, L. Lesniak, D.R. Lick, and C.E. Wall, editors, *Graph Theory with Applications to Algorithms and Computer Science*, pages 595–604. Wiley, 1985.

[1990] Nishizeki T. and K. Kashiwagi. On the 1.1 edge-coloring of multigraphs. *SIAM J. Discrete Math.* **3**, 391–410, 1990.

[1891] Petersen J. Die Theorie der regulären graphs. *Acta Math.* **15**, 193–220, 1891.

[1981] Plantholt M. The chromatic index of graphs with a spanning star. *J. Graph Theory* **5**, 5–13, 1981.

[1983] Plantholt M. On the chromatic index of graphs with large maximum degree. *Discrete Math.* **47**, 91–96, 1983.

[1991] Plantholt M. and S.K. Tipnis. Regular multigraphs of high degree are 1-factorizable. *J. London Math. Soc. (2)* **44**, 393–400, 1991.

[1990] Sevast'yanov S.V. Interval colorability of the edges of a bipartite graph (in Russian). *Metody Diskret. Analiz.* **50**, 61–72, 86, 1990.

[1979a] Seymour P.D. On multi-colourings of cubic graphs, and conjectures of Fulkerson and Tutte. *Proc. London Math. Soc. (3)* **38**, 423–460, 1979.

[1979b] Seymour P.D. Sums of circuits. In: J.A. Bondy and U.S.R. Murty, editors, *Graph Theory and Related Topics*, pages 341–355. Academic Press, 1979.

[1979c] Seymour P.D. Problem. In: J.A. Bondy and U.S.R. Murty, editors, *Graph Theory and Related Topics*, pages 367–368. Academic Press, 1979.

[1981b] Seymour P.D. On Tutte's extension of the four-colour problem. *J. Combin. Theory Ser. B* **31**, 82–94, 1981.

[1949] Shannon C.E. A theorem on coloring the lines of a network. *J. Math. Phys.* **28**, 148–151, 1949.

[1973] Szekeres G. Polyhedral decompositions of cubic graphs. *Bull. Austral. Math. Soc.* **8**, 367–387, 1973.

[1982] Thomason A.G. Cubic graphs with three Hamiltonian cycles are not always uniquely edge colorable. *J. Graph Theory* **6**, 219–221, 1982.

[1966] Tutte W.T. On the algebraic theory of graph colourings. *J. Combin. Theory* **1**, 15–50, 1966.

[1969] Tutte W.T. A geometrical version of the four color problem. In: R.C. Bose and T.A. Dowling, editors, *Combinatorial Mathematics and Its Applications*, pages 553–560. University of North Carolina Press, 1969.

[1976] Tutte W.T. Hamiltonian circuits. In: *Colloquio Internazionale sulle Teorie Combinatorie, Roma, Tomo I*, pages 193–199. Accademia Nazionale dei Lincei, 1976.

[1978] Tutte W.T. Colouring problems. *Math. Intelligencer* **1**, 72–75, 1978.

[1964] Vizing V.G. On an estimate of the chromatic class of a *p*-graph (in Russian). *Metody Diskret. Analiz.* **3**, 25–30, 1964.

[1965a] Vizing V.G. Critical graphs with given chromatic class (in Russian). *Metody Diskret. Analiz.* **5**, 9–17, 1965.

[1965b] Vizing V.G. The chromatic class of a multigraph (in Russian). *Kibernetika (Kiev)* no. 3, 29–39, 1965. English translation in *Cybernetics* **1**, 32–41.

[1968] Vizing V.G. Some unsolved problems in graph theory (in Russian). *Uspekhi Mat. Nauk* **23**, 117–134, 1968. English translation in *Russian Math. Surveys* **23**, 125–141.

[1989] Watkins J.J. Snarks. In: M.F. Capobianco, M. Guan, D.F. Hsu, and F. Tian, editors, *Graph Theory and Its Applications: East and West*, volume 576 of *Annals of the New York Academy of Sciences*, pages 606–622. New York Academy of Sciences, 1989.

[1980] Yap H.P. On the critical graph conjecture. *J. Graph Theory* **4**, 309–314, 1980.

[1993] Zhang C.-Q. Hamilton weights and unique 3-edge-colorings of cubic graphs. Manuscript, 1993.

13

Orientations and Flows

■ 13.1. TUTTE'S 5-FLOW CONJECTURE

A k-flow in an undirected graph G is an assignment to each edge of G of a direction and a value $1, 2, \ldots, k-1$ such that for each vertex x the sum of the values of edges directed into x is equal to the sum of the values of the edges directed out of x. Does every graph without bridges have a 5-flow? ■

This question is due to Tutte [1954]. Kilpatrick [1975], in a Ph.D. thesis, and Jaeger [1976a, 1979] proved independently that every graph without bridges has an 8-flow. Seymour [1981a] proved that it has a 6-flow (Theorem 33 in Chapter 1).

Steinberg [1984] proved Tutte's 5-flow conjecture for graphs embeddable on the projective plane. The similar result for orientable and nonorientable surfaces of Euler-characteristic at least -2 was obtained independently by M. Möller, Fouquet [1985], Jensen [1985], and Möller, Carstens, and Brinkmann [1988].

Not every graph without bridges has a 4-flow. The unique smallest bridge-less graph without a 4-flow is the Petersen graph. Tutte [1966] conjectured that if G is without a bridge and has no 4-flow, then G has a subgraph contractible to the Petersen graph (see Problem 13.2). The corresponding weaker statement with 4 replaced by 5 appears to be open.

An "apex graph" is a graph G with a vertex $v \in V(G)$ so that $G - v$ is planar. Tutte's 4-flow conjecture would imply that every bridge-less apex graph has a 4-flow. As pointed out by Seymour [1981b], this statement is equivalent to the following conjecture of H. Grötzsch: If G is a planar graph with $\Delta(G) \leq 3$, then G can be 3-edge-colored if and only if G has no subgraph all of whose vertices have degree 3 except precisely one vertex of degree 2 (see Problem 12.16). This conjecture implies the four-color theorem by the theorem of P.G. Tait (Theorem 26 in Chapter 1). The reduction methods used in proving the 5-flow conjecture for graphs embedded on low-genus surfaces can be applied to showing that an apex graph without bridges has a 5-flow.

k-flows and k-colorings are related by theorems of Tutte [1950]. Following Jensen [1985], we shall explain the close relation using results of Minty [1962, 1967] (see Theorem 31 in Chapter 1). In a directed graph $\vec{G}$ the "flow ratio" of a cycle (or an edge cut) is $\max\{m/n, n/m\}$, where n is the number of edges pointing in one direction and m the number of edges pointing in the opposite direction. If the edges of the cycle (or the edge cut) all point in the same direction, then the flow ratio is infinite. A graph G has a vertex k-coloring if and only if there exists an orientation $\vec{G}$ of G with flow ratio at most $k-1$ for every cycle (Theorem 31 in Chapter 1). Similarly, a graph has a k-flow if and only if there exists an orientation $\vec{G}$ of G with flow ratio at most $k-1$ for every edge cut (Minty [1967]). From the well-known duality between cycles and edge cuts in graphs (see, e.g., Chapter 2 of Seshu and Reed [1961], Chapter 12 of Bondy and Murty [1976], or Chapter 3 of Bryant and Perfect [1980]) the duality between k-colorings and k-flows follows. This relationship was implicit already in the work of Ghouila-Houri [1960].

In particular, if G is planar and G^* is its dual graph, then the following statements are equivalent:

(a) G has a vertex k-coloring,
(b) G^* has a k-flow,
(c) G^* has a face k-coloring.

Thus Tutte's 5-flow conjecture suggests a far-reaching generalization of the five-color theorem of Heawood [1890].

Tutte [1978] gave a very interesting introduction to difficult graph coloring problems and k-flow problems. Younger [1983] and Jaeger [1988] each gave an original and well-written survey of k-flows.

■ 13.2. TUTTE'S 4-FLOW CONJECTURE

A 4-flow in an undirected graph G is an assignment to each edge of G of a direction and a value 1, 2, or 3 such that for each vertex x the sum of the values of edges directed into x is equal to the sum of the values of the edges directed out of x. Is it true that if G has no bridge and no subgraph contractible to the Petersen graph, then G has a 4-flow? ■

The conjecture that the answer is affirmative is due to Tutte [1966]. The conjecture implies that every planar graph without bridges has a 4-flow, and this is equivalent to the four-color theorem.

Jaeger [1979] proved that every 4-edge-connected graph has a 4-flow. The proof relies on the interesting and useful fact that a graph has a k-flow if and only if it has a (nowhere zero) **A**-flow for any Abelian group **A** with k elements; that is, the value assigned to each edge is an element of $\mathbf{A}\setminus\{0\}$ (assuming **A** is additive). This can be proved by nonconstructive counting arguments using the principle of inclusion and exclusion, and it follows from the work of Tutte [1954]. (No constructive proof has been published so far as we know. However, the arguments given by Minty [1967] for the case $k = 4$ seem to provide a key to a constructive proof also in general.)

In particular, G has a 4-flow if and only if G has a $(\mathbf{Z}_2 \oplus \mathbf{Z}_2)$-flow. This is, in fact, the same as an edge cover $E(G) = E_1 \cup E_2$ so that E_1 and E_2 are disjoint unions of edge sets of cycles in G. Jaeger proved that every 4-edge-connected graph has such an edge cover, and he also proved that every 2-edge-connected graph G has a similar edge cover $E(G) = E_1 \cup E_2 \cup E_3$, hence a $(\mathbf{Z}_2 \oplus \mathbf{Z}_2 \oplus \mathbf{Z}_2)$-flow, hence an 8-flow. The 6-flow theorem of Seymour [1981a] (Theorem 33 in Chapter 1) was proved in the equivalent formulation that every 2-edge-connected graph has a $(\mathbf{Z}_2 \oplus \mathbf{Z}_3)$-flow.

It seems unknown whether the 4-flow conjecture is equivalent to its restriction to 3-regular graphs, that is, the statement that every bridge-less 3-regular graph without a subgraph contractible to the Petersen graph has a 4-flow. Note that a 3-regular graph has a 4-flow if and only if it has a $(\mathbf{Z}_2 \oplus \mathbf{Z}_2)$-flow, which again corresponds to a 3-edge coloring (see also Problem 12.15). A conjecture of Jaeger and Swart [1980], that every 4-edge-chromatic 3-regular graph contains a cycle of length at most 6, was recently disproved by M. Kochol, who applied to the Petersen graph a general construction, due to M. Kochol and M. Škoviera, of new graphs without a k-flow (where $k \leq 4$) from existing ones [personal communication from M. Škoviera in 1994]. The smallest example by Kochol of such a graph without cycles of length 6 or less has somewhere between 200 and 300 vertices.

Weaker forms of Tutte's 4-flow conjecture still imply the four-color theorem:

Conjecture 1. *If G has no bridge and no subgraph contractible to K_5, then G has a 4-flow.*

Conjecture 2. *If G has no bridge and $G - x$ is planar for some vertex x, then G has a 4-flow.*

Seymour [1981b] proved that Conjecture 2 is equivalent to a conjecture of H. Grötzsch (see Problem 12.16):

Conjecture 3. *If G is planar with maximum degree at most three, then G is 3-edge-colorable if and only if G has no subgraph in which one vertex has degree two and all others have degree three.*

At the Graph Minors Workshop held in Seattle in 1991, C.-Q. Zhang suggested another weaker form of Tutte's 4-flow conjecture as follows. Assume that G has no bridge and no subgraph contractible to the Petersen graph. Alspach and Zhang [1993] proved for G 3-regular, and Alspach, Goddyn, and Zhang [1993] proved for G in general that G has a cycle double cover (i.e., a collection of cycles $C_1, C_2, \ldots, C_m$ in G, possibly with repetitions, so that every edge of G is contained in exactly two cycles C_i and C_j where $i \neq j$). Denote by $\xi(G)$ the smallest possible number of colors in a coloring of the cycles of a cycle double cover of G so that no two cycles sharing an edge are colored the same. Equivalently, the cycles must be colored so that each color class corresponds to a subgraph of G with all degrees even. Then Tutte's 4-flow conjecture is equivalent to $\xi(G) \leq 3$. Zhang asked if there is a constant number k so that $\xi(G) \leq k$? Perhaps this can be proved at least for $k = 8$ or $k = 6$?

A conjecture was made by Szekeres [1973] and independently by Seymour [1979b] that every graph G without a bridge has a cycle double cover. A stronger version was formulated by Celmins [1984]: Let $\xi(G)$ be defined as above, defining $\xi(G) = \infty$ if G has no cycle double cover. Is it true that $\xi(G) \leq 5$ if G has no bridge? Jaeger [1985] wrote an in-depth survey of the cycle double cover conjecture, with emphasis in particular on the relation to integer flows. Also, a more recent paper by Fan [1992] explored this relationship.

■ 13.3. TUTTE'S 3-FLOW CONJECTURE

A 3-flow in an undirected graph G is an assignment to each edge of G of a direction and a value 1 or 2 such that for each vertex x the sum of the values of edges directed into x is equal to the sum of the values of the edges directed out of x. Is it true that if G has no bridge and no edge cut consisting of three edges, then G has a 3-flow? ■

The conjecture that the answer is affirmative was formulated by W.T. Tutte in 1972 (see the interesting survey by Steinberg [1993a]).

The existence of a k-flow is equivalent to the existence of a modulo k-flow (i.e., a k-flow in which the values and sums of values are calculated modulo k) as proved by Tutte [1950] (see also Younger [1983]). By reversing the direction of the edges of value 2 in a 3-flow, it follows that G has a 3-flow if and only if G has an orientation such that the in-degree minus the out-degree at each vertex is 0 modulo 3. For example, a 3-regular graph has a 3-flow if and only if it is bipartite.

An equivalent conjecture was presented in the section on unsolved problems in Bondy and Murty [1976]: Every 5-regular simple graph without bridges and edge cuts of size three allows an orientation so that each vertex has in-degree 1 or 4. That the 3-flow conjecture is implied by its restriction to 5-regular graphs seems first to have been observed by Tutte.

A planar graph G has a 3-flow if and only if its dual graph G^* has a vertex 3-coloring (see Problem 13.1). Since the 3-color problem for planar graphs is ***NP***-complete, as proved by M.R. Garey, D.S. Johnson and L.J. Stockmeyer (see Theorem 19 in Chapter 1), no good necessary and sufficient condition for the 3-color problem or for the 3-flow problem can be expected. Tutte's 3-flow conjecture suggests a possible sufficient condition for 3-flows.

For 2-flows there is, of course, a simple necessary and sufficient condition: G has a 2-flow if and only if all degrees in G are even.

The relation described above between 3-flows and 3-colorings shows that Tutte's 3-flow conjecture implies the theorem of H. Grötzsch (Theorem 10 in Chapter 1) that a planar graph G without triangles has a vertex 3-coloring.

A partial result was obtained by Steinberg and Younger [1989], who proved that the 3-flow conjecture is true for graphs embeddable on the projective plane. C. Thomassen [personal communication in 1993] remarked that the 3-flow conjecture for graphs embedded on the torus would follow from results of Thomassen [1993c], if the conjecture can be proved for every graph G for which $G - e$ is planar for some edge $e \in E(G)$.

A weaker conjecture by Jaeger [1988], that there exists an integer k such that every graph without an edge cut consisting of fewer than k edges has a 3-flow, is also open. Tutte's conjecture suggests this for $k = 4$. Lai and Zhang [1992] proved that if k is an integer such that G has no edge cut of fewer than k edges, and if $k \geq 4 \log_2 |V(G)|$, then G has a 3-flow.

■ 13.4. BOUCHET'S 6-FLOW CONJECTURE

A bidirection of a graph G is an orientation of each end of every edge of G (giving four possibilities for every edge: o——→——→o, o←——←——o, o——→←——o and o←——————→o). A bidirected k-flow on G is a bidirection of G and an assignment of a value $1, 2, \ldots, k - 1$ to each edge such that for each vertex x of G, the sum of the values of the edges whose ends at x are directed into x equals the sum of the values of the edges whose ends at x are directed out of x. Is it true that if G has a bidirected k-flow for some value of k and some bidirection of G, then G has a bidirected 6-flow with respect to the same bidirection? ■

The conjecture that the answer is affirmative is due to Bouchet [1983], who proved that this is so with 6 replaced by $6^3 = 216$. Moreover, Bouchet [1983] observed that the Petersen graph shows that the value 6 would be best possible.

A. Zyka from Czechoslovakia proved in his Ph.D. thesis that the conjecture is true with 6 replaced by 30. This was also done by J.-L. Fouquet. Khelladi [1987] proved the conjecture with 6 replaced by 18 for 4-connected graphs.

■ 13.5. JAEGER'S CIRCULAR FLOW CONJECTURE

An orientation of a graph is called a mod $(2p + 1)$-orientation ($p \geq 1$), if the out-degree of every vertex is congruent, modulo $2p + 1$, to its in-degree. Is it true for all $p \geq 1$, that every $4p$-edge-connected graph has a mod $(2p + 1)$-orientation? ■

Jaeger [1981] conjectured that the answer is affirmative and proved that G has a mod $(2p + 1)$-orientation if and only if G has a circular $(2p + 1)$-flow, defined as follows. For $k \geq 2$ a circular k-flow in G consists of an orientation of G together with an assignment to each edge of G of a k-vector c_{ij} for some i and j satisfying $1 \leq i, j \leq k$ and $|i - j| \in \{1, k - 1\}$, where the ith coordinate of c_{ij} is 1, the jth coordinate is -1, and all other coordinates are 0. Furthermore, for each vertex x of G, the (vector) sum of the values on the edges directed into x is equal to the sum of the values on the edges directed away from x.

Jaeger [1981] pointed out that the circular flow conjecture for $p = 1$ is equivalent to the 3-flow conjecture of W.T. Tutte (Problem 13.3), and that the conjecture for $p = 2$ implies the 5-flow conjecture of Tutte [1954] (Problem 13.1).

13.6. BERGE'S STRONG PATH PARTITION CONJECTURE

Let $\vec{G}$ be a directed graph and k an integer, where $1 \leq k \leq \max |V(\vec{P})|$ (maximum taken over all directed paths $\vec{P}$ in $\vec{G}$). A set $\mathcal{P}$ of disjoint directed paths $\vec{P}_1, \vec{P}_2, \ldots, \vec{P}_t$ in $\vec{G}$ that covers all vertices of $\vec{G}$ is called a path partition of $\vec{G}$. Such a $\mathcal{P}$ with a minimum value of $B_k(\mathcal{P}) = \sum_{i=1}^{t} \min\{k, |V(\vec{P})|\}$ is called k-optimal.

Is it true that for every k-optimal path partition $\mathcal{P}$ of $\vec{G}$ there exists a k-coloring of an induced subgraph of $\vec{G}$ such that the number of different colors on $\vec{P}_i$ is $\min\{k, |V(\vec{P}_i)|\}$ for every $\vec{P}_i$ in $\mathcal{P}$? ■

The conjecture that the answer is affirmative is due to C. Berge and is called the "strong path partition conjecture" (see Sridharan [1993a, 1993b]). Berge posed the conjecture in an attempt to unify the theorem of Gallai and Milgram [1960] and the theorem of B. Roy and T. Gallai (Theorem 32 in Chapter 1).

For $k = 1$ the conjecture has been proved by Linial [1978], generalizing the theorem of Gallai and Milgram [1960] that every directed graph $\vec{G}$ has a path partition with at most α paths, where α is the maximum number of independent vertices in $\vec{G}$.

For $k = \max |V(\vec{P})|$ the conjecture has been proved by Berge [1983], generalizing the theorem of Roy and Gallai.

The conjecture implies the weaker conjecture that every directed graph $\vec{G}$ has a path partition $\mathcal{P}$ with $B_k(\mathcal{P})$ of value at most the maximum number of vertices in an induced k-colorable subgraph of $\vec{G}$. This weaker statement, due to Linial [1981], has been named the "weak path partition conjecture" by C. Berge (see Sridharan [1993a, 1993b]). Sridharan [1993a, 1993b] proved the strong path partition conjecture for some special classes of graphs.

13.7. BERGE'S DIRECTED PATH-CONJECTURE

If G is a 4-chromatic graph and $\vec{G}$ is any orientation of G, does there exist a directed path $\vec{P}$ in $\vec{G}$ and a 4-coloring of G with each color class containing exactly one vertex from $\vec{P}$? ■

Berge [1980] conjectured that if $\vec{G}$ is any orientation of a graph G of chromatic number $k \geq 3$, then there exists a directed path $\vec{P}$ in $\vec{G}$ and a k-coloring of G such that each color class intersects $\vec{P}$ in exactly one vertex. The conjecture would strengthen the theorem of B. Roy and T. Gallai (Theorem 32 in Chapter 1), which states that every orientation of G contains a directed path with k vertices.

For $k = 3$ the conjecture is true, but for $k \geq 5$ it was disproved by Meyniel [1989]. Thus only the case $k = 4$ remains unsolved. The conjecture of Berge [1980] might still be true for all k-critical graphs with $k \geq 4$.

Gallai [1968] formulated a conjecture relating in a different way to the theorem of B. Roy and T. Gallai. He conjectured that given any k-critical graph G there exists an orientation of G with a unique directed path on k vertices. This conjecture was disproved by Youngs [1993] (see Problem 13.8).

■ 13.8. MINIMAL ORIENTATIONS OF CRITICAL GRAPHS

Let G be a k-critical graph. When does there exist a minimal orientation $\vec{G}$ of G, defined as an orientation $\vec{G}$ having exactly one directed path of length $k - 1$? What is the largest possible value $f(k)$ such that every k-critical graph of maximum degree at most $f(k)$ has a minimal orientation? It is known that $f(4) \leq 4$ and $f(5) \geq 5$. Does equality hold in either case? Does a minimal orientation always exist when G is without a triangle?

■

The first question was asked by Gallai [1968], who proved that every orientation of a k-chromatic graph has at least one directed path of length $k - 1$ (Theorem 32 in Chapter 1). This result was obtained independently by B. Roy.

The proof given by Gallai [1968] shows that if there exists an orientation with exactly one directed path of length $k - 1$, then there exists an acyclic such orientation $\vec{G}$. Assigning to each vertex x of $\vec{G}$ the length, $\varphi(x)$, of a longest directed path in $\vec{G}$ ending in x, one obtains a k-coloring φ of G with exactly one path in G of length $k - 1$ having its vertices colored $1, 2, \ldots, k$ in this order. Conversely, every such coloring φ gives an acyclic orientation of G with exactly one directed path of length $k - 1$ by directing the edge (i, j) from i to j when $\varphi(i) < \varphi(j)$. Thus Gallai's question can be rephrased in terms of colorings instead of orientations.

A breakthrough for Gallai's question was achieved by Youngs [1993], who gave the first examples of critical graphs without minimal orientations. The basic example of Youngs is 4-critical with 14 vertices and of maximum degree 5. Thus $f(4) \leq 4$. By joining G completely to a complete graph K_{k-4}, where $k \geq 5$, a k-critical graph without a minimal orientation and of maximum degree $k + 9$ is obtained. Thus $f(k) \leq k + 8$ for all $k \geq 5$. By joining G completely to a triangle-free $(k - 4)$-critical graph, where $k \geq 6$, one obtains a k-critical graph without minimal orientations (as will be explained below) and without complete subgraphs of size 6.

D. Hanson and B. Toft have noted that Youngs' 4-critical graph without a minimal orientation is the first member of a sequence consisting of k-critical graphs for all $k \geq 4$. Each graph has a 7-cycle C with vertices $x_1, x_2, \ldots, x_7$ in cyclic order, and for $k \geq 4$ it consists in addition of seven mutually disjoint copies $G_1, G_2, \ldots, G_7$ of the complete $(k - 3)$-graph, each of which is disjoint from C, with G_i completely joined to the vertices x_i, x_{i+1}, and x_{i+3} (indices modulo 7) for $i = 1, 2, \ldots, 7$. Hanson and Toft could show that also the second member of this sequence, with $k = 5$, on 21 vertices and of maximum degree 8, has no minimal orientation. This implies a bound of $f(5) \leq 7$. It is not clear if every graph in the sequence is without a minimal orientation, which would imply improved upper bounds on $f(k)$ for some small values of k.

Youngs [1993] proved for $k \geq 5$ that every k-critical graph of maximum degree k has a minimal orientation. Hence $f(k) \geq k$ for $k \geq 5$. It is not known if $f(5) = 5$. Only few k-critical graphs of maximum degree k are known for $k \geq 5$, and it has been conjectured that they do not exist at all for $k \geq 9$ (see Problem 4.8). It is open if every 4-critical graph of maximum degree 4 has a minimal orientation.

If disjoint graphs G_1 and G_2 are k_1-critical and k_2-critical, respectively, and if G is the complete join of G_1 and G_2, obtained by joining all vertices of G_1 to all vertices of G_2 by edges, then G is $(k_1 + k_2)$-critical. Gallai [1968] noted that if both G_1 and G_2 have minimal orientations, then G has a minimal orientation. D. Hanson and B. Toft noted that the converse is also true. This follows from the observation, due to Las Vergnas [1973] and easily proved by induction, that the complete join of two directed paths that start from vertices x and y, with an arbitrary direction assigned to each edge joining the paths, always contains a directed Hamilton path (a path containing all vertices of the two paths) starting in x or in y. Taking the complete join of p copies of Youngs' basic 4-critical example without a minimal orientation, one thus obtains a $4p$-critical graph on $14p$ vertices in which every orientation has at least 2^p directed paths of length $4p - 1$. Thus the minimum number of directed paths of length $k - 1$ in a k-critical graph can grow exponentially with k and with the number of vertices in the graph.

Bondy [1976] proved the following stronger version of the theorem of Roy and Gallai: If $\vec{H}$ is any strongly connected orientation of a k-chromatic multigraph H (i.e., for every pair of vertices x and y there is a directed path from x to y in $\vec{H}$), where $k \geq 2$, then $\vec{H}$ contains a directed cycle of length at least k. This was conjectured by M. Las Vergnas (see the collection of open problems in Bondy and Murty [1976]). The theorem of Roy and Gallai can be proved from this statement by joining a new vertex to each vertex of the given graph by two edges, one in each direction. For the theorem of Bondy, similar questions to those given above may be raised, replacing "orientation" by "strongly connected orientation" and "directed path of length $k - 1$" by "directed cycle of length at least k." But we are not aware of any work done in this direction.

13.9. ALON–TARSI ORIENTATIONS AND CHROMATIC NUMBER

Let $\vec{G}$ be an orientation of a graph G. An Eulerian subgraph of $\vec{G}$ is a directed subgraph $\vec{H}$ such that the in-degree of every vertex in $\vec{H}$ is equal to its out-degree in $\vec{H}$. A subgraph $\vec{H}$ is called "even" or "odd" according to the parity of $|E(\vec{H})|$. Define AT(G) to be the smallest number k so that there exists an orientation $\vec{G}$ with the number of even spanning Eulerian subgraphs different from the number of odd spanning Eulerian subgraphs, and so that every vertex of G has out-degree at most $k - 1$ in $\vec{G}$.

Does there exist a polynomial algorithm to evaluate AT(G) for any given graph G? Does there exist a polynomial algorithm to color a graph G with at most AT(G) colors (or with lists of colors of size at most AT(G) at the vertices)? ■

Alon and Tarsi [1992] used algebraic methods to prove that

$$\chi(G) \leq \chi_\ell(G) \leq \mathrm{AT}(G)$$

(Theorem 34 in Chapter 1) and remarked that it would be interesting to find a nonalgebraic proof of the inequality $\chi(G) \leq \mathrm{AT}(G)$. They noted that a polynomial algorithm exists when restricting to acyclic orientations $\vec{G}$—in such an orientation there is exactly one even spanning Eulerian subgraph, namely the edge-less graph, and no odd spanning Eulerian subgraph. The acyclic orientations of G provide the upper bound

$$\mathrm{AT}(G) \leq \mathrm{col}(G),$$

where col(G) denotes the coloring number of G (i.e., the largest number k such that there exists a subgraph of G with minimum degree $k - 1$). It is easy to determine the number col(G) and to find a coloring of the vertices of G with col(G) colors (see Theorem 12 and the discussion of it in Chapter 1).

Another upper bound for $\chi_\ell(G)$, call it R(G), is obtained by considering only orientations $\vec{G}$ without odd directed cycles. Again, such a $\vec{G}$ has no odd spanning Eulerian subgraph. Hence $\chi_\ell(G) \leq \mathrm{AT}(G) \leq R(G) \leq \mathrm{col}(G)$. The inequality $\chi_\ell(G) \leq R(G)$ can also be proved by a direct argument. As noted by Alon and Tarsi, it is not difficult to decide if $\vec{G}$ has an odd directed cycle, by looking at odd powers of the adjacency matrix of $\vec{G}$, or by checking if G has a subgraph corresponding to a strongly connected component of $\vec{G}$ and containing an odd cycle (see Theorem 6.14 of Harary, Norman, and Cartwright [1965]). A theorem of Richardson [1953] implies that if $\vec{G}$ has no odd directed cycle, then $\vec{G}$ has a "kernel," that is, an independent set S of vertices so that every vertex in $\vec{G}$ not belonging to S has at least one out-neighbor in S. J.A. Bondy, R. Boppana, and A. Siegel noted (see Alon and Tarsi [1992]) that Richardson's theorem can be used directly to color G from lists of size $\Delta^+(\vec{G}) + 1$ in polynomial time, where $\Delta^+(\vec{G})$ denotes the maximum out-degree of the vertices in $\vec{G}$ (apply Richardson's theorem to the induced subgraph on the set of vertices containing color 1 in their lists, and color the vertices belonging to its kernel with color 1; continue with the next color, etc.). However, it does not seem clear how to determine the number R(G) polynomially.

Two important theorems on list coloring were obtained by determining the value of R(G) for G belonging to a given class of graphs. Alon and Tarsi [1992] proved that every bipartite planar graph is 3-choosable (see also Problem 2.13), and F. Galvin proved the list-coloring conjecture (see Section 1.9) for all bipartite graphs, in particular settling a famous conjecture of J. Dinitz (see also Problem 12.20).

BIBLIOGRAPHY

[1992] Alon N. and M. Tarsi. Colorings and orientations of graphs. *Combinatorica* **12**, 125–134, 1992.

[1993] Alspach B., L. Goddyn, and C.-Q. Zhang. Graphs with the circuit cover property. Manuscript, 1993. To appear in *Trans. Amer. Math. Soc.*

[1993] Alspach B. and C.-Q. Zhang. Cycle covers of cubic multigraphs. *Discrete Math.* **111**, 11–17, 1993.

[1980] Berge C. Problem. In: M. Deza and I.G. Rosenberg, editors, *Combinatorics 79, part II*, volume 8–9 of *Annals of Discrete Mathematics*, page 297. North-Holland, 1980.

[1983] Berge C. Path partitions in directed graphs. In: C. Berge, D. Bresson, P. Camion, J.F. Maurras, and F. Sterboul, editors, *Combinatorial Mathematics*, volume 17 of *Annals of Discrete Mathematics*, pages 59–63. North-Holland, 1983.

[1976] Bondy J.A. Diconnected orientations and a conjecture of Las Vergnas. *J. London Math. Soc. (2)* **14**, 277–282, 1976.

[1976] Bondy J.A. and U.S.R. Murty. *Graph Theory with Applications*. Macmillan, 1976.

[1983] Bouchet A. Nowhere-zero integral flows on bidirected graphs. *J. Combin. Theory Ser. B* **34**, 279–292, 1983.

[1980] Bryant V. and H. Perfect. *Independence Theory in Combinatorics*. Chapman and Hall, 1980.

[1984] Celmins U.A. *On cubic graphs that do not have an edge-3-colouring*. Ph.D. thesis, University of Waterloo, 1984.

[1992] Fan G. Integer flows and cycle covers. *J. Combin. Theory Ser. B* **54**, 113–122, 1992.

[1985] Fouquet J.-L. Conjecture du 5-flot pour les graphes presque planaires. In: *Séminaire de Mathématiques Discrètes et Applications, Grenoble*, 1985.

[1968] Gallai T. On directed paths and circuits. In: P. Erdős and G. Katona, editors, *Theory of Graphs*, pages 115–118. Academic Press, 1968.

[1960] Gallai T. and A.N. Milgram. Verallgemeinerung eines graphentheoretischen Satzes von Rédei. *Acta Sci. Math. (Szeged)* **21**, 181–186, 1960.

[1960] Ghouila-Houri A. Sur l'existence d'un flot ou d'une tension prenant ses valeurs dans un groupe abélien. *C.R. Acad. Sci. Paris* **250**, 3931–3933, 1960.

[1965] Harary F., R.Z. Norman, and D. Cartwright. *Structural Models: An Introduction to the Theory of Directed Graphs*. Wiley, 1965.

[1890] Heawood P.J. Map colour theorem. *Quart. J. Pure Appl. Math.* **24**, 332–338, 1890.

[1976a] Jaeger F. On nowhere-zero flows in multigraphs. In: C. St. J.A. Nash-Williams and J. Sheehan, editors, *Proc. Fifth British Combinatorial Conference 1975*, pages 373–378. Utilitas Mathematics Publications Inc., 1976.

[1979] Jaeger F. Flows and generalized coloring theorems in graphs. *J. Combin. Theory Ser. B* **26**, 205–216, 1979.

[1981] Jaeger F. On circular flows in graphs. In: A. Hajnal, L. Lovász, and V.T. Sós, editors, *Finite and Infinite Sets*, volume 37 of *Colloquia Mathematica Societatis János Bolyai*, pages 391–402. North-Holland, 1981.

[1985] Jaeger F. A survey of the cycle double cover conjecture. In: B. Alspach and C.D. Godsil, editors, *Cycles in Graphs*, volume 27 of *Annals of Discrete Mathematics*, pages 1–12. North-Holland, 1985.

[1988] Jaeger F. Nowhere-zero flow problems. In: L.W. Beineke and R.J. Wilson, editors, *Selected Topics in Graph Theory*, volume 3, pages 71–95. Academic Press, 1988.

[1980] Jaeger F. and T. Swart. Conjecture 1. In: M. Deza and I.G. Rosenberg, editors, *Combinatorics 79, part II*, volume 8–9 of *Annals of Discrete Mathematics*, page 305. North-Holland, 1980.

[1985] Jensen T.R. *Tutte's k-flow problems* (in Danish). Master's thesis, Odense University, 1985.

[1987] Khelladi A. Nowhere-zero integral chains and flows in bidirected graphs. *J. Combin. Theory Ser. B* **43**, 95–115, 1987.

[1975] Kilpatrick P.A. *Tutte's first colour-cycle conjecture*. Ph.D. thesis, Cape Town, 1975.

[1992] Lai H.-J. and C.-Q. Zhang. Nowhere-zero 3-flows of highly connected graphs. *Discrete Math.* **110**, 179–183, 1992.

[1973] Las Vergnas M. Sur les circuits dans les sommes complétées de graphes orientés. *Cahiers Centre Études Rech. Opér.* **15**, 231–244, 1973.

[1978] Linial N. Covering digraphs by paths. *Discrete Math.* **23**, 257–272, 1978.

[1981] Linial N. Extending the Greene–Kleitman theorem to directed graphs. *J. Combin. Theory Ser. A* **30**, 331–334, 1981.

[1989] Meyniel H. About colorings, stability and paths in directed graphs. *Discrete Math.* **74**, 149–150, 1989.

[1962] Minty G.J. A theorem on n-coloring the points of a linear graph. *Amer. Math. Monthly* **63**, 623–624, 1962.

[1967] Minty G.J. A theorem on three-coloring the edges of a trivalent graph. *J. Combin. Theory* **2**, 164–167, 1967.

[1988] Möller M., H.G. Carstens, and G. Brinkmann. Nowhere-zero flows in low genus graphs. *J. Graph Theory* **12**, 183–190, 1988.

[1953] Richardson M. Solutions of irreflexive relations. *Ann. of Math.* **58**, 573–580, 1953.

[1961] Seshu S. and M.B. Reed. *Linear Graphs and Electrical Networks*. Addison–Wesley, 1961.

[1979b] Seymour P.D. Sums of circuits. In: J.A. Bondy and U.S.R. Murty, editors, *Graph Theory and Related Topics*, pages 341–355. Academic Press, 1979.

[1981a] Seymour P.D. Nowhere-zero 6-flows. *J. Combin. Theory Ser. B* **30**, 130–135, 1981.

[1981b] Seymour P.D. On Tutte's extension of the four-colour problem. *J. Combin. Theory Ser. B* **31**, 82–94, 1981.

[1993a] Sridharan S. On the Berge's strong path partition conjecture. *Discrete Math.* **112**, 289–293, 1993.

[1993b] Sridharan S. On the strong path partition conjecture of Berge. *Discrete Math.* **117**, 265–270, 1993.

[1984] Steinberg R. Tutte's 5-flow conjecture for the projective plane. *J. Graph Theory* **8**, 277–289, 1984.

[1993a] Steinberg R. The state of the three color problem. In: J. Gimbel, J.W. Kennedy, and L.V. Quintas, editors, *Quo Vadis, Graph Theory?* volume 55 of *Annals of Discrete Mathematics*, pages 211–248. North-Holland, 1993.

[1989] Steinberg R. and D.H. Younger. Grötzsch's theorem for the projective plane. *Ars Combin.* **28**, 15–31, 1989.

[1973] Szekeres G. Polyhedral decompositions of cubic graphs. *Bull. Austral. Math. Soc.* **8**, 367–387, 1973.

[1993c] Thomassen C. Grötzsch's 3-color theorem and its counterparts for the torus and the projective plane. Manuscript, 1993. To appear in *J. Combin. Theory Ser. B*.

[1950] Tutte W.T. On the imbedding of linear graphs in surfaces. *Proc. London Math. Soc. (2)* **51**, 474–483, 1950.

[1954] Tutte W.T. A contribution to the theory of chromatic polynomials. *Canad. J. Math.* **6**, 80–91, 1954.

[1966] Tutte W.T. On the algebraic theory of graph colourings. *J. Combin. Theory* **1**, 15–50, 1966.

[1978] Tutte W.T. Colouring problems. *Math. Intelligencer* **1**, 72–75, 1978.

[1983] Younger D.H. Integer flows. *J. Graph Theory* **7**, 349–357, 1983.

[1993] Youngs D.A. Minimal orientations of colour critical graphs. Manuscript, 1993. To appear in *Combinatorica*.

14

Chromatic Polynomials

■ 14.1. COEFFICIENTS OF CHROMATIC POLYNOMIALS

Let $P(G,k) = k^n + a_1k^{n-1} + a_2k^{n-2} + \cdots + a_{n-1}k + a_n$ be the chromatic polynomial of G. That is, $P(G,k)$ is the number of different k-colorings of G with colors $1, 2, \ldots, k$. Is it true that the absolute values of the coefficients form a unimodal sequence, that is,

$$1 \leq |a_1| \leq |a_2| \leq \cdots \leq |a_{i-1}| \leq |a_i| \geq |a_{i+1}| \geq \cdots \geq |a_{n-1}|$$
$$\geq |a_n| = 0$$

for some i with $1 \leq i \leq n$? ■

Attention to this problem was drawn by Read [1968]. As an equivalent reformulation one may ask: Is the statement "$|a_i| > |a_{i+1}|$ and $|a_{i+1}| < |a_{i+2}|$" false for all i?

Hoggar [1974] conjectured more strongly that the inequality $a_ia_{i+2} < a_{i+1}^2$ holds for all i with $1 \leq i \leq n-2$. A sequence of numbers satisfying this inequality is said to have the strong logarithmic concavity property, implying that it is unimodal in absolute value. Hoggar observed that the class of polynomials with coefficients satisfying strong logarithmic concavity is closed under multiplication, and that this is not the case for the class of polynomials with coefficients that are unimodal in absolute value. From the definition of chromatic polynomials it is clear that $P(G,k) \cdot P(H,k)$ is the chromatic polynomial of the disjoint union of the graphs G and H; hence the class of chromatic polynomials is closed under multiplication. Read and Tutte [1988] remarked that this makes the question by Hoggar [1974] seem the more natural of the two.

The strong logarithmic concavity, and hence the unimodality property, has been verified for the chromatic polynomials of all graphs on less than 10 vertices, as reported by Read and Tutte [1988].

For accounts of related properties of the coefficients of chromatic polynomials see Meredith [1972], Wilf [1976], Farrell [1980], Brenti [1992], and the classical paper by Whitney [1932b].

■ 14.2. CHARACTERIZATION OF CHROMATIC POLYNOMIALS

Is it possible to find a set of necessary and sufficient algebraic conditions for a polynomial to be the chromatic polynomial of some graph? ■

The question was asked by Read [1968], who proved that the chromatic polynomial $P(G, k)$ of a graph G on n vertices satisfies the following necessary conditions:

(i) The degree of $P(G, k)$ is n,
(ii) the coefficient of k^n is 1,
(iii) the coefficient of k^{n-1} is $-|E(G)|$,
(iv) the constant term is 0,
(v) the terms alternate in sign, and
(vi) $P(G, \lambda) \leq \lambda(\lambda - 1)^{n-1}$ for any positive integer λ, if G is connected.

However, Read [1968] also noted that these conditions are not sufficient for characterizing chromatic polynomials among all polynomials.

Read and Tutte [1988] remarked that for any given polynomial, at least it is possible by (i) and (iii) above to determine the number of vertices and edges a graph must have, if it exists, in order to have the given chromatic polynomial. What remains would be to generate all such graphs and calculate their chromatic polynomials, which is finite process, but certainly very impractical.

Some additional necessary conditions are obtained by studying the zeros of $P(G, k)$. It follows from (ii), (iv), and (v) that $k = 0$ is a zero, that $P(G, \lambda)$ is nonzero for any real number $\lambda < 0$, and that all rational zeros must be integers. Since $P(G, k)$ is the number of k-colorings of G, it follows that $P(G, k)$ has no rational zeros other than $0, 1, 2, \ldots, \chi(G) - 2, \chi(G) - 1$. It can be proved (see Tutte [1974a]) that $P(G, \lambda)$ is nonzero for all real numbers $0 < \lambda < 1$ (for further results along these lines, see Problem 14.5).

Lehmer [1985] gave a complete list of all chromatic polynomials of graphs on at most 6 vertices, and a random sample of chromatic polynomials of graphs on 10 vertices.

■ 14.3. CHROMATIC UNIQUENESS

Let $P(G, k)$ denote the chromatic polynomial of G. G is called chromatically unique if $P(G, k) = P(H, k)$ implies that H is isomorphic to G. What is a necessary and sufficient condition for G to be chromatically unique? ■

This question was asked by Chao and Whitehead [1978]. It is a natural variation of the question about chromatic equivalence (Problem 14.4) posed by Read [1968]: What is a necessary and sufficient condition for two graphs to have the same chromatic polynomial?

Chao and Whitehead [1978] proved that all cycles with at least 3 vertices and all θ-graphs with at least 4 vertices are chromatically unique. A θ-graph is the union of two cycles with one edge in common. Loerinc [1978] proved that generalized θ-graphs are chromatically unique; these are formed by joining two vertices by three paths that are pairwise disjoint except for their endpoints. It is clear that complete graphs and edge-less graphs are chromatically unique.

Chao and Whitehead [1978, 1979] asked for the values of $n \geq 4$ for which the wheel W_n, consisting of one vertex joined completely to the vertices of a cycle of length $n - 1$, is chromatically unique, and they showed that W_n is chromatically unique for $n = 4, 5$, but not for $n = 6$. Xu and Li [1984] proved that every W_n with n odd is chromatically unique, and that W_8 is not. Li and Whitehead [1992] showed that W_{10} is chromatically unique. The results for W_8 and W_{10} were obtained independently by R.C. Read (see Read and Tutte [1988]). For W_n with n even and $n \geq 12$ the question of chromatic uniqueness is open. Tutte [1970a] gave a general formula for $P(W_n, k)$ for $n \geq 4$.

G is called an H-homeomorph, if G can be obtained from H by a series of subdivisions of edges. Whitehead and Zhao [1984a] gave explicit lists of infinite families of chromatically unique K_4-homeomorphs. They also described infinite families of pairs of chromatically equivalent K_4-homeomorphs; additional such families were recently found by Whitehead [1993]. Weiming [1987] proved that almost all K_4-homeomorphs are chromatically unique. The proof is based on a lemma by Chao and Zhao [1983], stating that if G is a K_4-homeomorph and G' is a graph such that $P(G', k) = P(G, k)$, then G' is also a K_4-homeomorph. There is no known characterization of the chromatically unique K_4-homeomorphs.

Chao and Novacky [1982] considered the graphs $T(n, k)$, for $n \geq k \geq 2$, introduced by Turán [1941, 1954], and proved that any such graph is chromatically unique. $T(n, k)$ has n vertices in k parts of sizes that differ by at most 1, each part consisting of an independent set, and all pairs of vertices from different parts being adjacent. Xu [1991] showed that every complete bipartite graph $K_{m,n}$ with $m \geq n \geq 2$ is chromatically unique, as conjectured by Salzberg, López, and Giudici [1986]. It seems open if complete k-partite graphs for $k \geq 3$ in general are chromatically unique.

Chia [1986] showed that every connected chromatically unique graph G has at most two blocks. If G has exactly two blocks, each block is vertex-transitive and chromatically unique. Moreover, if H is a vertex-transitive 2-connected chromatically unique graph, then every connected graph with blocks H and K_2 is chromatically unique. The latter result was conjectured by Whitehead and Zhao [1984b]. Read [1987] called a graph "weakly chromatically unique" if it is chromatically unique, but the graph obtained by adding a new isolated vertex is not. He then proved a stronger form of the conjecture of Whitehead and Zhao [1984b] by showing that G is weakly chromatically unique if and only if G consists of two blocks H and K_2, where H is a chromatically unique vertex-transitive graph. More results related to connectivity can be found in Read [1987]. Chia [1986] conjectured that all connected vertex-transitive graphs are chromatically unique. This conjecture was disproved by Liu and Li [1991].

Some additional results and open problems can be found in the survey by Bari and Kahn [1989] and in the excellent expository paper by Koh and Teo [1990].

The known results on chromatic uniqueness *"were obtained by painstaking detective work—by carefully wringing from a chromatic polynomial every drop of information about the graph, until it could be shown that there was only the one possibility."* (Read and Tutte [1988]). Results concerning relations between chromatic polynomial and graph structure can be found in Eisenberg [1970], Meredith [1972], Farrell [1980], Whitehead and Zhao [1984b], and Whitehead [1989]. From this point of view, the problem is closely related to the problem of characterizing chromatic polynomials (Problem 14.2): Is there a practical method to determine for a given polynomial if it is the chromatic polynomial of some graph?

■ 14.4. CHROMATIC EQUIVALENCE

What is a necessary and sufficient condition for two graphs to have the same chromatic polynomial? ■

This question was asked by Read [1968], who noted, for example, that all trees with the same number of vertices are chromatically equivalent, and observed that more nontrivial examples exist as well.

Some variations of the question were stated in Read and Tutte [1988]: Given two graphs, is there a way of recognizing that they are chromatically equivalent without actually working out the chromatic polynomial for either of them? And, if given one graph, is there a way of constructing different graphs that have the same chromatic polynomial?

If G is the union of two smaller graphs G_1 and G_2 that intersect in a complete graph $K_r = G_1 \bigcap G_2$, then (see, e.g., Read [1968])

$$P(G,k) = \frac{P(G_1,k) \cdot P(G_2,k)}{k(k-1)(k-2)\cdots(k-r+1)}.$$

In general, it is possible to form nonisomorphic graphs starting from the same G_1, G_2, and r, hence such graphs are chromatically equivalent. For example, all connected graphs with the same collection of blocks are chromatically equivalent, one particular case being all trees of the same size. More general examples arise from the class of triangulated graphs (also called chordal graphs or rigid circuit graphs) introduced by Hajnal and Surányi [1958] and characterized by Dirac [1961], defined as the class of graphs for which every cycle of length at least 4 has a diagonal (i.e., an edge joining two nonconsecutive vertices of the cycle). Dirac showed that triangulated graphs are precisely the graphs that can be built up, starting from a single vertex, by successively introducing one new vertex and joining it to the vertices of a complete subgraph in the existing graph. Knowing the sizes of the complete subgraphs used in the construction of a triangulated graph G is sufficient to determine $P(G,k)$ by the formula given above: Let m_r $(0 \leq r \leq \omega)$ be the number of complete r-graphs used to construct G. Then

$$P(G,k) = k^{m_0}(k-1)^{m_1}(k-2)^{m_2}\cdots(k-\omega)^{m_\omega}.$$

Many different triangulated graphs have the same sequence $m_0, m_1, m_2, \ldots, m_\omega$; thus they are all chromatically equivalent to G. Dmitriev [1980] gave infinitely many examples showing that a graph that is not triangulated can be chromatically equivalent to a triangulated graph. The smallest example found by Dmitriev is obtained from the complete graph K_6 by subdividing an edge by one new vertex (this particular example was also found earlier and independently by Read [1975]). A nontrivial result obtained by Dmitriev [1982] states that if H has chromatic polynomial

$$P(H,k) = k(k-1)(k-2)\cdots(k-\omega+1)(k-\omega)^{m_\omega},$$

then H is in fact a triangulated graph. Borodin and Dmitriev [1991] proved the similar result that if

$$P(H,k) = k(k-1)(k-2)\cdots(k-\omega+1)(k-\omega)^{m_\omega}(k-j),$$

where $0 \le j < \omega$, then H must also be triangulated. Since the triangulated graphs have chromatic polynomials of a particularly simple form, it seems an interesting problem to characterize graphs that are chromatically equivalent to triangulated graphs. A somewhat similar problem was solved by Wakelin and Woodall [1992].

For graphs on n vertices, $n \ge 3$, any G and H must be chromatically equivalent if they have identical collections of isomorphism types for their n vertex-deleted subgraphs. This was proved by Tutte [1967, 1979]. The reconstruction conjecture suggests that G and H are in fact isomorphic (see Bondy and Hemminger [1977]). Tutte's result can also be expressed by saying that the chromatic polynomial is "reconstructible". The chromatic number is thus reconstructible, but it seems unknown if the edge-chromatic number is.

Akiyama and Harary [1981] asked whether there exists a graph G that is not isomorphic to its own complement $\overline{G}$, but is chromatically equivalent to $\overline{G}$. This question was answered affirmatively by Koh and Teo [1991].

A certain operation of "rotor flipping" was introduced by Brooks, Smith, Stone, and Tutte [1940], and studied also by Tutte [1974b, 1980] and Lee [1975]. A rotor R of a graph G is a subgraph with a rotational symmetry θ such that the vertices of attachment for R (i.e., the vertices of G incident both to elements of $E(R)$ and to elements of $E(G) - E(R)$) form an orbit under the action of θ. The rotor R is "flipped" by replacing R in G by its reflection. Read and Tutte [1988] remarked that it is not known if rotor flipping can change the chromatic polynomial of a planar graph. Foldes [1978] proved that rotor flipping may change the chromatic polynomial of a nonplanar graph. It has been pointed out by J.H. Przytycki [personal communication in 1994] that results obtained by Jones [1992] on invariants in knot theory can be translated into partial results on the effects of rotor flipping on the chromatic polynomial of a planar graph, thus providing certain sufficient conditions that the chromatic polynomial will remain unchanged following the operation. The connections with knot theory were first described by Anstee, Przytycki, and Rolfsen [1989] and developed further by Traczyk [1989].

■ 14.5. ZEROS OF CHROMATIC POLYNOMIALS

What numbers can be zeros of the chromatic polynomial of a graph G? What is the answer if G is restricted to be planar? What if G is restricted to be a plane triangulation? ■

Read [1968] asked this question for graphs G in general. It follows from basic results (see Problem 14.2) that $P(G, \lambda) = 0$ for $\lambda = 0, 1, 2, \ldots, \chi(G) - 2, \chi(G) - 1$, and that these are the only rational solutions, but in general not all the solutions. Moreover, from the observation that the coefficients of $P(G, k)$ alternate in sign follows that $P(G, \lambda) \neq 0$ for all $\lambda < 0$. Tutte [1974a] showed that $P(G, \lambda) \neq 0$ for $0 < \lambda < 1$, which has been extended to a slightly more general result by Woodall [1992a].

Woodall [1977] proved that if $m, n \geq 2$ have different parities, then $P(K_{m,n}, \lambda) = 0$ has a solution in the interval $1 < \lambda < 2$, where $K_{m,n}$ denotes the complete bipartite graph with parts of size m and n. Woodall [1977] also proved that for fixed $m \geq 2$, and any integer i with $2 \leq i \leq m/2$, $P(K_{m,n}, \lambda) = 0$ has real solutions λ arbitrarily close to i, for n chosen large enough.

Jackson [1993] proved that $P(G, \lambda)$ never has a zero in the interval $1 < \lambda \leq 32/27$, and that the value $32/27$ is best possible. It seems unknown if there exists any zero-free interval of the form $\lambda_1 < \lambda < \lambda_2$, where $32/27 < \lambda_1 < \lambda_2$. Jackson [1993] conjectured that if G is 3-connected and not bipartite, then $P(G, \lambda) \neq 0$ for all λ in the interval $1 < \lambda < 2$. This is true if G is a plane triangulation (Birkhoff and Lewis [1946]), and it has been verified by computer search for general graphs G on at most nine vertices by G.F. Royle, as reported by Jackson [1993].

The statement "$P(G, 4) > 0$ for every planar graph G" is equivalent to the four-color theorem, and remains equivalent when G is restricted to the class of plane triangulations. In an attempt to settle the four-color problem, Birkhoff and Lewis [1946] studied the zeros of chromatic polynomials of plane triangulations. They proved that $P(G, \lambda) > 0$ holds for every plane triangulation G and all $\lambda \geq 5$, and formulated a conjecture that would imply the truth of the corresponding statement with 5 replaced by 4, and moreover with $P(G, \lambda)$ replaced by any derivative $P^{(i)}(G, \lambda)$ for $1 \leq i \leq |V(G)|$. The conjecture of Birkhoff and Lewis [1946] has remained unsolved. It is not even known if there exists a planar graph G with $P(G, \lambda) = 0$ for any λ in the interval $4 < \lambda < 5$. Possible extensions and generalizations of the Birkhoff–Lewis conjecture have been considered by S. Beraha, Queens College, City University of New York [personal communication in 1994].

Woodall [1992b] proved that the unique real zero ($\simeq 2.547\ldots$) of the polynomial $t^3 - 9t^2 + 29t - 32$ is the smallest noninteger real zero of the chromatic polynomial of any plane triangulation. This polynomial divides the chromatic polynomial of the octahedron graph. Woodall [1977] had conjectured this result, and he also conjectured that if G is a plane triangulation, then $P(G, \lambda) \neq 0$ in the interval $2.678\ldots < \lambda < 3$. The latter conjecture has been disproved by Woodall, who instead conjectured that there are no zero-free intervals between $2.547\ldots$ and 3 [personal communication from B. Jackson in 1994].

14.6. BERAHA CONJECTURE

Let n be a positive integer and define $B_n = 2 + 2\cos(2\pi/n)$. Is it true for every $\varepsilon > 0$ that there exists a plane triangulation G such that the chromatic polynomial $P(G, k)$ of G has a zero λ in the interval $B_n - \varepsilon < \lambda < B_n + \varepsilon$? ■

The question was asked by Beraha [1975] in his Ph.D. thesis, and it was mentioned by Tutte [1974a].

The number $B_n = 2 + 2\cos(2\pi/n)$ is occasionally referred to as the nth Beraha number. The first few such numbers are $4, 0, 1, 2, \tau^2, 3, \ldots$, where $\tau = (1 + \sqrt{5})/2$ (the "golden ratio"). This sequence satisfies $\lim_{n\to\infty} B_n = 4$; hence the truth of the conjecture implies the existence of plane triangulations having chromatic polynomials with zeros arbitrarily close to 4. Beraha and Kahane [1979] showed that such triangulations indeed exist, and they pointed out the connection with the four-color problem. The four-color theorem is equivalent to the statement that the chromatic polynomial of any planar graph has no zero equal to 4.

It is clear that the conjecture is true for $B_2 = 0, B_3 = 1, B_4 = 2$, and $B_6 = 3$, by the existence of 4-chromatic planar graphs. Beraha, Kahane, and Reid [1973] proved that the conjecture holds for B_7 and B_{10}. For $B_5 = \tau^2$, Tutte [1970a] proved that the chromatic polynomial of a plane triangulation G is small at τ^2:

$$|P(G, \tau^2)| \leq \tau^{5-|V(G)|}.$$

On the other hand, Tutte [1970b] also proved that $P(G, \tau^2)$ is nonzero for every plane triangulation G. Hence to prove the conjecture for B_5, one has to prove the existence of an infinite family of plane triangulations with zeros tending to τ^2. This was done by Beraha, Kahane, and Weiss [1980]. The conjecture appears to be open for $n = 8$, 9 and $n \geq 11$.

14.7. CHESSBOARD PROBLEM

Find a general formula for the chromatic polynomial $P(X_{m,n}, \lambda)$ of the graph $X_{m,n}$ associated with an $m \times n$ "chessboard"—the graph with vertices at the integer lattice points (x, y) with $1 \leq x \leq n$, $1 \leq y \leq m$, and in which each vertex is joined to the four or fewer vertices at distance 1 from it. ■

The problem was stated in the survey article by Read and Tutte [1988], adding that: *"This is an easy question to ask, but without doubt a fiendishly difficult one to answer."* However, they also added that finding an answer might be within the bounds of possibility.

General formulas for the chromatic polynomials of certain infinite families of graphs were found by Biggs, Damerell, and Sands [1972]. Read and Tutte [1988] mentioned additional cases where such formulas have been found.

■ 14.8. COEFFICIENTS FOR HYPERGRAPHS

Given a hypergraph $H = (V, \mathcal{E})$, denote by $P(H, k)$ the number of different k-colorings of H using colors $1, 2, \ldots, k$ (two k-colorings are considered different if they assign different colors to some element of V). Then

$$P(H,k) = \sum_{m=1}^{|V|} b_m \binom{k}{m} m! = \sum_{m=1}^{|V|} a_m k^m,$$

where b_m is the number of partitions of V into m nonempty pairwise disjoint sets $X_1, X_2, \ldots, X_m$ such that no X_i contains an edge of H. Is it true that the numbers $b_1, b_2, \ldots, b_{|V|}$ form a unimodal sequence (i.e., a sequence that is first nondecreasing and then nonincreasing)? ■

The question was asked by Chvátal [1974b], who remarked that the only existing result in this direction seemed to be the inequality

$$b_{m+1} \geq b_m \cdot \frac{2^{|V|/m}}{m+1},$$

which can be obtained by an easy counting argument described by Chvátal [1970a] in the special case of graphs.

In the restricted case when H is a graph, the corresponding question for the sequence $|a_1|, |a_2|, \ldots, |a_{|V|}|$ was asked by Read [1968] (see Problem 14.1). For the sequence $|b_1|, |b_2|, \ldots, |b_{|V|}|$ the problem seems open even with this restriction, and it has been studied by Brenti [1992], who gave several interesting variations, and by Wakelin [1993].

Stanley [1993] introduced a noteworthy generalization of the chromatic polynomial in terms of a homogeneous symmetric function, and investigated problems on the coefficients arising from its diffferent expansions.

BIBLIOGRAPHY

[1981] Akiyama J. and F. Harary. A graph and its complement with specified properties. VII. A survey. In: G. Chartrand, Y. Alavi, D.L. Goldsmith, L. Lesniak-Foster, and D.R. Lick, editors, *The Theory and Applications of Graphs., Proc. 4th International Graph Theory Conference, Kalamazoo, 1980*, pages 1–12. Wiley, 1981.

[1989] Anstee R.P., J.H. Przytycki, and D. Rolfsen. Knot polynomials and generalized mutation. *Topology Appl.* **32**, 237–249, 1989.

[1989] Bari R. and S.Z. Kahn. Chromatic equivalence and chromatic uniqueness. In: R. Kulli, editor, *Recent Studies in Graph Theory*, pages 1–13. Vishwa International Publications, 1989.

[1975] Beraha S. *Infinite non-trivial families of maps and chromials*. Ph.D. thesis, Johns Hopkins University, 1975.

[1979] Beraha S. and J. Kahane. Is the four-color conjecture almost false? *J. Combin. Theory Ser. B* **27**, 1–12, 1979.

[1973] Beraha S., J. Kahane, and R. Reid. B_7 and B_{10} are limit points of chromatic zeros. *Notices Amer. Math. Soc.* **20**, 45, 1973.

[1980] Beraha S., J. Kahane, and N.J. Weiss. Limits of chromatic zeroes of some families of maps. *J. Combin. Theory Ser. B* **28**, 52–65, 1980.

[1972] Biggs N.L., R.M. Damerell, and D.A. Sands. Recursive families of graphs. *J. Combin. Theory Ser. B* **12**, 123–131, 1972.

[1946] Birkhoff G.D. and D.C. Lewis. Chromatic polynomials. *Trans. Amer. Math. Soc.* **60**, 355–451, 1946.

[1977] Bondy J.A. and R.L. Hemminger. Graph reconstruction—a survey. *J. Graph Theory* **1**, 227–268, 1977.

[1991] Borodin O.V. and I.G. Dmitriev. Characterization of chromatically rigid polynomials (in Russian). *Sibirsk. Mat. Zh.* **32**, 22–27, 1991. Translation in *Siberian Math. J.* **32**, 17–21, 1991.

[1992] Brenti F. Expansions of chromatic polynomials and log-concavity. *Trans. Amer. Math. Soc.* **332**, 729–756, 1992.

[1940] Brooks R.L., C.A.B. Smith, A.H. Stone, and W.T. Tutte. The dissection of rectangles into squares. *Duke Math. J.* **7**, 312–340, 1940.

[1982] Chao C.Y. and G.A. Novacky. On maximally saturated graphs. *Discrete Math.* **41**, 139–143, 1982.

[1978] Chao C.Y. and E.G. Whitehead. On chromatic equivalence of graphs. In: Y. Alavi and D.R. Lick, editors, *Graph Theory and Applications*, volume 642 of *Lecture Notes in Mathematics*, pages 121–131. Springer-Verlag, 1978.

[1979] Chao C.Y. and E.G. Whitehead. Chromatically unique graphs. *Discrete Math.* **27**, 171–177, 1979.

[1983] Chao C.Y. and L.-C. Zhao. Chromatic polynomials of a family of graphs. *Ars Combin.* **15**, 111–129, 1983.

[1986] Chia G.L. A note on chromatic uniqueness of graphs. *J. Graph Theory* **10**, 541–543, 1986.

[1970a] Chvátal V. A note on coefficients of chromatic polynomials. *J. Combin. Theory* **9**, 95–96, 1970.

[1974b] Chvátal V. Problem. In: C. Berge and D. Ray-Chaudhuri, editors, *Hypergraph Seminar*, volume 411 of *Lecture Notes in Mathematics*, page 281. Springer-Verlag, 1974.

[1961] Dirac G.A. On rigid circuit graphs. *Abh. Math. Sem. Univ. Hamburg* **25**, 71–76, 1961.

[1980] Dmitriev I.G. Weakly cyclic graphs with integral chromatic number (in Russian). *Metody Diskret. Analiz.* **34**, 3–7, 1980.

[1982] Dmitriev I.G. Characterization of a class of k-trees (in Russian). *Metody Diskret. Analiz.* **38**, 9–18, 1982.

[1970] Eisenberg B. *On the coefficients of the chromatic polynomial of a graph*. Ph.D. thesis, Adelphi University, 1970.

[1980] Farrell E.J. On chromatic coefficients. *Discrete Math.* **29**, 257–264, 1980.

[1978] Foldes S. The rotor effect can alter the chromatic polynomial. *J. Combin. Theory Ser. B* **25**, 237–239, 1978.

[1958] Hajnal A. and J. Surányi. Über die Auflösung von Graphen in vollständige Teilgraphen. *Ann. Univ. Sci. Budapest Eötvös Sect. Math.* **1**, 113–121, 1958.

[1974] Hoggar S.G. Chromatic polynomials and logarithmic concavity. *J. Combin. Theory Ser. B* **16**, 248–254, 1974.

[1993] Jackson B. A zero-free interval for chromatic polynomials of graphs. *Combin. Probab. Comput.* **2**, 325–336, 1993.

[1992] Jones V.F.R. Commuting transfer matrices and link polynomials. *Internat. J. Math.* **3**, 205–212, 1992.

[1990] Koh K.M. and K.L. Teo. The search for chromatically unique graphs. *Graphs Combin.* **6**, 259–285, 1990.

[1991] Koh K.M. and K.L. Teo. Chromatic equivalence of a graph and its complement. *Bull. Inst. Combin. Appl.* **3**, 81–82, 1991.

[1975] Lee L.A. *On chromatically equivalent graphs*. Ph.D. thesis, The George Washington University, 1975.

[1985] Lehmer D.H. The chromatic polynomial of a graph. *Pacific J. Math.* **118**, 463–469, 1985.

[1992] Li N.-Z. and E.G. Whitehead. The chromatic uniqueness of W_{10}. *Discrete Math.* **104**, 197–199, 1992.

[1991] Liu R.Y. and N.-Z. Li. Chromatic equivalence of connected vertex-transitive graphs (in Chinese). *Math. Appl.* **4**, 50–53, 1991 (English summary).

[1978] Loerinc B. Chromatic uniqueness of the generalized θ-graph. *Discrete Math.* **23**, 313–316, 1978.

[1972] Meredith C.H.J. Coefficients of chromatic polynomials. *J. Combin. Theory Ser. B* **13**, 14–17, 1972.

[1968] Read R.C. An introduction to chromatic polynomials. *J. Combin. Theory* **4**, 52–71, 1968.

[1975] Read R.C. Review. *Math. Rev.* **50** (# 6906), 1975.

[1987] Read R.C. Connectivity and chromatic uniqueness. *Ars Combin.* **23**, 209–218, 1987.

[1988] Read R.C. and W.T. Tutte. Chromatic polynomials. In: L.W. Beineke and R.J. Wilson, editors, *Selected Topics in Graph Theory*, volume 3, pages 15–42. Academic Press, 1988.

[1986] Salzberg P.M., M.A. López, and R.E. Giudici. On the chromatic uniqueness of bipartite graphs. *Discrete Math.* **58**, 286–294, 1986.

[1993] Stanley R.P. A symmetric function generalization of the chromatic polynomial of a graph. Manuscript, 1993.

[1989] Traczyk P. A note on rotant links. Technical report, Warsaw University, 1989.

[1941] Turán P. On an extremal problem in graph theory (in Hungarian). *Mat. Fiz. Lapok.* **48**, 436–452, 1941. English translation in: P. Erdős, editor, *Collected Papers of Paul Turán*, Volume 1, pages 231–240, Akadémiai Kiadó Budapest, 1990.

[1954] Turán P. On the theory of graphs. *Colloq. Math.* **3**, 19–30, 1954.

[1967] Tutte W.T. On dichromatic polynomials. *J. Combin. Theory* **2**, 301–320, 1967.

[1970a] Tutte W.T. On chromatic polynomials and the golden ratio. *J. Combin. Theory* **9**, 289–296, 1970.

[1970b] Tutte W.T. More about chromatic polynomials and the golden ratio. In: R. Guy, H. Hanani, N.W. Sauer, and J. Schönheim, editors, *Combinatorial Structures and Their Applications*, pages 439–453. Gordon and Breach, 1970.

[1974a] Tutte W.T. Chromials. In: C. Berge and D. Ray-Chaudhuri, editors, *Hypergraph Seminar*, volume 411 of *Lecture Notes in Mathematics*, pages 243–266. Springer-Verlag, 1974.

[1974b] Tutte W.T. Codichromatic graphs. *J. Combin. Theory Ser. B* **16**, 168–174, 1974.

[1979] Tutte W.T. All the king's horses (a guide to reconstruction). In: J.A. Bondy and U.S.R. Murty, editors, *Graph Theory and Related Topics*, pages 15–33. Academic Press, 1979.

[1980] Tutte W.T. Rotors in graph theory. In: J. Srivastava, editor, *Combinatorial Mathematics: Optimal Designs and Their Applications*, volume 6 of *Annals of Discrete Mathematics*, pages 343–347. North-Holland, 1980.

[1993] Wakelin C.D. The chromatic polynomial relative to the complete graph basis. Manuscript 1993. Submitted to *J. Graph Theory*.

[1992] Wakelin C.D. and D.R. Woodall. Chromatic polynomials, polygon trees, and outerplanar graphs. *J. Graph Theory* **16**, 459–466, 1992.

[1987] Weiming L. Almost every K_4 homeomorph is chromatically unique. *Ars Combin.* **23**, 13–36, 1987.

[1989] Whitehead E.G. Chromatic polynomials and the structure of graphs. In: *Graph Theory and Its Applications: East and West*, volume 576 of *Annals of the New York Academy of Sciences*, pages 630–632. New York Academy of Sciences, 1989.

[1993] Whitehead E.G. Chromaticity of K_4 homeomorphs. Manuscript, 1993.

[1984a] Whitehead E.G. and L.-C. Zhao. Chromatic uniqueness and chromatic equivalence in K_4-homeomorphs. *J. Graph Theory* **8**, 355–364, 1984.

[1984b] Whitehead E.G. and L.-C. Zhao. Cutpoints and the chromatic polynomial. *J. Graph Theory* **8**, 371–377, 1984.

[1932b] Whitney H. The coloring of graphs. *Ann. of Math.* **33**, 688–718, 1932.

[1976] Wilf H.S. Which polynomials are chromatic? In: B. Segre, editor, *Colloquio Internazionale sulle Teorie Combinatorie, Roma, Tomo I*, pages 247–256. Accademia Nazionale dei Lincei, 1976.

[1977] Woodall D.R. Zeros of chromatic polynomials. In: P.J. Cameron, editor, *Surveys in Combinatorics: Proc. Sixth British Combinatorial Conference*, pages 199–223. Academic Press, 1977.

[1992a] Woodall D.R. An inequality for chromatic polynomials. *Discrete Math.* **101**, 327–331, 1992.

[1992b] Woodall D.R. A zero-free interval for chromatic polynomials. *Discrete Math.* **101**, 333–341, 1992.

[1991] Xu S.-J. The chromatic uniqueness of complete bipartite graphs. *Discrete Math.* **94**, 153–159, 1991.

[1984] Xu S.-J. and N.-Z. Li. The chromaticity of wheels. *Discrete Math.* **51**, 207–212, 1984.

15

Hypergraphs

■ 15.1. ERDŐS' PROPERTY B

Let $\mathcal{F}$ be a family of sets called edges. $\mathcal{F}$ is said to have property B if there is a subset S of $\bigcup \mathcal{F}$ such that every edge has nonempty intersection with S and with $\overline{S}$, where $\overline{S}$ denotes the complement of S in $\bigcup \mathcal{F}$. What is the minimum number $m(n)$ of edges in a family $\mathcal{F}$ of n-sets not having property B? ■

In terms of hypergraph coloring, this problem asks for the minimum number of edges possible in a 3-chromatic n-uniform hypergraph. It is due to Erdős [1963, 1964], who gave the upper bound

$$m(n) < n^2 2^{n+1}.$$

Beck [1978] established the lower bound

$$n^{\frac{1}{3}-\varepsilon} 2^n \leq m(n) \text{ for all } \varepsilon > 0 \text{ and all } n > n(\varepsilon).$$

A short proof of Beck's result was given by Spencer [1981].

Erdős [1963] gave the so far only known exact values $m(2) = 3$ and $m(3) = 7$. Based on work of Abbott and Hanson [1969], the upper bound $m(4) \leq 23$ was proved by Seymour [1974a] and Toft [1975a]. Moreover Aizley and Selfridge [1977] announced the lower bound $m(4) \geq 19$. The recursive upper bound

$$m(2k+1) \leq (2k+1) \cdot m(2k-1) + 2^{2k} \text{ for all } k \geq 1,$$

giving the still best known bound $m(5) \leq 51$, was found by Abbott and Hanson [1969]. Toft [1975a] obtained the bound

$$m(2k+2) \leq (2k+2) \cdot m(2k) + 2^{2k+1} + \frac{1}{2}\binom{2k+2}{k+1} \text{ for all } k \geq 1.$$

The results of a computer search carried out by Exoo [1990] were interpreted by the experimenter as evidence that the correct value of $m(4)$ is, in fact, 23. Exoo [1990] searched for improved upper bounds for $m(n)$ also for other small values of n. However, his constructions did not improve any of the known bounds obtained from the recursive formulas given above.

It seems unknown if the edges of a minimum family of n-sets not having property B always have pairwise nonempty intersection, or at least have so for n not too large. It is the case for $n = 2$ and 3, but seems unlikely for $n \geq 4$.

The minimum number of vertices in a 3-chromatic n-uniform hypergraph is obviously $2n - 1$, attained by taking all n-sets of the $2n - 1$ vertices as edges. Seymour [1974c] proved with an elegant linear algebra argument that the number of edges in any 3-chromatic hypergraph is at least as big as the number of vertices, and he called a 3-chromatic hypergraph with equal numbers of edges and vertices "square" (see also Problem 15.9). The graph K_3 and the finite projective plane of order 2 (also called the "Fano configuration," having seven vertices and seven edges) are the unique 2- and 3-uniform 3-chromatic hypergraphs with minimum edge numbers. They are both square. Does this behavior persist for larger values of n? For all values of n? Perhaps giving an indication that this situation might change for $n \geq 4$, Thomassen [1992] has proved that every critical 3-chromatic square hypergraph has either an edge of size at most 3 or a vertex of degree at most 3.

Property B was introduced and studied by Bernstein [1908] and Miller [1937] in the infinite version. A more general property $B(s)$, due to Erdős, has been studied in detail by Abbott and Liu [1979, 1981] (see Problem 15.2).

■ 15.2. PROPERTY $B(s)$

Let $\mathcal{F}$ be a family of n-sets called edges. $\mathcal{F}$ is said to have property $B(s)$, $s \leq n$, if there is a subset S of the set $\bigcup \mathcal{F}$ such that $0 < |F \cap S| < s$ for every $F \in \mathcal{F}$. Let $m(n, s)$ denote the size of a smallest family $\mathcal{F}$ of n-sets not having property $B(s)$. What is the behavior of $m(n, s)$? Determine the exact value of $m(n, s)$ for any n and s where $n \geq s \geq 4$. Is it true for all $s \geq 4$ that the inequality $m(n, s) < m(s, s)$ holds for some n? ■

The problem of determining or estimating $m(n, s)$ was raised by Erdős and Hajnal [1961] and studied by Abbott and Liu [1979, 1981]. It is not difficult to see that $m(n, 2) = 3$ when n is even and $m(n, 2) = 4$ when n is odd. Moreover, it is known that $m(n, 3) = 7$ when n is divisible by 3 or 4. No exact value of $m(n, 4)$ seems to be known (see Abbott and Liu [1988]). The case of $m(4, 4)$ is a long-standing open problem (see Problem 15.1). Seymour [1974a] and Toft [1975a] obtained the bound $m(4, 4) \leq 23$, based on a construction by Abbott and Hanson [1969].

The last question was asked by Abbott and Liu [1987], who in solving a problem of P. Erdős had shown that the inequality $m(n, s) \geq m(s, s)$ does not hold in general. In particular they proved that $m(n, s) < m(s, s)$ for s sufficiently large and n large compared to s.

The corresponding question for $s \leq 3$ has a negative answer, as shown by Abbott and Liu [1979]. For $s = 4$ the answer is affirmative, since the bound $m(4,4) \geq 19$ has been announced by Aizley and Selfridge [1977], and $m(8,4) \leq 15$ was proved by Abbott and Liu [1987]. The property $B(4)$ was treated in detail by Abbott and Liu [1988], and Exoo [1990]. They proved $m(n,4) \leq 29$ except possibly for the eight values $n = 19, 23, 29, 31, 37, 38,$ 46 and 47. For $s = 5$ the answer seems unknown.

Property $B(n)$ of a family $\mathcal{F}$ of n-sets is equivalent to 2-colorability of $\mathcal{F}$ and is also called property B. Thus $m(n,n)$ equals the function $m(n)$ due to Erdős [1963, 1964] (see Problem 15.1). Property B was introduced by Bernstein [1908] and Miller [1937].

■ 15.3. FINITE PROJECTIVE PLANES

A finite projective plane of order $n \geq 1$ has $n^2 + n + 1$ points and $n^2 + n + 1$ lines all of size $n + 1$, with each point on exactly $n + 1$ lines, each pair of points on exactly one line, and each pair of lines having exactly one point in common. Does there exist a constant s for which the points of every finite projective plane of order $n \geq 3$ can be colored with colors 1 and 2 such that the number of points of color 1 in each line lies strictly between 0 and s? ■

A finite projective plane of order n may be thought of as an $(n + 1)$-uniform $(n + 1)$-regular hypergraph $H(n)$ with each pair of vertices contained together in exactly one edge, and each pair of edges having exactly one vertex in common. For $n = 1$ and $n = 2$ the hypergraph $H(n)$ is K_3 and the Fano configuration, respectively, each of which is 3-chromatic. For $n \geq 3$ it is easy to see that $H(n)$ is 2-colorable (e.g., consider three edges that do not share a common vertex, and for each vertex of $H(n)$ color it 1 if it is contained in precisely one of these edges; otherwise, color it 2). The stronger property asked for above, that there exists a set S of vertices such that $0 < |E \cap S| < s$ for all edges E, is called property $B(s)$. For an n-uniform hypergraph the property $B(n)$ is simply the property B to be 2-colorable (see Problem 15.1).

The property $B(s)$ was proposed by Erdős and Hajnal [1961] as a stronger form of property B (see also Abbott and Liu [1979]). The question above is due to Erdős and was stated by Erdős, Silverman, and Stein [1983], who proved with probabilistic methods that $H(n)$ has property $B(c \log n)$ for $c > 2e$ and n large enough. In addition, they gave a constructive proof that every $H(n)$ has property $B(n - c'\sqrt{n})$, where $c' > 0$ is a constant.

It is not known for which values of n there exist projective planes of order n. All known planes are of prime power order. A special type, the projective plane $PG(2, p^t)$ of order $n = p^t$, where p is a prime, can be constructed from the 3-dimensional vector-space $GF(p^t)^3$ over the finite field $GF(p^t)$ with p^t elements, letting the points and lines of $PG(2, p^t)$ be the 1- and 2-dimensional subspaces, respectively (see, e.g., the authoritative book by Hall [1986]). For $PG(2, p^t)$ and $p \geq 3$ Abbott and Liu [1985] replaced $2e$ in the result of Erdős, Silverman, and Stein [1983] by the smaller constant $2/\log 2$. Moreover, Boros [1988] proved that the projective plane $PG(2, p^t)$ has property $B(p+2)$ for all $p \geq 3$, thus extending an earlier result by Bruen and Fischer [1974] that $PG(2, 3^t)$ has property $B(5)$.

15.4. STEINER TRIPLE SYSTEMS

A Steiner triple system (STS) is a pair $(V, \mathcal{E})$, where V is a set of $n \geq 3$ vertices and $\mathcal{E}$ is a set of 3-subsets of V such that each 2-subset of V is a subset of exactly one of the sets in $\mathcal{E}$. $(V, \mathcal{E})$ is k-chromatic if V can be partitioned into k sets none of which contains an element of $\mathcal{E}$ as a subset, but cannot be partitioned into fewer than k such sets. For which pairs (n, k) do k-chromatic Steiner triple systems of order n exist? ■

The chromatic numbers of Steiner triple systems were first studied by Rosa [1970a, 1970b]. It has been known since Kirkman [1847] that a STS of order n exists if and only if $n \equiv 1$ or 3 (mod 6). Rosa [1970a, 1970b] proved that a 2-chromatic STS exists only if $n = 3$, and that a 3-chromatic STS exists of every possible order $n \geq 7$. The non-2-colorability was generalized to Steiner $(t+1)$-systems (where 3 and 2 are replaced by $t+1$ and t, respectively) with t even, by Gionfriddo and Lo Faro [1993].

De Brandes, Phelps, and Rödl [1982] proved that a 4-chromatic STS of order n exists for every $n \geq 25$ with $n \equiv 1$ or 3 (mod 6), except possibly for $n = 39$, 43, or 45. They conjectured that a 4-chromatic STS exists for each of these three values.

For every $k \geq 3$ there exists a number n_k such that for every $n \equiv 1$ or 3 (mod 6) with $n \geq n_k$ there exists a k-chromatic STS of order n. This was proved by de Brandes, Phelps, and Rödl [1982]. By the results mentioned above, it is known that $n_3 = 7$ and $25 \leq n_4 \leq 49$. De Brandes, Phelps, and Rödl [1982] gave as a general upper bound

$$n_k \leq ck^2 \log k,$$

where $c > 0$ is a constant. Phelps and Rödl [1986] proved that this inequality gives the right order of magnitude for n_k.

Let $n \equiv 1$ or 3 (mod 6) be fixed, and let

$$C(n) = \{k : \text{ there exists a } k\text{-chromatic STS of order } n\}.$$

De Brandes, Phelps, and Rödl [1982] conjectured that $C(n)$ is an interval.

Concerning the computational complexity of determining the chromatic number of a Steiner triple system, Phelps and Rödl [1984] proved that deciding 14-colorability of a STS is *NP*-complete, hence the existence of a polynomial algorithm to determine the chromatic number of any given STS is not to be expected. De Brandes, Phelps and Rödl [1982] asked if it is a difficult problem to decide 3-colorability of a Steiner triple system.

15.5. STEINER QUADRUPLE SYSTEMS

A Steiner quadruple system (SQS) is a pair $(V, \mathcal{E})$, where V is a set of $n \geq 4$ vertices and $\mathcal{E}$ a set of 4-subsets of V such that each 3-subset of V is a subset of exactly one of the sets in $\mathcal{E}$. $(V, \mathcal{E})$ is k-colorable if V can be partitioned into k sets none of which contains an element of $\mathcal{E}$ as a subset. Does a 2-colorable SQS on n vertices exist for every $n \equiv 2$ or 4 (mod 6) except $n = 14$? ■

The question was asked by Phelps [1991]. Hanani [1960] proved that a Steiner quadruple system of order $n \geq 4$ exists if and only if $n \equiv 2$ or 4 (mod 6). Phelps and Rosa [1980] proved that no SQS of order $n = 14$ is 2-colorable.

Two-colorable Steiner quadruple systems were first studied by Doyen and Vandensavel [1971], who proved that a 2-colorable SQS of order n exists for every $n \equiv 4$ or 8 (mod 12). Phelps and Rosa [1980] proved the existence for order $n = 2 \cdot 5^a \cdot 13^b \cdot 17^c$ for all $a, b, c \geq 0$. Phelps [1991] found a 2-colorable SQS of order $n = 22$; hence the smallest open cases are $n = 30$ and $n = 38$. Gionfriddo and Lo Faro [1993] proved that if a 2-coloring exists, then the two-color classes must be of equal size. Gionfriddo, Milici, and Tuza [to appear] obtained a necessary and sufficient condition for 2-colorability, from which the equal size of the color classes also follows.

Colbourn, Colbourn, Phelps, and Rödl [1982] gave a polynomial algorithm for deciding 2-colorability of any SQS.

15.6. MINIMUM-WEIGHT 3-CHROMATIC HYPERGRAPHS

Let the weight of an edge F in a hypergraph H be $2^{-|F|}$, and let the weight $w(H)$ of H be the sum of the weights of all the edges of H. Define $f(n)$ as the infimum of all possible weights of 3-chromatic hypergraphs in which every edge has size at least n.

What is the order of magnitude of $f(n)$? Is $f(n)$ equal to the weight of an n-uniform 3-chromatic hypergraph? If not, what are the values of c within the interval $1 \leq c < 2$ that will ensure this to be the case with weight $c^{-|F|}$ (instead of $2^{-|F|}$) for each edge F? Is $c = 1$ the only such value? ■

Erdős [1963] introduced $f(n)$, proved $1/2 \leq f(n)$, and conjectured $f(n) \to \infty$ as $n \to \infty$ (this conjecture was also mentioned in the very interesting paper on 3-chromatic hypergraphs by Erdős and Lovász [1975]). From the upper bound $m(n) < n^2 2^{n+1}$ of Erdős [1964] for the minimum number $m(n)$ of edges in a 3-chromatic n-uniform hypergraph (see Problem 15.1) it follows that $f(n) \leq m(n)/2^n < 2n^2$, and moreover that if the weight is changed to $c^{-|F|}$, with $c > 2$, for each edge F (instead of $2^{-|F|}$), then the corresponding function f must be identically 0.

Erdős and Selfridge [1973] described a generalization of the inequality $1/2 \leq f(n)$ in an interesting game-theoretic setting. In a so-called positional game played on H two players alternately select a previously unselected vertex of H. The goal of each player is to collect all the vertices of some edge of H. Erdős and Selfridge [1973] proved that if $w(H) < 1/2$, then the second player has a strategy that forces a draw. This implies that H has a 2-coloring in which the sizes of the color classes differ by at most one. Such equipartite colorings of hypergraphs have been studied by Berge and Sterboul [1977]. Positional games in general, and in particular higher-dimensional versions of Tic-tac-toe, were studied by Hales and Jewett [1963], Erdős and Selfridge [1973], and Beck [1981] (see also the book by Berge [1989], where Chapter 4 contains a treatment of hypergraph coloring and, as a special case, positional games).

Beck [1978] proved Erdős' conjecture that $f(n) \to \infty$ as $n \to \infty$. Indeed,

$$^{7f(n)+100}2 \geq n,$$

where $^{k}2$ denotes a "stack" of 2's of height k defined inductively by $^{0}2 = 1$ and $^{k+1}2 = 2^{(^{k}2)}$.

As mentioned, we have the bound $f(n) \leq m(n)/2^n$ by the definition of $m(n)$. Beck [1978] stated that he could not decide if strict inequality holds, and thus the second question above was raised.

■ 15.7. POSITIONAL GAMES

A positional game POS(H) on a (finite) hypergraph H is a "Tic-tac-toe" type of game played by two players. The players alternately select a previously unselected vertex of H. The goal of each player is to collect all the vertices of some edge of H. Describe sufficient conditions under which the second player has a strategy to force at least a draw: for example, conditions of the type $\Delta(H) \leq f(n)$, where $\Delta(H)$ is the maximum degree of H and $f(n)$ is some function of the minimum edge size n of H. The question may also be posed for restricted classes of hypergraphs, such as n-uniform hypergraphs, or "linear" hypergraphs (where any two edges share at most one vertex). ■

Since in the game POS(H) it cannot be a disadvantage to start, and since its length is finite, there is always a strategy for the first player to force at least a draw. There may also be a strategy for the second player to force at least a draw (as, indeed, is the case for the usual game of Tic-tac-toe). A necessary condition for the second player to have a strategy to force at least a draw is that H has a 2-coloring in which the sizes of the color classes differ by at most one. Such equipartite 2-colorings have been studied by Berge and Sterboul [1977] (see also Berge [1989]).

If the number of edges in an n-uniform hypergraph is strictly less than 2^{n-1}, then the second player can force a draw, as proved by Erdős and Selfridge [1973], who also proved that this is best possible. The extremal situations in connection with the theorem of Erdős and Selfridge were investigated by Lu [1992]. For linear hypergraphs the corresponding number of edges is of the order of magnitude 4^n, as conjectured by P. Erdős and proved by Beck [1981].

Erdős and Lovász [1975] proved that if H is n-uniform and $\Delta(H) \le 2^{n-1}/(4n)$, then H is 2-colorable. This was the first application of the now famous Lovász local lemma (see Chapter 4 in Graham, Rothschild, and Spencer [1990], or Chapter 5 in Alon and Spencer [1992]). The problem above asks for a similar theorem with the stronger conclusion that the second player can force a draw. The problem was raised by Beck [1981], who asked as one of the most interesting problems on positional games: For which order of magnitude of $\Delta(H)$ for an n-uniform hypergraph H is there a strategy for the second player to force a draw?

Hales and Jewett [1963] noted that if all the edges $E_1, E_2, \dots, E_m$ of H have size at least n, and if $\Delta(H) \le n/2$, then H has distinct vertices $x_1, y_1, x_2, y_2, \dots, x_m, y_m$ such that $x_i, y_i \in E_i$ for all i (by a standard argument from matching theory). The second player consequently has a simple strategy for forcing a draw: whenever the first player chooses some x_i, the reply is to choose the corresponding y_i, and vice versa.

Beck and Csirmaz [1982] gave an excellent overview of many variations and general results on positional games, including infinite versions. Some of their problems with particular relevance to hypergraph coloring were answered by Galvin [1990].

15.8. TIC-TAC-TOE

The (k, n) version of the positional game Tic-tac-toe, for positive integers k and n, is played on a "board" consisting of a hypergraph $H(k, n)$ defined as follows. The vertices of $H(k, n)$ are the k^n integer coordinate points $(a_1, a_2, \dots, a_n)$ with $1 \le a_i \le k$ for all i, and the edges of $H(k, n)$ are all possible sets of k of these points on a straight line in $\mathbf{R}^n$. The hypergraph $H(k, n)$ is called the n-dimensional cube of side k.

In Tic-tac-toe on $H(k, n)$ two players alternately select a previously unselected vertex of $H(k, n)$. The goal of each player is to collect all the vertices of some edge of $H(k, n)$.

Determine for which pairs (k, n) the hypergraph $H(k, n)$ has a 2-coloring. Determine for which pairs (k, n) the hypergraph $H(k, n)$ has an equipartite 2-coloring, that is, a 2-coloring such that the color classes differ in size by at most one. Determine for which pairs the second player has a strategy to force at least a draw when playing Tic-tac-toe on $H(k, n)$.

As weaker problems, but still extremely difficult in terms of finding exact answers, or just orders of magnitude, consider the following three questions. For each n, what is the largest value $\rho(n)$ of k for which $H(k, n)$ is not 2-colorable? For each n, what is the largest value $\sigma(n)$ of k for which $H(k, n)$ is not equipartite 2-colorable? For each n, what is the largest value

$\tau(n)$ of k for which the second player in Tic-tac-toe on $H(k, n)$ cannot force a draw?

Can the second player force a draw when the number of vertices is at least twice the number of edges? ■

For a general description of positional games on a hypergraph, see Problem 15.7. Tic-tac-toe for higher dimensions has been studied by Gardner [1959], Hales and Jewett [1963], Erdős and Selfridge [1973], Paul [1978], and Beck [1981], among others. A fine survey was given by Patashnik [1980], including a description of a computer investigation of the case $H(4, 3)$ of 3-dimensional Tic-tac-toe of side 4, also called "Qubic."

A necessary condition for the second player to have a strategy to force at least a draw is that $H(k, n)$ have an equipartite 2-coloring. Of course, the first player might have a winning strategy even if $H(k, n)$ has an equipartite 2-coloring. For Tic-tac-toe there is only one known example of this kind, namely the game of "Qubic" as analyzed by Patashnik [1980].

The k-uniform hypergraph $H(k, n)$ has $((k+2)^n - k^n)/2$ edges (see Moser [1948] for an elegant four-line proof, and also Paul [1978] and Patashnik [1980]). Thus the final question above asks if the second player can force a draw when $k \geq 2 \cdot (2^{1/n} - 1)^{-1}$ ($\approx 2n/\log 2$). The question is due to Hales and Jewett [1963] and was answered in the affirmative by Beck [1981] for $n \geq 100$. The background for the question is the simple observation that if there is a pair of vertices in each edge, such that all the pairs are pairwise disjoint, then the second player has an easy strategy to force a draw. For example the game "Five-in-a-row" on a 5×5 board (i.e., Tic-tac-toe on $H(5, 2)$) allows such a "Hales–Jewett pairing". The game "Four-in-a-row" on a 4×4 board has not enough room for a Hales–Jewett pairing, but the second player can force a draw nevertheless. Indeed, the second player can force a draw when playing Four-in-a-row on a 5×5 board (see Chapter 22 in Berlekamp, Conway and Guy [1982] for these and many other interesting results and variations on Tic-tac-toe; for example, a proof that "Nine-in-a-row" on an infinite 2-dimensional board has a draw strategy for the second player).

As for the functions ρ, σ, and τ, of course $\rho(n) \leq \sigma(n) \leq \tau(n)$. Hales and Jewett [1963] proved that $\rho(n) \longrightarrow \infty$ as $n \longrightarrow \infty$, and $\rho(n) \leq n$ for all n. They also established an exponential upper bound on $\tau(n)$, namely $\tau(n) \leq 3^n - 2$. Beck [1981] proved that

$$\tau(n) < (\log 3/\log 2)n + 4\sqrt{n \log n} + 4 \text{ for } n \geq 100.$$

Paul [1975, 1979] proved that $\sigma(n) \leq n - 1$ for $n \geq 4$, $\rho(n) \leq n - 2$ for $n \geq 8$, and $\rho(n) \leq n - 3$ for $n \geq 15$. The result on "Qubic" by Patashnik [1980] referred to above shows that $\tau(3) \geq 4$. While $\rho(2) = 2$, $\rho(3) = 3$, and $\rho(4) = 3$, the exact values of $\rho(n)$ remain unknown for $n \geq 5$. Graham, Wen-Ching Winnie Li, and Paul [1981a, 1981b] proved that $\rho(n)/n \longrightarrow 0$ as $n \longrightarrow \infty$.

The result that $\rho(n) \longrightarrow \infty$ as $n \longrightarrow \infty$ can also be expressed as follows. For a given k there exists a least integer $HJ(k)$ such that the hypergraph $H(k, n)$ has chromatic number at least 3 for all $n \geq HJ(k)$. This is a Ramsey theory formulation of the

Hales–Jewett theorem. (To be precise, the function $HJ(k)$ is usually defined in terms of a subhypergraph $C(k, n)$ of $H(k, n)$, where an edge consists of points $x_1, x_2, \ldots, x_k$ from $\{1, 2, \ldots, k\}^n$ so that in each coordinate j either the x_{ij} are constant, or $x_{ij} = i$. See Graham, Rothschild, and Spencer [1990]. We do not know if this makes any significant difference for the corresponding function, but we expect not.) Graham, Rothschild, and Spencer [1990] said that "*... the Hales–Jewett theorem strips van der Waerden's theorem of its unessential elements and reveals the heart of Ramsey theory. It provides a focal point from which many results can be derived and acts as a cornerstone for much of the more advanced work.*" (The theorem of van der Waerden is described in Problem 15.12.) The slow growth of $\rho(n)$ is equivalent to a fast growth of $HJ(k)$. The best upper bound for $HJ(k)$ so far has been obtained by Shelah [1988] (see Graham, Rothschild, and Spencer [1990] for a very fine exposition of Shelah's extraordinary achievement and its implications). Graham, Rothschild, and Spencer [1990] mentioned that the senior author (R. Graham) has a standing offer of \$1000 for a proof (or disproof) that $HJ(k) \leq {}^{ck}2$ for some constant $c > 0$ (defining ${}^{n}m$ inductively by ${}^{0}m = 1$ and ${}^{n+1}m = m^{({}^{n}m)}$ for $n, m \geq 0$). In the same way that $\rho(n)$ corresponds to $HJ(k)$, so does the function $\tau(n)$ correspond to a function $GHJ(k)$. Concerning this function Graham, Rothschild, and Spencer [1990] remarked that "*... the exact nature of GHJ remains a puzzle.*"

■ 15.9. SQUARE HYPERGRAPHS

A hypergraph H with an equal number of vertices and edges, say with $|V(H)| = |E(H)| = v$, is called a square hypergraph of size v.

Does there exist a polynomial algorithm to decide for a square hypergraph if it is critical 3-chromatic?

Determine for all integers $n \geq 2$ the least possible size $w(n)$ of a critical 3-chromatic square hypergraph in which every edge has size at least n. ■

Square hypergraphs may at first sight seem rather esoteric. However, Seymour [1974c] showed that they are extremely interesting combinatorial objects. Seymour [1974c] proved with a simple linear algebra argument that a critical 3-chromatic hypergraph H satisfies $|E(H)| \geq |V(H)|$. Moreover, he proved that if $|E(H)| = |V(H)|$, then the dual hypergraph H^* is also a critical 3-chromatic square hypergraph. (The dual hypergraph H^* has $E(H)$ as vertex set and an edge for each vertex x of H consisting of all the edges in H containing x.) This implies that the second problem posed above is equivalent to the following: Determine for all integers $n \geq 2$ the least possible size $w(n)$ of a 3-chromatic square hypergraph in which all vertices have degree at least n.

Even more strikingly, Seymour [1974c] characterized the critical 3-chromatic square hypergraphs in terms of directed graphs. He proved that given a critical 3-

chromatic square hypergraph H of size v, the collection of edges $E_1, E_2, \ldots, E_v$ has a system of distinct representatives $x_1, x_2, \ldots, x_v$ where $x_i \in E_i$ for $1 \leq i \leq v$. Define a directed graph $\vec{G}(H)$ on the vertex set $\{x_1, x_2, \ldots, x_v\}$ of H by joining each x_i to all the vertices of $E_i - x_i$ with edges directed away from x_i. Then $\vec{G}(H)$ is strongly connected and without directed cycles of even length. Conversely, from any strongly connected directed graph $\vec{G}$ without even cycles, define a hypergraph $H(\vec{G})$ on the vertex set $V(\vec{G})$ by letting each vertex x correspond to an edge consisting of x itself together with all the vertices in $\vec{G}$ that can be reached by one edge directed away from x. Then $H(\vec{G})$ is a critical 3-chromatic square hypergraph, as shown by Seymour [1974c].

Thus to ask for a polynomial algorithm to decide if a square hypergraph is critical 3-chromatic is equivalent to ask for a polynomial algorithm to decide if a directed graph is without directed cycles of even length. This difficult problem has been attacked by Thomassen [1985, 1986, 1992] and Seymour and Thomassen [1987]. The related problem of deciding if a square hypergraph has chromatic number at least 3 seems very hard; in fact, it is equivalent to the ***NP***-complete problem (see Theorem 19 in Chapter 1) of deciding if a given hypergraph is 2-colorable. Indeed, by adding new edges and new vertices to any given hypergraph it is easy to produce a square hypergraph that is 2-colorable if and only if the given one is.

In a paper on qualitative matrix theory, Thomassen [1986] investigated so-called sign nonsingular square matrices. These are nonsingular (i.e., with nonzero determinants) real matrices for which changes of nonzero elements to other values of the same signs do not destroy nonsingularity. Using the aforementioned result of Seymour [1974c] it can be seen that the (0,1)-valued incidence vectors of a critical 3-chromatic square hypergraph of size v form the rows of a (0,1)-valued sign nonsingular irreducible matrix of rank v, and vice versa (see Thomassen [1986]). (A sign nonsingular matrix is said to be irreducible if it cannot be written, after permuting rows and columns, as

$$\begin{bmatrix} M_1 & M_2 \\ 0 & M_3 \end{bmatrix}$$

where M_1 and M_3 are nonempty square matrices.) This gives yet another interesting equivalence between critical 3-chromatic square hypergraphs and a seemingly different type of structure.

Concerning the function w it is not clear that $w(n)$ exists for all values of n. Indeed Lovász [1975] asked about this, formulated as a problem on even directed cycles. Thomassen [1985] proved constructively, in terms of directed graphs without even cycles, that $w(n)$ exists for all n and satisfies $2^{n-1} \leq w(n) \leq 4^n$. Thus $w(n)$ grows exponentially with n. The only exact values that are known seem to be $w(2) = 3$ and $w(3) = 7$.

Thomassen [1992] proved that every critical 3-chromatic square hypergraph either has an edge of size at most 3 or a vertex of degree at most 3. This result was again obtained as a corollary of an interesting result on directed graphs without even cycles.

■ 15.10. SIZE OF 3-CHROMATIC UNIFORM CLIQUES

A hypergraph is said to be a clique if any two edges have a nonempty intersection. What are the possible values of the number of edges and the number of vertices of n-uniform 3-chromatic cliques? ■

This question is due to Erdős and Lovász [1975], who noted that every n-uniform clique is 3-colorable. A result of Całczyńska-Karłowicz [1964] implies that there is only a finite number of n-uniform cliques for any fixed n. Thus numbers $\min|\mathcal{F}|$, $\min|\bigcup\mathcal{F}|$, $M(n) = \max|\mathcal{F}|$, and $N(n) = \max|\bigcup\mathcal{F}|$ exist for all n, where the minima and maxima are taken over all 3-chromatic n-uniform cliques $\mathcal{F}$. For the maxima Erdős and Lovász [1975] established the following inequalities:

$$n!(e-1) \leq M(n) \leq n^n$$

$$\frac{1}{2}\binom{2n-2}{n-1} + 2n - 2 \leq N(n) \leq \frac{n}{2}\binom{2n-1}{n-1}.$$

Tuza [1985] gave improved bounds on $N(n)$:

$$2\binom{2n-4}{n-2} + 2n - 4 \leq N(n) \leq \binom{2n-1}{n-1} + \binom{2n-4}{n-2}$$

and conjectured (see Erdős [1984]) that this lower bound is best possible for $n \geq 4$. The special values $N(3) = 7$ and $N(4) = 16$ were determined by Hanson and Toft [1983]. For the function M, Tuza [to appear] obtained the improved upper bound

$$M(n) \leq (1 - 1/e + o(1))n^n,$$

where $o(1) \to 0$ as $n \to \infty$.

In an n-uniform clique with the weaker property that no set of $n-1$ vertices intersects all edges, the minimum possible number of edges is bounded by $c \cdot n$ for some constant c. This deep result was obtained by Kahn [1994a] (see also Kahn [1994b]).

■ 15.11. MONOCHROMATIC SUM-SETS

For $n > k > 0$ and for every k-subset $S \subseteq \{1, 2, \ldots, n\}$ let $P(S)$ denote the set of all numbers that can be written as a sum without repetitions of elements in S; that is, $P(S) = \{\sum_{i\in I} i : I \subseteq S\}$. Let $\mathcal{F}_{n,k}$ be the family of those sets $P(S)$ that are contained in $\{1, 2, \ldots, n\}$,

$$\mathcal{F}_{n,k} = \{P(S) \subseteq \{1, 2, \ldots, n\} : S \subseteq \{1, 2, \ldots, n\} \text{ and } |S| = k\}.$$

For every positive k, if there exists a number n so that the hypergraph $\mathcal{F}_{n,k}$ is not 2-colorable, let $F(k)$ denote the smallest such n. What is the order of magnitude of $F(k)$? ■

It follows from a theorem attributed to J. Folkman that $F(k)$ exists for all $k > 0$ (see Graham, Rothschild, and Spencer [1990]). The problem of determining an upper bound on $F(k)$ was studied by Taylor [1981a], who obtained

$$F(k) \leq {}^{4k-3}3,$$

where ${}^{n}m$ denotes a "stack" of m's of height n, defined inductively by ${}^{0}m = 1$ and ${}^{n+1}m = m^{{}^{n}m}$ for $n, m \geq 0$.

A lower bound was proved by Erdős and Spencer [1989]:

$$F(k) \geq 2^{ck^2/\log k}$$

for a constant $c > 0$. No better bounds seem known.

■ 15.12. ARITHMETIC PROGRESSIONS

For $n > k > 0$ let $\mathcal{W}_{n,k}$ denote the family of subsets of $\{1, 2, \ldots, n\}$ consisting of arithmetic progressions of k terms, that is, subsets $\{a, a+b, a+2b, \ldots, a+(k-1)b\}$ with $a, b > 0$ and $a + (k-1)b \leq n$. For $k > 0$ let $W(k)$ denote the smallest n (it exists) such that in any 2-coloring of the integers $1, 2, \ldots, n$, some element of $\mathcal{W}_{n,k}$ is monochromatic. Does $W(k)$ satisfy

$$W(k) \leq {}^{ck}2$$

for some constant $c > 0$ (where ${}^{n}m$ is defined inductively by ${}^{0}m = 1$ and ${}^{n+1}m = m^{{}^{n}m}$ for $n, m \geq 0$)? ■

The question was asked by R.L. Graham (see Shelah [1988]), who conjectured that the answer is affirmative. Graham offered \$1000 for a proof of this conjecture.

The theorem of van der Waerden [1927] states that $W(k)$ exists for all $k > 0$, as conjectured by I. Schur (see A. Brauer's introduction to *I. Schur—Gesammelte Abhandlungen*, Springer–Verlag, 1973). Other proofs were given by Hales and Jewett [1963], Graham and Rothschild [1971], and Furstenberg [1982], without producing a "reasonable" upper bound for $W(k)$. The first such bound was proved by Shelah [1988], who was in fact awarded \$500 by Graham for the improved bound, half of the original prize for settling the conjecture completely (Graham, Rothschild, and Spencer [1990] reported that the original offer of \$1000 still stands, nevertheless). Shelah's bound can be formulated in terms of the Ackermann hierarchy $f_0, f_1, f_2, \ldots$

of functions from the set of nonnegative integers to itself, defined inductively as follows:

$$f_0(x) = x + 1, \text{ and}$$
$$f_{i+1}(x) = f_i^{(x)}(x) \text{ for } i \geq 0,$$

where $f^{(x)}$ denotes the function f iterated x times. Expressed in terms of these functions, Shelah [1988] obtained the upper bound $W(k) \leq f_4(c'k)$ for a constant $c' > 0$, thus substantially improving the previously known bounds. These involved functions similar to the Ackermann function $f_\infty(x) = f_x(x)$, which grows very rapidly compared with any function $f_k(x)$ for fixed $k > 0$. (Graham, Rothschild, and Spencer [1990] included a proof and a discussion of Shelah's result in their monograph on Ramsey theory.) The Graham conjecture would imply that $W(k) \leq f_3(ck)$ for constant $c > 0$.

The best known lower bounds for $W(k)$ grow exponentially. Graham, Rothschild, and Spencer [1990] remarked that this is perhaps the right order of magnitude. Prömel and Voigt [1990] obtained stronger versions of van der Waerden's theorem formulated in terms of hypergraph coloring. Alon, Caro, and Tuza [1989] investigated a "totally multicolored" version of the problem, where each color is used at most a constant number of times, and where the aim is to find arithmetic progressions all terms of which are colored differently.

■ 15.13. UNPROVABILITY

For positive integers p, n, and N, with $p, n < N$, let $\mathcal{P}_{p,n,N}$ be the family of sets where each set consists of all p-subsets of a set Y, for all sets $Y \subseteq \{1, 2, \ldots, N\}$ with $|Y| \geq n$ and $|Y| \geq \min Y$. Let $r_{n,p}(k)$ be the smallest N, if one exists, such that $\mathcal{P}_{p,n,N}$ is not k-colorable. Is it possible to prove from the Peano axioms of arithmetic that $r_{p+1,p}(2)$ exists for all $p > 0$? ■

The problem was mentioned by Paris [1990] and by Loebl and Nešetřil [1991]. Paris and Harrington [1977] proved that $r_{n,p}(k)$ exists for all integers $n, k, p > 0$ and also proved that the existence cannot be proved from Peano's axioms. Thus Paris and Harrington gave the first example of a natural combinatorial statement that is unprovable. The first example of a true statement that can be formulated but not proved in Peano arithmetic was given by Gödel [1931].

Paris [1990] reported that the result of Paris and Harrington [1977] was later strengthened, in one direction by Paris and Harrington, who showed that the existence for all $k > 0$ of $r_{k+1,k}(k)$ is unprovable, and in another direction by J. Quinsey, who proved that the existence for all $p > 0$ of $r_{p+1,p}(3)$ is unprovable in Peano arithmetic.

Ketonen and Solovay [1981] suggested that the existence of functions such as $\gamma(p) = r_{p+1,p}(2)$ might be shown unprovable by showing that $\gamma(p)$ grows faster than any function describable in ordinary Peano arithmetic (see also the interesting paper by Erdős and Mills [1981]).

15.14. THE DIRECT PRODUCT OF HYPERGRAPHS

Given two hypergraphs $H = (X, \mathcal{E})$ and $H' = (Y, \mathcal{F})$, with $\mathcal{E} = (E_i : i \in I)$, $\mathcal{F} = (F_j : j \in J)$, their direct product is a hypergraph $H \times H'$ with vertex set $X \times Y$ and with edges $E_i \times F_j$ for $(i, j) \in I \times J$. Given $p, q > 0$ let $f(p, q)$ denote the smallest chromatic number $\chi(H \times H')$ of any direct product where $\chi(H) = p$ and $\chi(H') = q$. Is it true that

$$f(p,q) \longrightarrow \infty \text{ as } p, q \longrightarrow \infty?$$

■

This problem was posed by Berge and Simonovits [1974]. They remarked that the problem of finding a good estimate for $f(p, q)$ seems to be difficult.

Berge and Simonovits [1974] proved that the maximum chromatic number of a direct product $H \times H'$, with $\chi(H) = p$ and $\chi(H') = q$, is attained by the product $K_p^2 \times K_q^2$, where K_m^r denotes the hypergraph consisting of all r-subsets of a set of size m. Sterboul [1974] proved among other similar results

$$\lim_{m \to \infty} \frac{\chi(K_m^2 \times K_m^2)}{\sqrt{m}} = 1.$$

The chromatic numbers of direct products of the form $K_m^r \times K_n^s$ were also studied by Erdős and Rado [1956] and by Chvátal [1969].

15.15. MAXIMAL COMPLETE SUBGRAPHS IN PERFECT GRAPHS

Let G be a graph and let $\mathcal{F}(G)$ denote the hypergraph with vertex set $V(G)$ whose edge set is the family of subsets of $V(G)$ consisting of the vertex sets of the maximal complete subgraphs of G. Does there exist some constant k so that it is always possible to k-color the hypergraph $\mathcal{F}(G)$ if G is a perfect graph? ■

This question was asked by Duffus, Sands, Sauer, and Woodrow [1991], who proved that $\mathcal{F}(G)$ can be 2-colored if G is a comparability graph, that is, if $V(G)$ is the set of elements of a partially ordered set P with two elements x and y of P joined by an edge in G if and only if x and y are comparable in P. The comparability graphs form a well-known class of perfect graphs (see, e.g., Berge [1975]). Duffus, Kierstead, and Trotter [1991] proved that if $\overline{G}$ is the complement of a comparability graph G, then $\mathcal{F}(\overline{G})$ is 3-colorable. By the perfect graph theorem of L. Lovász (Theorem 24 in Chapter 1) $\overline{G}$ is perfect if G is perfect.

For graphs G in general, there is no constant k such that $\mathcal{F}(G)$ can always be colored using at most k colors. This follows from constructions of triangle-free graphs with arbitrarily high chromatic numbers (Theorem 21 in Chapter 1).

15.16. COLORING TRIANGULABLE MANIFOLDS

For a finite simplicial complex K of dimension $m > 1$, define H_K as the hypergraph having the $(m-2)$-dimensional simplexes of K as its vertex set, and having, for every $(m-1)$-dimensional simplex in K, an edge containing precisely the vertices that form the faces of the simplex.

For $m > 1$, let X be any "triangulable" m-dimensional space; that is, X is homeomorphic to the geometric realization $|K|$ of a finite simplicial complex K. K is then said to triangulate X. Define the (possibly infinite) chromatic number $\chi(X)$ of X as

$$\chi(X) = \sup_K \chi(H_K) \text{ where } K \text{ triangulates } X.$$

Do there exist an $m \geq 3$ and two m-dimensional (closed, Hausdorff) triangulable manifolds X and Y with $\chi(X) \neq \chi(Y)$? ■

This question was asked by Sarkaria [1983a],[1] who anticipated a negative answer. It should be noted that all manifolds of dimension $m \leq 3$, but not all manifolds of dimension $m > 3$, are triangulable (see Massey [1991] for results and references).

The answer to the corresponding question for $m = 2$ is affirmative since every 2-dimensional manifold can be triangulated, and since the bound given by P.J. Heawood for any surface **S** of Euler characteristic $\varepsilon(\mathbf{S}) < 2$,

$$\chi(\mathbf{S}) \leq \left\lfloor \frac{1}{2}\left(7 + \sqrt{49 - 24\varepsilon(\mathbf{S})}\right) \right\rfloor$$

(see Theorem 2 in Chapter 1) is met with equality, except in the case of the Klein bottle with $\chi(\text{Klein bottle}) = 6$ (see Theorem 3 in Chapter 1). Hence there exist 2-dimensional manifolds of all chromatic numbers larger than 5.

For dimension $m \geq 3$ Sarkaria [1982] proved, using the four-color theorem, that every closed triangulable manifold X has chromatic number $\chi(X) \leq 4$. Sarkaria [1983a] also gave a short proof of $\chi(X) \leq 6$, using ideas of Grünbaum [1970b] and without using the four-color theorem.

[1] We are grateful to T. Bier for making us aware of K.S. Sarkaria's work.

Sarkaria [1981, 1983b, 1987] introduced and studied more general chromatic numbers $\chi_i(X)$ for $0 \le i < m$ defined as follows. Given a triangulation K of X, $\chi_i(K)$ is defined as the chromatic number of the hypergraph having the i-dimensional simplexes of K as its vertices, with edges corresponding to the $(i + 1)$-dimensional simplexes of K, and with hypergraph incidence defined by simplicial containment. Let

$$\chi_i(X) = \sup_K \chi_i(K) \text{ where } K \text{ triangulates } X.$$

Among other results, Sarkaria proved that $\chi_{m-1}(X) = 2$ for all $m \ge 2$, and that $\chi_i(X)$ is infinite if $m > 2i + 2$, and he conjectured that $\chi_i(X)$ is finite if $m \le 2i + 2$. Grünbaum [1970b] studied similar hypergraphs. However, he considered the different coloring problem of coloring the vertices subject to the condition that no two vertices belonging to the same edge may be colored alike.

Sarkaria [1981, 1987] also studied the chromatic number of pseudo-manifolds. In two dimensions these are obtained from surfaces by identifying finite sets of single points (for the general definition, see, e.g., Massey [1991]). He thus formulated a generalization of the empire problem of P.J. Heawood (Problem 3.1) to higher dimensions.

■ 15.17. BERGE'S CONJECTURE ON EDGE-COLORING

Let the closure of a hypergraph H, denoted by $\hat{H}$, be defined as the hypergraph with the same vertex set as H, such that E is an edge in $\hat{H}$ if and only if E is a nonempty subset of some edge in H. In particular, $\hat{H}$ in general contains loops (i.e., edges of size 1). H is said to be linear if the intersection of every pair of edges in H contains at most one vertex. If H is linear, is it true that

$$\chi'(\hat{H}) = \Delta(\hat{H}),$$

where the edge-chromatic number χ' is the smallest number of colors that can be used to color the edges so that two edges receive different colors if they have nonempty intersection, and where Δ denotes the maximum degree? ■

This question was asked by Berge [1990], who conjectured that the answer is affirmative.

Berge's conjecture would generalize the theorem of V.G. Vizing (Theorem 29 in Chapter 1) that a simple graph G can be edge-colored with $\Delta(G) + 1$ colors. Vizing's theorem follows by observing that a simple graph is always linear, and that $\hat{G}$ is obtained from G by adding one loop for every vertex, hence $\Delta(\hat{G}) = \Delta(G) + 1$. (Note that the "degree" of a vertex means precisely the number of edges containing it—thus a loop is *not* counted twice, as it is mostly practiced for graphs.)

Stein [1983] proved $\Delta(\hat{H}) = \Delta_0(\hat{H})$ for every linear hypergraph H, where Δ_0 denotes the maximum number of edges with pairwise nonempty intersection. In general, it is clear that $\chi' \geq \Delta_0 \geq \Delta$. Thus for Berge's conjecture it is sufficient to prove the seemingly weaker statement $\chi'(\hat{H}) \leq \Delta_0(\hat{H})$. The conjecture that $\Delta_0(\hat{H}) = \Delta(\hat{H})$ holds for the closure of every hypergraph H is an unsolved conjecture of Chvátal [1974a].

Partial results on the conjectures of Berge and Chvátal were obtained by Gionfriddo and Tuza [1994].

BIBLIOGRAPHY

[1969] Abbott H.L. and D. Hanson. On a combinatorial problem of Erdős. *Canad. Math. Bull.* **12**, 823–829, 1969.

[1979] Abbott H.L. and A. Liu. On property $B(s)$. *Ars Combin.* **7**, 255–260, 1979.

[1981] Abbott H.L. and A. Liu. On property $B(s)$ II. *Discrete Math.* **37**, 135–141, 1981.

[1985] Abbott H.L. and A. Liu. Property $B(s)$ and projective planes. *Ars Combin.* **20**, 217–220, 1985.

[1987] Abbott H.L. and A. Liu. On a problem of Erdős concerning property B. *Combinatorica* **7**, 215–219, 1987.

[1988] Abbott H.L. and A. Liu. On property $B(4)$ of families of sets. *Ars Combin.* **26**, 59–68, 1988.

[1977] Aizley P. and J.L. Selfridge. Abstract. *Notices Amer. Math. Soc.* **24**, A–452, 1977.

[1989] Alon N., Y. Caro, and Z. Tuza. Sub-Ramsey numbers for arithmetic progressions. *Graphs Combin.* **5**, 303–306, 1989.

[1992] Alon N. and J.H. Spencer. *The Probabilistic Method.* Wiley, 1992 (with an appendix on open problems written by P. Erdős).

[1978] Beck J. On 3-chromatic hypergraphs. *Discrete Math.* **24**, 127–137, 1978.

[1981] Beck J. On positional games. *J. Combin. Theory Ser. A* **30**, 117–133, 1981.

[1982] Beck J. and L. Csirmaz. Variations on a game. *J. Combin. Theory Ser. A* **33**, 297–315, 1982.

[1975] Berge C. Perfect graphs. In: D.R. Fulkerson, editor, *Studies in Graph Theory, Part 1*, volume 11 of *MAA Studies in Mathematics*, pages 1–22. The Mathematical Association of America, 1975.

[1989] Berge C. *Hypergraphs. Combinatorics of Finite Sets*, volume 45 of *North-Holland Mathematical Library*. North-Holland, 1989.

[1990] Berge C. On two conjectures to generalize Vizing's theorem. In: M. Gionfriddo, editor, *Le Matematiche*, volume XLV, pages 15–23, 1990.

[1974] Berge C. and M. Simonovits. The coloring numbers of the direct product of two hypergraphs. In: C. Berge and D. Ray-Chaudhuri, editors, *Hypergraph Seminar*, volume 411 of *Lecture Notes in Mathematics*, pages 21–33. Springer-Verlag, 1974.

[1977] Berge C. and F. Sterboul. Equipartite colorings in graphs and hypergraphs. *J. Combin. Theory Ser. B* **22**, 97–113, 1977.

[1982] Berlekamp E.R., J.H. Conway, and R.K. Guy. *Winning Ways, Vol. 1 and 2.* Academic Press, 1982.

[1908] Bernstein F. Zur Theorie der trigonometrische Reihen. *Leipz. Ber.* **60**, 325–328, 1908.

[1988] Boros E. $PG(2, p^s)$, $p > 2$, has property $B(p + 2)$. *Ars Combin.* **25**, 111–113, 1988.

[1974] Bruen A. and J.C. Fischer. Blocking sets and complete arcs. *Pacific J. Math.* **53**, 73–84, 1974.

[1964] Całczyńska-Karłowicz M. Theorem on families of finite sets. *Bull. Acad. Polon. Sci. Sér. Sci. Math. Astronom. Phys.* **12**, 87–89, 1964.

[1969] Chvátal V. On finite polarized partition relations. *Canad. Math. Bull.* **12**, 321–326, 1969.

[1974a] Chvátal V. Intersecting families of edges in hypergraphs having the hereditary property. In: C. Berge and D. Ray-Chaudhuri, editors, *Hypergraph Seminar*, volume 411 of *Lecture Notes in Mathematics*, pages 61–66. Springer-Verlag, 1974.

[1982] Colbourn C.J., M.J. Colbourn, K.T. Phelps, and V. Rödl. Coloring Steiner quadruple systems. *Discrete Appl. Math.* **4**, 103–111, 1982.

[1982] De Brandes M., K.T. Phelps, and V. Rödl. Coloring Steiner triple systems. *SIAM J. Algebraic Discrete Methods* **3**, 241–249, 1982.

[1971] Doyen J. and S. Vandensavel. Nonisomorphic Steiner quadruple systems. *Bull. Soc. Math. Belg.* **23**, 393–410, 1971.

[1991] Duffus D., H.A. Kierstead, and W.T. Trotter. Fibres and ordered set coloring. *J. Combin. Theory Ser. A* **58**, 158–164, 1991.

[1991] Duffus D., B. Sands, N.W. Sauer, and R.E. Woodrow. Two-colouring all two-element maximal antichains. *J. Combin. Theory Ser. A* **57**, 109–116, 1991.

[1963] Erdős P. On a combinatorial problem. *Nordisk Mat. Tidskr.* **11**, 5–10, 1963.

[1964] Erdős P. On a combinatorial problem II. *Acta Math. Acad. Sci. Hungar.* **15**, 445–447, 1964.

[1984] Erdős P. On some problems in graph theory, combinatorial analysis and combinatorial number theory. In: B. Bollobás, editor, *Graph Theory and Combinatorics*, pages 1–17. Academic Press, 1984.

[1961] Erdős P. and A. Hajnal. On a property of families of sets. *Acta Math. Acad. Sci. Hungar.* **12**, 87–123, 1961.

[1975] Erdős P. and L. Lovász. Problems and results on 3-chromatic hypergraphs and some related questions. In: A. Hajnal, R. Rado, and V.T. Sós, editors, *Infinite and Finite Sets*, volume 10 of *Colloq. Math. Soc. János Bolyai*, pages 609–627. North-Holland, 1975.

[1981] Erdős P. and G. Mills. Some bounds for the Ramsey–Paris–Harrington numbers. *J. Combin. Theory Ser. A* **30**, 53–70, 1981.

[1956] Erdős P. and R. Rado. A partition calculus in set theory. *Bull. Amer. Math. Soc.* **62**, 427–489, 1956.

[1973] Erdős P. and J.L. Selfridge. On a combinatorial game. *J. Combin. Theory Ser. B* **14**, 298–301, 1973.

[1983] Erdős P., R. Silverman, and A. Stein. Intersection properties of families containing sets of nearly the same size. *Ars Combin.* **15**, 247–259, 1983.

[1989] Erdős P. and J.H. Spencer. Monochromatic sumsets. *J. Combin. Theory Ser. A* **50**, 162–163, 1989.

[1990] Exoo G. On constructing hypergraphs without property B. *Ars Combin.* **30**, 3–12, 1990.

[1982] Furstenberg H. *Recurrence in ergodic theory and combinatorial number theory*. Princeton University Press, 1982.

[1990] Galvin F. Hypergraph games and the chromatic number. In: A. Baker, B. Bollobás, and A. Hajnal, editors, *A Tribute to Paul Erdős*, pages 201–206. Cambridge University Press, 1990.

[1959] Gardner M. Ticktacktoe. In: *The Scientific American Book of Mathematical Puzzles and Diversions*, pages 37–46. Simon and Schuster, 1959. Republished as *Hexaflexagons and Other Mathematical Diversions* (with a new afterword and a new bibliography), The University of Chicago Press, 1988.

[1993] Gionfriddo M. and G. Lo Faro. 2-colourings in $S(t, t+1, v)$. *Discrete Math.* **111**, 263–268, 1993.

[to appear] Gionfriddo M., S. Milici, and Z. Tuza. Blocking sets in SQS($2v$). To appear in *Combin. Probab. Comput.*

[1994] Gionfriddo M. and Z. Tuza. On conjectures of Berge and Chvátal. *Discrete Math.* **124**, 79–86, 1994.

[1931] Gödel K. Über formal unentscheidbare Sätze der Principia Mathematica und verwandter Systeme I. *Monatsh. Math. Phys.* **38**, 173–198, 1931.

[1971] Graham R.L. and B.L. Rothschild. Ramsey's theorem for n-parameter sets. *Trans. Amer. Math. Soc.* **159**, 257–292, 1971.

[1990] Graham R.L., B.L. Rothschild, and J.H. Spencer. *Ramsey Theory*, 2nd ed. Wiley, 1990 (first edition, 1980).

[1981a] Graham R.L., Wen-Ching Winnie Li and J.L. Paul. Monochromatic lines in partitions of $\mathbf{Z}^n$. In: K.L. McAvaney, editor, *Combinatorial Mathematics VIII*, number 884 in *Lecture Notes in Mathematics*, pages 35–48. Springer-Verlag, 1981.

[1981b] Graham R.L., Wen-Ching Winnie Li, and J.L. Paul. Homogeneous collinear sets in partitions of $\mathbf{Z}^n$. *J. Combin. Theory Ser. A* **31**, 21–32, 1981.

[1970b] Grünbaum B. Polytopes, graphs and complexes. *Bull. Amer. Math. Soc.* **76**, 1131–1201, 1970.

[1963] Hales A.W. and R.I. Jewett. Regularity and positional games. *Trans. Amer. Math. Soc.* **106**, 222–229, 1963.

[1986] Hall M. *Combinatorial Theory*, 2nd ed. Wiley, 1986 (first edition, Blaisdell 1967).

[1960] Hanani H. On quadruple systems. *Canad. J. Math.* **12**, 145–157, 1960.

[1983] Hanson D. and B. Toft. On the maximum number of vertices in n-uniform cliques. *Ars Combin.* **16-A**, 205–216, 1983.

[1994a] Kahn J. On a problem of Erdős and Lovász. II. $n(r) = O(r)$. *J. Amer. Math. Soc.* **7**, 125–143, 1994.

[1994b] Kahn J. Recent results on some not–so–recent hypergraph matching and covering problems. In: P. Frankl, Z. Füredi, G.O.H. Katona, and D. Miklós, editors, *Extremal Problems for Finite Sets*, volume 3 of *Bolyai Society Mathematical Studies*. János Bolyai Mathematical Society, 1994.

[1981] Ketonen J. and R. Solovay. Rapidly growing Ramsey functions. *Ann. of Math. (2)* **113**, 267–314, 1981.

[1847] Kirkman T.P. On a problem in combinations. *Cambridge and Dublin Math. J.* **2**, 191–204, 1847.

[1991] Loebl M. and J. Nešetřil. Fast and slow growing. In: A.D. Keedwell, editor, *Surveys in Combinatorics: Proc. 13th British Combinatorial Conference*, pages 119–160. Cambridge University Press, 1991.

[1975] Lovász L. Problem 2. In: M. Fiedler, editor, *Recent Advances in Graph Theory, Proc. Symposium, Prague, June 1974*, page 541. Academia Praha, 1975.

[1992] Lu X. A characterization on n-critical economical generalized tic-tac-toe games. *Discrete Math.* **110**, 197–203, 1992.

[1991] Massey W.S. *A Basic Course in Algebraic Topology*. Springer-Verlag, 1991.

[1937] Miller E.W. On a property of families of sets. *C.R. Soc. Sci. Varsovie* **30**, 31–38, 1937.

[1948] Moser L. Solution to Problem E 773. *Amer. Math. Monthly* **55**, 99, 1948.

[1990] Paris J. Combinatorial statements independent of arithmetic. In: J. Nešetřil and V. Rödl, editors, *Mathematics of Ramsey Theory*, pages 232–245. Springer-Verlag, 1990.

[1977] Paris J. and L. Harrington. A mathematical incompleteness in Peano arithmetic. In: J. Barwise, editor, *Handbook of Mathematical Logic*, volume 90 of *Studies in Logic and the Foundations of Mathematics*, pages 1133–1142. North-Holland, 1977.

[1980] Patashnik O. Qubic: $4 \times 4 \times 4$ tic-tac-toe. *Math. Mag.* **53**, 202–216, 1980.

[1975] Paul J.L. The q-regularity of lattice point paths in $\mathbf{R}^n$. *Bull. Amer. Math. Soc.* **81**, 492–494 and 1136, 1975.

[1978] Paul J.L. Tic-tac-toe in n dimensions. *Math. Mag.* **51**, 45–49, 1978.

[1979] Paul J.L. Partitioning the lattice points in $\mathbf{R}^n$. *J. Combin. Theory Ser. A* **26**, 238–248, 1979.

[1991] Phelps K.T. A class of 2-chromatic SQS(22). *Discrete Math.* **97**, 333–338, 1991.

[1984] Phelps K.T. and V. Rödl. On the algorithmic complexity of coloring simple hypergraphs and Steiner triple systems. *Combinatorica* **4**, 79–88, 1984.

[1986] Phelps K.T. and V. Rödl. Steiner triple systems with minimum independence number. *Ars Combin.* **21**, 167–172, 1986.

[1980] Phelps K.T. and A. Rosa. 2-chromatic Steiner quadruple systems. *European J. Combin.* **1**, 253–258, 1980.

[1990] Prömel H.-J. and B. Voigt. A sparse Gallai–Witt theorem. In: R. Bodendiek and R. Henn, editors, *Topics in Combinatorics and Graph Theory*, pages 747–755. Physica Heidelberg, 1990.

[1970a] Rosa A. On the chromatic number of Steiner triple systems. In: R. Guy, H. Hanani, N.W. Sauer, and J. Schönheim, editors, *Combinatorial Structures and Their Applications*, pages 369–371. Gordon and Breach, 1970.

[1970b] Rosa A. Steiner triple systems and their chromatic number. *Acta Fac. Rerum Natur. Univ. Comenian. Math. Publ.* **24**, 159–174, 1970.

[1981] Sarkaria K.S. On coloring manifolds. *Illinois J. Math.* **25**, 464–469, 1981.

[1982] Sarkaria K.S. A four color theorem for manifolds of dimension ≥ 3. *Abstracts Amer. Math. Soc.*, 1982.

[1983a] Sarkaria K.S. Addendum to "On coloring manifolds". *Illinois J. Math.* **27**, 612–613, 1983.

[1983b] Sarkaria K.S. On neighborly triangulations. *Trans. Amer. Math. Soc.* **277**, 213–239, 1983.

[1987] Sarkaria K.S. Heawood inequalities. *J. Combin. Theory Ser. A* **46**, 50–78, 1987.

[1974a] Seymour P.D. A note on a combinatorial problem of Erdős and Hajnal. *J. London Math. Soc. (2)* **8**, 681–682, 1974.

[1974c] Seymour P.D. On the two-colouring of hypergraphs. *Quart. J. Math. Oxford Ser. (2)* **25**, 303–312, 1974.

[1987] Seymour P.D. and C. Thomassen. Characterization of even directed graphs. *J. Combin. Theory Ser. B* **42**, 36–45, 1987.

[1988] Shelah S. Primitive recursive bounds for van der Waerden numbers. *J. Amer. Math. Soc.* **1**, 683–697, 1988.

[1981] Spencer J.H. Coloring n-sets red and blue. *J. Combin. Theory Ser. A* **30**, 112–113, 1981.

[1983] Stein P. Chvátal's conjecture and point-intersections. *Discrete Math.* **43**, 321–323, 1983.

[1974] Sterboul F. On the chromatic number of the direct product of hypergraphs. In: C. Berge and D. Ray-Chaudhuri, editors, *Hypergraph Seminar*, volume 411 of *Lecture Notes in Mathematics*, pages 165–174. Springer-Verlag, 1974.

[1981a] Taylor A. Bounds for the disjoint union theorem. *J. Combin. Theory Ser. A* **30**, 339–344, 1981.

[1985] Thomassen C. Even cycles in directed graphs. *European J. Combin.* **6**, 85–89, 1985.

[1986] Thomassen C. Sign-nonsingular matrices and even cycles in directed graphs. *Linear Algebra and Appl.* **75**, 27–41, 1986.

[1992] Thomassen C. The even cycle problem for directed graphs. *J. Amer. Math. Soc.* **5**, 217–229, 1992.

[1975a] Toft B. On colour-critical hypergraphs. In: A. Hajnal, R. Rado, and V.T. Sós, editors, *Infinite and Finite Sets*, volume 10 of *Colloquia Mathematica Societatis János Bolyai*, pages 1445–1457. North-Holland, 1975.

[1985] Tuza Z. Critical hypergraphs and intersecting set-pair systems. *J. Combin. Theory Ser. B* **39**, 134–145, 1985.

[to appear] Tuza Z. Inequalities for minimal covering sets in set systems of given rank. To appear in *Discrete Math.*

[1927] van der Waerden B.L. Beweis einer Baudetschen Vermutung. *Nieuw. Arch. Wisk.* **15**, 212–216, 1927.

16

Infinite Chromatic Graphs

16.1. SPARSE SUBGRAPHS OF HIGH CHROMATIC NUMBER

Is it true for every integer $g \geq 4$ that every $\aleph_0$-chromatic graph contains an $\aleph_0$-chromatic subgraph in which every cycle has length at least g?

Is it true for every infinite cardinal $\kappa > \aleph_0$ that every κ-chromatic graph contains a triangle-free κ-chromatic subgraph?

Is there any infinite cardinal κ for which it is true for every integer $g \geq 4$ that every κ-chromatic graph G contains an $\aleph_0$-chromatic subgraph in which every cycle has length at least g? ■

It is an easy consequence of the theorem of de Bruijn and Erdős [1951] (Theorem 1 in Chapter 1) that a graph G has infinite chromatic number if and only if G contains an infinite sequence $\{G_k\}$ of pairwise disjoint finite graphs such that $\chi(G_k) \geq k$. The union of the graphs G_k is $\aleph_0$-chromatic, and therefore the study of $\aleph_0$-chromatic graphs reduces in a sense to finite graphs.

Indeed, it follows from the preceding observation that an affirmative answer to the first question stated above for a value g_0 of g is equivalent to an affirmative answer to Problem 7.5 for g_0 and all k. To see this, assume first that the answer to Problem 7.5 is negative for some value k_0 of k. Let G be the disjoint union of finite j-chromatic counterexamples G_j for infinitely many j. Then G is $\aleph_0$-chromatic without a k_0-chromatic subgraph in which every cycle has length at least g_0. But then G is a counterexample to the first question above. Conversely, suppose that the answer to Problem 7.5 is affirmative for g_0 and all k. For an $\aleph_0$-chromatic graph G consider an infinite sequence $\{G_j\}$ of disjoint finite subgraphs of G with G_j being j-chromatic. Then there exists an infinite sequence $\{H_k\}$ of disjoint subgraphs of the graphs from $\{G_j\}$ so that H_k is k-chromatic and every cycle in each H_k has length at least g_0. The union H of all the graphs H_k is an $\aleph_0$-chromatic subgraph of G in which every cycle has length at least g_0.

The first question is due to P. Erdős and A. Hajnal, and it has been solved in the affirmative for $g = 4$ by V. Rödl (see Problem 7.5).

If $\aleph_0$ in the first question is replaced by any larger cardinal, then the answer is negative for $g \geq 5$. This follows from the result of Erdős and Hajnal [1966] that every graph of chromatic number strictly larger than κ, where $\kappa \geq \aleph_0$, contains the complete bipartite graph K_{n,κ^+} as a subgraph for every positive integer n (κ^+ denotes the least cardinal satisfying $\kappa^+ > \kappa$) and hence contains 4-cycles (see also the comprehensive survey on infinite graphs by Thomassen [1983]). However, the problem remains open for $g = 4$. This gives rise to the second question above, due to Erdős and Hajnal, who conjectured that the answer is affirmative (see Erdős [1975]).

The existence of triangle-free κ-chromatic graphs for all infinite cardinals κ was first proved by Erdős and Rado [1959]. The order of such graphs may, in fact, be just κ (see Erdős and Rado [1960]). More generally, the existence for every integer g of κ-chromatic graphs with all odd cycles of length at least g was proved by Erdős and Hajnal [1966].

The third question above is a weaker version of the first question. Again it is due to P. Erdős and A. Hajnal, who conjectured that the answer is affirmative (see Erdős [1975]).

■ 16.2. INFINITE CHROMATIC SUBGRAPHS

Let m and n be cardinals where $m > n$, and let G be a graph of chromatic number m. Is it true that G has a subgraph of chromatic number n? ■

The question was asked by Galvin [1973], who remarked that for $n \leq \aleph_0$ and any $m > n$ the answer is affirmative by the theorem of de Bruijn and Erdős [1951] (Theorem 1 in Chapter 1), and he asked in particular if the answer is affirmative for $m = \aleph_2$ and $n = \aleph_1$. Komjáth [1988] proved that the affirmative answer with $m = \aleph_2$ and $n = \aleph_1$ is consistent with the peculiar statement $2^{\aleph_0} = \aleph_3$, but he remarked that it is not known if it is consistent with the generalized continuum hypothesis.

If one asks for an induced subgraph of G with chromatic number n, the question is also open. However, and perhaps surprisingly, Galvin proved (see Erdős [1981]) that with this stronger conclusion, the affirmative answer for $m = \aleph_2$ and $n = \aleph_1$ implies the continuum hypothesis.

■ 16.3. ALMOST BIPARTITE SUBGRAPHS

Is it true that for every $f(n) \longrightarrow \infty$ there is always a graph of chromatic number $\aleph_0$ every n-vertex subgraph of which can be made bipartite by deleting at most $f(n)$ edges? ■

The question was asked by Erdős, Hajnal, and Szemerédi [1982]. In particular, they asked if the statement holds for $f(n) = \log^{(i)} n$, the i times iterated logarithm, for any fixed $i \geq 1$. Rödl [1982] solved the corresponding question for r-uniform hypergraphs in the affirmative for all $r \geq 3$.

Erdős, Hajnal, and Szemerédi [1982] pointed out that the problem is of finite character. In this connection they mentioned a construction due to L. Lovász, giving for some constant c and for every $k \geq 2$ a finite graph G_k of chromatic number $k+2$, such that every subgraph H can be made bipartite by deleting at most $c|V(H)|^{1-1/k}$ edges. We do not know if this is best possible. The union of an infinite sequence of graphs G_k with $k \to \infty$ shows that for the particular function $f(n) = cn$ the answer is affirmative. Lovász's construction is not given in detail in the paper by Erdős, Hajnal, and Szemerédi [1982]. For $k = 2$ it is due to Gallai [1963a] and is formed by completing an $n \times n$ grid (with n even) to a square tessellation of the Klein bottle. The construction for larger values of k seems to be a generalization of this (but we do not know exactly how it is defined). Other constructions with similar properties can be derived from further results of Lovász [1983b].

The answer to the question for every $f(n) \to \infty$, obtained by replacing "chromatic number $\aleph_0$" with "uncountable chromatic number" is negative. Erdős, Hajnal, and Szemerédi [1982] proved that every graph G with uncountable chromatic number contains a subgraph H with n vertices, for infinitely many n, such that the number of edges that must be removed from H to make the graph bipartite is at least $d \cdot n$, for a constant $d > 0$ depending on G.

The question was mentioned by Erdős [1992], who offered \$100 for a solution.

16.4. LARGE FINITE *n*-CHROMATIC SUBGRAPHS

Is there a function $f : \mathbf{Z}_+ \to \mathbf{Z}_+$ for which every graph of uncountable chromatic number contains an n-chromatic graph of at most $f(n)$ vertices for all $n \geq 1$? ■

The question is due to Erdős [1975]. Erdős, Hajnal, and Szemerédi [1982] proved that if such a function $f(n)$ exists, then $f(n)$ must tend to infinity faster than any function obtained by iterating the exponential function a fixed number of times.

The answer to the corresponding question for graphs of chromatic number $\aleph_0$ is that no such $f(n)$ exists. In fact, for any function $f : \mathbf{Z}_+ \to \mathbf{Z}_+$ there exists a graph G_f of chromatic number $\chi(G_f) = \aleph_0$ so that for all $n \geq 3$, every subgraph $H \subseteq G_f$ on $f(n)$ vertices has chromatic number $\chi(H) < n$. G_f can be constructed as follows: For each $k \geq 3$ let G_k denote a k-chromatic graph on n_k vertices for which every subgraph of G_k on at most $n_k/2$ vertices is 3-colorable, where $n_3 \geq 2f(3)$ and $n_k \geq \max\{2f(k), n_{k-1}\}$ for $k \geq 4$. The existence of such graphs G_k for all $k \geq 3$ follows from a result of Erdős [1962]. Finally, let G_f be the disjoint union of all the graphs G_k.

The problem was mentioned by Erdős [1992], who offered \$250 for a solution.

16.5. TREES IN TRIANGLE-FREE GRAPHS

If G is a triangle-free graph of infinite chromatic number, does G contain every finite tree as an induced subgraph? ■

The question was asked by Gyárfás [1975], who conjectured that the answer is affirmative. To prove this for all infinite chromatic numbers it suffices to prove it for the chromatic number $\aleph_0$ (this follows from the theorem of de Bruijn and Erdős [1951], see Theorem 1 in Chapter 1 and the discussion in Problem 16.1).

The existence of triangle-free κ-chromatic graphs for all infinite cardinals κ was first proved by Erdős and Rado [1959]. The order of such graphs may, in fact, be just κ (see Erdős and Rado [1960]). More generally, the existence for all integers g of κ-chromatic graphs in which all odd cycles have length at least g was proved by Erdős and Hajnal [1966].

Gyárfás, Szemerédi, and Tuza [1980] proved, among other results, that the conjecture is true for every graph G without a 4-cycle (such a G is automatically $\aleph_0$-colorable by a result of Erdős and Hajnal [1966]).

■ 16.6. UNAVOIDABLE CLASSES OF FINITE SUBGRAPHS

Let A be a class of finite graphs. What properties must A have that for every infinite cardinal κ there should be a graph G of chromatic number κ every finite subgraph of which should be in A? In particular, does this hold when A is the set of finite subgraphs of any graph with chromatic number $\aleph_1$? ■

The question was asked by Erdős [1975], motivated by the "in particular" question due to Taylor [1971]. Erdős [1985] gave an excellent overview of this and related problems.

Erdős and Hajnal [1966] proved that A must contain all bipartite graphs. Erdős, Hajnal, and Shelah [1974] proved among other results that A must contain all cycles of length at least g for some g.

If A_1 and A_2 are two classes of finite graphs with the property described above, is $A_1 \cap A_2$ also such a class? This question was asked by Erdős [1985].

Erdős [1985] asked what the answer would be if "finite (sub)graph" is replaced throughout by "countable (sub)graph." Perhaps if A is the set of countable subgraphs of any graph with chromatic number $\aleph_2$, there exists for every cardinal κ some graph of chromatic number κ so that all its countable subgraphs belong to A.

The problems above have also been studied by Hajnal and Komjáth [1984].

■ 16.7. 4-CHROMATIC SUBGRAPHS

Let G_1 and G_2 be two graphs of uncountable chromatic number. Is there a 4-chromatic graph G which is a subgraph of both G_1 and G_2? ■

The question was asked by Erdős [1985].

The similar statement for 2-chromatic graphs is true. Erdős and Hajnal [1966] proved that any graph of uncountable chromatic number in fact contains all finite bipartite graphs as subgraphs. The similar statement for 3-chromatic graphs is also true by a theorem of Erdős, Hajnal, and Shelah [1974], that every graph of uncountable chromatic number has odd cycles of all sufficiently large lengths.

Erdős [1990b] remarked that it is not impossible that 4 can be replaced by $\aleph_0$. A. Hajnal proved that 4 cannot be replaced by $\aleph_1$ (see Erdős [1985]).

■ 16.8. AVOIDING 5-CYCLES AND LARGE BIPARTITE SUBGRAPHS

Does there exist a graph with uncountable chromatic number not containing a 5-cycle and not containing a complete bipartite subgraph $K_{\aleph_0,\aleph_0}$? ■

This question is due to P. Erdős and A. Hajnal (see Erdős [1975]).

Hajnal [1971] proved that there is a graph with chromatic number $\aleph_1$ that does not contain a triangle and also no $K_{\aleph_0,\aleph_0}$.

Erdős and Hajnal [1966] proved that every graph with uncountable chromatic number contains the complete bipartite graph $K_{n,\aleph_1}$ as a subgraph for every $n > 0$.

■ 16.9. CONNECTIVITY OF SUBGRAPHS

Is it true that every graph of chromatic number at least $\aleph_1$ has a subgraph that cannot be disconnected by deletion of a finite set of vertices? ■

The question is due to Erdős and Hajnal [1966], who proved that there is a graph of chromatic number $\aleph_1$ every subgraph of which can be disconnected by the deletion of a countable set of vertices (see also Erdős [1975]).

Komjáth [1986] proved that every graph of uncountable chromatic number contains an n-connected subgraph of uncountable chromatic number for every finite number n. Komjáth [1988] also proved that if one asks for a subgraph of uncountable chromatic number and of infinite connectivity, then both the positive answer and the negative answer are consistent. Komjáth [1988] added that maybe the affirmative answer to the question of Erdős and Hajnal is provably true.

The variation of the problem when "deletion of a finite set of vertices" is replaced by "deletion of a finite set of edges" has been attacked by R. Diestel and C. Thomassen. This also remains an open problem, and it seems a difficult one [personal communication from C. Thomassen in 1993].

■ 16.10. SET OF ODD CYCLE LENGTHS

Let G be a graph of chromatic number $\aleph_0$. Denote by $n_1 < n_2 < n_3 < \cdots$ the sequence of odd integers n_i for which G contains a cycle of length n_i.

Is it true that

$$\sum_{i>0} \frac{1}{n_i} = \infty? \qquad \blacksquare$$

The question is due to P. Erdős and A. Hajnal (see Erdős [1975]). They conjectured that the answer is affirmative.

Erdős [1990b] suggested that perhaps the upper density

$$\limsup_N \frac{|\{i : n_i \leq N\}|}{N}$$

is in fact positive.

Erdős [1994] asked if there is a constant $c > 0$ so that the inequality

$$\limsup_N \frac{1}{\log N} \sum_{n_i < N} \frac{1}{n_i} \geq c$$

holds, and asked if perhaps it holds with $c = \frac{1}{2}$. Moreover, if $m_1 < m_2 < m_3 < \cdots$ is the sequence of integers m_i for which G has cycles of length m_i, is it then true that

$$\limsup_N \frac{1}{\log N} \sum_{m_i < N} \frac{1}{m_i} = 1?$$

A similar conjecture of Erdős concerning the set of both even and odd cycle lengths in G was studied and solved by Gyárfás, Komlós, and Szemerédi [1984], with further results obtained by Gyárfás, Prömel, Szemerédi, and Voigt [1985] (see also Bollobás [1986]).

See Erdős and Hajnal [1985] and Erdős [1994] for discussions of these and other related problems.

■ 16.11. UNAVOIDABLE CYCLE LENGTHS

Call an infinite sequence of integers $u_1 < u_2 < u_3 < \cdots$ "unavoidable" if every graph with infinite chromatic number must contain a cycle of length u_j for some j. Is it possible to characterize all such unavoidable sequences? More specifically, is it true that any given sequence $u_1 < u_2 < u_3 < \cdots$ is unavoidable if and only if it satisfies

$$\sum_{j=1}^{\infty} \frac{1}{u_j} = \infty?$$

Can it at least be determined if the sequence $(2^n : n \geq 2)$ is unavoidable?

■

The first question was discussed by P. Erdős and P. Mihók at a party given by J. Nešetřil in Prague in 1982. They considered in particular the sequence $(2^n : n \geq 2)$, the status of which seems to remain open (see Erdős and Hajnal [1985]). The possible complete characterization of unavoidable sequences was suggested by Mihók [1983a] in his doctoral thesis.

The suggestion of Mihók is partially supported by a result of Bollobás [1977], implying that every infinite arithmetic progression is unavoidable.

■ 16.12. COLORING NUMBER

The coloring number col(G) of a graph G is the smallest cardinal κ such that for some well-ordering $<$ of the vertex set of G, the degree of every vertex x in the subgraph induced by the vertices y with $y \leq x$ is strictly less than κ. For cardinals m and n, where $m > n$, when does $\text{col}(G) = m$ imply that G contains a subgraph G' (or an induced subgraph) of coloring number n? Is this always the case when m and n are finite and $m = n + 1$? ■

This question was asked by E.C. Milner in connection with the International Conference on Combinatorics in Keszthely, Hungary, in July 1993, held in celebration of the eightieth birthday of P. Erdős.

For $\text{col}(G) = n + 1$, when n is an integer, there is not necessarily a finite subgraph G' of G with $\text{col}(G') = n$. Erdős and Hajnal [1966] proved that if $\text{col}(G') \leq n$ for all finite subgraphs G' of G, then $\text{col}(G) \leq 2n - 2$, and in fact this is the best possible. In particular, the theorem of de Bruijn and Erdős [1951] (Theorem 1 in Chapter 1) does not hold when replacing the chromatic number by the coloring number. It does not hold even if "all finite subgraphs" is replaced by "all countable subgraphs" (we are grateful to E.C. Milner [private communication in 1994] for drawing our attention to this fact stemming from the paper by Erdős and Hajnal [1966]). Related problems in the infinite case have been studied by Erdős and Hajnal [1966, 1971] and Shelah [1975a, 1975b].

The corresponding question for the chromatic number instead of the coloring number has been raised by Galvin [1973] (Problem 16.2). This question has a positive answer for the case of $n \leq \aleph_0$ by the theorem of de Bruijn and Erdős.

■ 16.13. DIRECT PRODUCT

The direct product $G \times H$ of two graphs G and H is the graph with vertex set $V(G \times H) = V(G) \times V(H)$ and an edge $(x, y)(x', y') \in E(G \times H)$ if and only if $xx' \in E(G)$ and $yy' \in E(H)$.

Does there exist a pair of graphs G and H with $\chi(G) = \chi(H) = \aleph_{k+2}$ and $\chi(G \times H) \leq \aleph_k$ for some $k \geq 0$?

Does there exist a pair of graphs G and H for which $\chi(G) = \chi(H) = \aleph_\omega$ and $\chi(G \times H) < \aleph_\omega$, where ω denotes the first infinite ordinal? ■

Both problems were mentioned by Erdős [1985].

A. Hajnal proved that $\chi(G \times H)$ is infinite if $\chi(G)$ and $\chi(H)$ are both infinite (see Erdős [1985]). Moreover, Hajnal [1985] proved that if G has infinite chromatic number and H has finite chromatic number, then $\chi(G \times H) = \chi(H)$.

Hajnal [1985] proved that there are graphs G and H with $\chi(G) = \chi(H) = \aleph_{k+1}$ and $\chi(G \times H) = \aleph_k$ for all $k \geq 0$.

It follows from results by L. Sokoup and by R. Laver and M. Foreman that the following statement is consistent, but not provable in Zermelo–Fraenkel set theory extended with the generalized continuum hypothesis: There are graphs G_1, G_2 with $\chi(G_i) = |V(G_i)| = \aleph_2$ $(i = 1, 2)$ such that $\chi(G_1 \times G_2) = \aleph_0$ (see Hajnal [1985]).

For finite graphs it is an open problem due to S.T. Hedetniemi, whether there exists a pair of graphs G and H for which $\chi(G \times H) < \min\{\chi(G), \chi(H)\}$ (Problem 11.1).

■ 16.14. PARTITION PROBLEM OF GALVIN AND HAJNAL

If the chromatic number of G is uncountable, does there exist a partition $V(G) = \bigcup_\alpha X_\alpha$ of $V(G)$ into uncountably many subsets, so that every induced subgraph $G[X_\alpha]$ has uncountable chromatic number? ■

The question was raised by Hajnal [1970]. It is a stronger version of a conjecture by F. Galvin proved by Hajnal [1970]: If $\chi(G) > \aleph_0$, then there exists a partition of $V(G)$ into two parts so that the subgraph induced by each part has uncountable chromatic number.

The problem was studied by Komjáth [1990], who proved that a negative answer is consistent, assuming that the existence of measurable cardinals is consistent (see a textbook on large cardinals, e.g., Drake [1974] or Jech [1978]). Thus a proof of the affirmative answer to Hajnal's question would also be a proof of the nonexistence of measurable cardinals. This would solve a classical open problem in set theory.[1]

■ 16.15. SMALL SUBGRAPHS OF LARGE CHROMATIC NUMBER

Is it consistent that if $|V(G)| = \aleph_{\omega+1}$ and $\chi(G) = \aleph_1$, then G has a subgraph H with $|V(H)| = \aleph_\omega$ and $\chi(H) = \aleph_1$ (ω denotes the first infinite ordinal)? ■

[1] It is, however, generally believed that the existence of measurable cardinals is consistent, even though the consistency itself is known to be unprovable on the basis of Zermelo–Fraenkel set theory.

The corresponding question with "consistent" replaced with "true" is due to P. Erdős and A. Hajnal (see Erdős [1985]). This version of the problem was also asked by Shelah [1975a]. However, it was proved by Komjáth [1988] that it is in fact consistent that there exists a graph with $\aleph_{\omega+1}$ vertices and chromatic number $\aleph_1$, such that every subgraph whose vertex set has strictly smaller cardinality is of countable chromatic number. Komjáth noted that the question above remains open.

Many similar questions were treated by Erdős and Hajnal [1968] and discussed by Erdős [1985].

BIBLIOGRAPHY

[1977] Bollobás B. Cycles modulo *k*. *Bull. London Math. Soc.* **9**, 97–98, 1977.

[1986] Bollobás B. *Extremal Graph Theory with Emphasis on Probabilistic Methods*. Number 62 in *CBMS Regional Conference Series in Mathematics*. American Mathematical Society, 1986.

[1951] de Bruijn N.G. and P. Erdős. A colour problem for infinite graphs and a problem in the theory of relations. *Nederl. Akad. Wetensch. Proc. Ser. A* **54**, 371–373, 1951. (*Indag. Math.* **13**).

[1974] Drake F.R. *Set Theory*. North-Holland, 1974.

[1962] Erdős P. On circuits and subgraphs of chromatic graphs. *Mathematika* **9**, 170–175, 1962.

[1975] Erdős P. Problems and results on finite and infinite combinatorial analysis. In: A. Hajnal, R. Rado, and V.T. Sós, editors, *Infinite and Finite Sets*, volume 10 of *Colloquia Mathematica Societatis János Bolyai*, pages 403–424. North-Holland, 1975.

[1981] Erdős P. On the combinatorial problems which I would most like to see solved. *Combinatorica* **1**, 25–42, 1981.

[1985] Erdős P. Problems and results on chromatic numbers in finite and infinite graphs. In: Y. Alavi, G. Chartrand, L. Lesniak, D.R. Lick, and C.E. Wall, editors, *Graph Theory with Applications to Algorithms and Computer Science*, pages 201–213. Wiley, 1985.

[1990b] Erdős P. Some of my favourite unsolved problems. In: A. Baker, B. Bollobás, and A. Hajnal, editors, *A Tribute to Paul Erdős*, pages 467–479. Cambridge University Press, 1990.

[1992] Erdős P. On some of my favourite problems in various branches of combinatorics. In: J. Nešetřil and M. Fiedler, editors, *Fourth Czechoslovakian Symposium on Combinatorics, Graphs and Complexity*, volume 51 of *Annals of Discrete Mathematics*, pages 69–79. North-Holland, 1992.

[1994] Erdős P. Some of my recent problems in combinatorial number theory, geometry and combinatorics. In: Y. Alavi and A. Schwenk, editors, *Graph Theory, Combinatorics, and Algorithms: Proc. 7th International Graph Theory Conference, Kalamazoo, 1992*. Wiley, 1994.

[1966] Erdős P. and A. Hajnal. On chromatic number of graphs and set-systems. *Acta Math. Acad. Sci. Hungar.* **17**, 61–99, 1966.

[1968] Erdős P. and A. Hajnal. On chromatic numbers of infinite graphs. In: *Theory of Graphs, Proc. Colloquium, Tihany, Hungary, 1966*, pages 83–98. Academic Press, 1968.

[1971] Erdős P. and A. Hajnal. Unsolved problems in set theory. In: *Axiomatic Set Theory*, volume XIII of *Proc. Symposia in Pure Mathematics*, pages 17–48. American Mathematical Society, 1971.

[1985] Erdős P. and A. Hajnal. Chromatic number of finite and infinite graphs and hypergraphs. *Discrete Math.* **53**, 281–285, 1985.

[1974] Erdős P., A. Hajnal, and S. Shelah. On some general properties of chromatic numbers. In: Á. Császár, editor, *Topics in Topology*, volume 8 of *Colloquia Mathematica Societatis János Bolyai*, pages 243–255. North-Holland, 1974.

[1982] Erdős P., A. Hajnal, and E. Szemerédi. On almost bipartite large chromatic graphs. In: A. Rosa, G. Sabidussi, and J. Turgeon, editors, *Theory and Practice of Combinatorics*, volume 12 of *Annals of Discrete Mathematics*, pages 117–123. North-Holland, 1982.

[1959] Erdős P. and R. Rado. Partition relations connected with the chromatic number of a graph. *J. London Math. Soc.* **34**, 63–72, 1959.

[1960] Erdős P. and R. Rado. A construction of graphs without triangles having preassigned order and chromatic number. *J. London Math. Soc.* **35**, 445–448, 1960.

[1963a] Gallai T. Kritische Graphen I. *Publ. Math. Inst. Hungar. Acad. Sci.* **8**, 165–192, 1963.

[1973] Galvin F. Chromatic numbers of subgraphs. *Period. Math. Hungar.* **4**, 117–119, 1973.

[1975] Gyárfás A. On Ramsey covering numbers. In: A. Hajnal, R. Rado, and V.T. Sós, editors, *Infinite and Finite Sets*, volume 10 of *Colloquia Mathematica Societatis János Bolyai*, pages 801–816. North-Holland, 1975.

[1984] Gyárfás A., J. Komlós, and E. Szemerédi. On the distribution of cycle lengths in graphs. *J. Graph Theory* **4**, 441–462, 1984.

[1985] Gyárfás A., H.-J. Prömel, E. Szemerédi, and B. Voigt. On the sum of reciprocals of cycle lengths in sparse graphs. *Combinatorica* **5**, 41–52, 1985.

[1980] Gyárfás A., E. Szemerédi, and Z. Tuza. Induced subtrees in graphs of large chromatic number. *Discrete Math.* **30**, 235–244, 1980.

[1970] Hajnal A. On some combinatorial problems involving large cardinals. *Fund. Math.* **69**, 39–53, 1970.

[1971] Hajnal A. A negative partition relation. *Proc. Natl. Acad. Sci. U.S.A* **68**, 142–144, 1971.

[1985] Hajnal A. The chromatic number of the product of two $\aleph_1$-chromatic graphs can be countable. *Combinatorica* **5**, 137–139, 1985.

[1984] Hajnal A. and P. Komjáth. What must and what need not be contained in a graph of uncountable chromatic number. *Combinatorica* **4**, 47–52, 1984.

[1978] Jech T. *Set Theory*. Academic Press, 1978.

[1986] Komjáth P. Connectivity and chromatic number of infinite graphs. *Israel J. Math.* **56**, 257–266, 1986.

[1988] Komjáth P. Consistency results on infinite graphs. *Israel J. Math.* **61**, 285–294, 1988.

[1990] Komjáth P. A Galvin–Hajnal conjecture on uncountably chromatic graphs. In: A. Baker, B. Bollobás, and A. Hajnal, editors, *A Tribute to Paul Erdős*, pages 313–316. Cambridge University Press, 1990.

[1983b] Lovász L. Self-dual polytopes and the chromatic number of distance graphs on the sphere. *Acta Sci. Math. (Szeged)* **45**, 317–323, 1983.

[1983a] Mihók P. *On graphs critical with respect to their characteristics* (in Czech). Ph.D. thesis, University of Šafárik, Czechoslovakia, 1983.

[1982] Rödl V. Nearly bipartite graphs with large chromatic number. *Combinatorica* **2**, 377–389, 1982.

[1975a] Shelah S. A compactness theorem for singular cardinals, free algebras, Whitehead problem and transversals. *Israel J. Math.* **21**, 319–349, 1975.

[1975b] Shelah S. Notes on partition calculus. In: A. Hajnal, R. Rado, and V.T. Sós, editors, *Infinite and Finite Sets*, volume 10 of *Colloquia Mathematica Societatis János Bolyai*, pages 1257–1276. North-Holland, 1975.

[1971] Taylor W. Atomic compactness and elementary equivalence. *Fund. Math.* **71**, 103–112, 1971.

[1983] Thomassen C. Infinite graphs. In: L.W. Beineke and R.J. Wilson, editors, *Selected Topics in Graph Theory*, volume 2, pages 129–160. Academic Press, 1983.

17

Miscellaneous Problems

■ 17.1. LIST-COLORING BIPARTITE GRAPHS

Determine the smallest number $n(k)$ of vertices in a bipartite graph G of list-chromatic number exceeding k, for $k \geq 2$. That is, no matter how lists of colors of size k are assigned to the vertices of G, it is impossible to (properly) vertex-color G such that every vertex receives a color from its list. ■

The problem is due to Erdős, Rubin, and Taylor [1979].

Let $m(k)$ denote the minimum number of edges possible in a 3-chromatic k-uniform hypergraph, that is, the minimum number of k-sets in a family of sets not having property B (see Problem 15.1). Erdős, Rubin, and Taylor [1979] proved that

$$m(k) + 2 \leq n(k) \leq 2m(k).$$

It is known that $m(3) = 7, m(4) \leq 23, m(5) \leq 51$, and that in general $m(k) < k^2 2^{k+1}$ (see Problem 15.1). However, the order of magnitude of $m(k)$, and hence of $n(k)$, has not been determined.

Erdős, Rubin, and Taylor [1979] observed that $n(2) = 6 = 2m(2)$, and they remarked "*. . . although it is most likely that $n(3) = 14$, it would be quite a surprise if $n(k) = 2m(k)$ were to persist for large k.*" Indeed, Hanson, MacGillivray, and Toft [1994] confirmed the exact value of $n(3) = 14$. Moreover, they noted that $n(k) \leq kn(k-2) + 2^k$, which does not improve $n(k) \leq 2m(k)$ when using the best known bounds for $m(k)$ for k odd (see Abbott and Hanson [1969]). It is, however, an improvement for k even. In particular, using this bound one obtains $n(4) \leq 40$ and $n(6) \leq 304$, compared to the best known bounds $2m(4) \leq 46$ and $2m(6) \leq 360$ (see Seymour [1974a] and Toft [1975a]). List-chromatic numbers of complete bipartite graphs with one side much larger than the other were obtained by Hoffman and Johnson [1993].

For graphs of chromatic numbers at least 3 the corresponding problem has not yet been studied, as far as we know.

17.2. LIST-COLORING THE UNION OF GRAPHS

Let G and H be two graphs on the same set of vertices. If G is k-choosable and H is l-choosable, is their union kl-choosable? ■

As pointed out by N. Alon [personal communication in 1994] this question is a special case of a question of Erdős, Rubin, and Taylor [1979]. A graph is $(a : b)$-choosable if for any assignment of lists of length a to the vertices of the graph it is possible to choose a b-subset of the list for each vertex while keeping the chosen b-subsets disjoint for adjacent vertices. A conjecture of Erdős, Rubin, and Taylor [1979] states that if G is $(a : b)$-choosable then it is also $(am : bm)$-choosable for every positive integer m. If true, then G being k-choosable (i.e., $(k : 1)$-choosable) would imply that G is $(kl : l)$-choosable. This together with H being l-choosable easily implies that $G \cup H$ is kl-choosable.

The existence of a function $f(k, l)$ such that $G \cup H$ is $f(k, l)$-choosable was proved by N. Alon [personal communication in 1994]. It follows from the theorem of Alon [1993], stating that if a graph is k-choosable, then there is an upper bound $d(k)$ on its average degree depending only on k. The value of $d(k)$ obtained by Alon (by a probabilistic proof) is $4\binom{k^4}{k}\log\left(2\binom{k^4}{k}\right)$, which is superexponential in k.

A closely related question is the following. Let G be k-choosable and let $G[I_2]$ denote the graph obtained from G by replacing each vertex by two independent vertices. Find a best possible upper bound for $\chi_\ell(G[I_2])$ in terms of k. If G is bipartite, is $G[I_2]$ then k^2-choosable? Note that in the latter case $G[I_2]$ is the union of two graphs each consisting of two disjoint copies of G. Of course G and $G[I_2]$ have the same chromatic number, and the question here is how much their list-chromatic numbers can differ.

Alon [1993] noted that his result mentioned above, combined with a polynomial algorithm for the coloring number (see Section 1.3), implies the existence of a function g and a polynomial algorithm that for a given input graph G finds a number s, such that the list-chromatic number $\chi_\ell(G)$ lies between s and $g(s)$. Alon [1993] remarked that no similar result seems known for the usual chromatic number (Problem 10.2).

17.3. COCHROMATIC NUMBER

The cochromatic number $z(G)$ of a graph G is the smallest number of sets into which the vertex set $V(G)$ can be partitioned so that each set is either independent or induces a complete graph. Is there a constant $c > 0$ with the property that if $\chi(G) = n$ then G has a subgraph H where $z(H) \geq c \cdot n/\log n$? ■

The question was asked by J. Gimbel at the Julius Petersen Graph Theory Conference at Hindsgavl, Denmark, in July 1990 (see Gimbel [1992]).

Erdős, Gimbel, and Straight [1990] considered the problem of bounding the cochromatic number $z(G)$ in terms of $\chi(G)$ when the size $\omega(G)$ of a largest complete subgraph in G is bounded by a constant. They proved that if $\omega(G) < t$, then $z(G) > \chi(G) - r(t,t)$, where the Ramsey number $r(s,t)$ denotes the smallest positive integer n such that every graph with n vertices contains either a complete s-graph or an independent set of t vertices. Lesniak and Straight [1977] proved that if $\omega(G) < 3$, then $z(G) = \chi(G)$, except for $G = K_2$. Erdős, Gimbel, and Straight [1990] proved that if $\omega(G) < 4$, then $z(G) \geq \chi(G) - 1$, except for $G = K_3$, and they conjectured that if $\omega(G) < 5$ and $z(G) > 3$, then $z(G) \geq \chi(G) - 2$. The conjecture is best possible, since there exist graphs G that satisfy $\omega(G) = 4$, $\chi(G) = 6$, and $z(G) = 3$, as shown by Erdős, Gimbel, and Straight [1990].

The cochromatic number was first introduced by Lesniak and Straight [1977], and it has also been studied by Straight [1979, 1980], Gimbel [1984, 1986], Broere and Burger [1986, 1989], Gimbel and Straight [1987], Erdős, Gimbel, and Kratsch [1992], and Erdős and Gimbel [1993].

■ 17.4. STAR CHROMATIC NUMBER

Let G be a graph, and let k and d be positive integers with $k \geq 2d$. A (k,d)-coloring of G is an assignment $\varphi : V(G) \longrightarrow \{1, 2, \ldots, k\}$ of colors to the vertices of G in such a way that $d \leq |\varphi(u) - \varphi(v)| \leq k - d$ holds for every edge uv. The star chromatic number is defined as

$$\chi^*(G) = \inf\{k/d : G \text{ has a } (k,d)\text{-coloring}\}.$$

Is it possible to characterize the 3-colorable graphs G that satisfy $\chi^*(G) = \chi(G)$? Is it possible to characterize the graphs G that are critical and satisfy $\chi^*(G) = \chi(G)$? Is there a triangle-free 5-critical such example? A graph G is called *-perfect if $\chi^*(H) = \chi(H)$ for every induced subgraph H of G. Is such a graph perfect? Is every Berge graph *-perfect? ■

The star chromatic number was introduced and studied by Vince [1988], who noted that the Petersen graph is an example with $\chi^* = \chi = 3$, and that the Mycielski graph is a critical graph with $\chi^* = \chi$. A usual k-coloring is equivalent to a $(k,1)$-coloring as defined above, and thus $\chi^* \leq \chi$ holds. In fact $\omega \leq \chi^* \leq \chi$, where ω denotes the size of a largest complete subgraph.

Vince [1988] showed that in the definition of $\chi^*(G)$, the "inf" can be replaced by "min" taken over all $k \leq |V(G)|$.

The star chromatic number was studied also by Bondy and Hell [1990], by Zhu [1992], by Guichard [1993], who proved that it is in general ***NP***-hard to decide if $\chi^*(G) = \chi(G)$, and by Abbott and Zhou [1993a, 1994], who proved the existence of triangle-free graphs G with $\chi^*(G) = \chi(G) = k$ for all $k \geq 3$, and asked the third question above.

A perfect graph is *-perfect, and a *-perfect graph is a Berge graph, hence the affirmative answers to both of the last two questions would be equivalent to an affirmative answer to the strong perfect graph conjecture (see Section 1.6, Problem 8.1, and also Problem 8.2).

17.5. HARMONIOUS CHROMATIC NUMBER

A harmonious k-coloring of a graph G is a coloring of the vertices of G with k colors so that adjacent vertices receive different colors, and for all i, j with $1 \leq i < j \leq k$ at most one edge joins a vertex of color i to a vertex of color j. The harmonious chromatic number $\chi_H(G)$ is the smallest k such that G has a harmonious k-coloring.

Let $n, t > 0$ be integers and let $T_{n,t}$ denote the complete n-ary tree with t levels; that is, the vertices of $T_{n,t}$ are partitioned into t levels so that level 1 contains a single vertex, each vertex in level l, for $1 < l \leq t$, is adjacent to exactly one vertex in level $l - 1$, and each vertex in level m, for $1 \leq m < t$, is adjacent to n vertices in level $m + 1$. What is the harmonious chromatic number of $T_{n,t}$? ■

The harmonious chromatic number was introduced by Frank, Harary, and Plantholt [1982], and independently by Hopcroft and Krishnamoorthy [1983] in a slightly different version, allowing that two adjacent vertices are colored with the same color. It was studied in the version above by Lee and Mitchem [1987] and by Mitchem [1989], who mentioned that the value of $\chi_H(T_{n,t})$ does not seem known in general. Mitchem [1989] solved the simplest nontrivial case by proving that $\chi_H(T_{n,3}) = \left\lceil \frac{3}{2}(n+1) \right\rceil$.

The harmonious chromatic number was also studied by Miller and Pritikin [1991] and by Kundrík [1992].

17.6. ACHROMATIC NUMBER

Let G be a graph. The achromatic number $\psi(G)$ is defined as the largest number k for which the vertices of G can be colored with k colors so that for all $1 \leq i < j \leq k$ at least one edge of G joins a vertex of color i to a vertex of color j. Is it possible to characterize all graphs with $\binom{n}{2}$ edges and achromatic number n for $n \geq 2$? ■

The achromatic number was introduced and first studied by Harary, Hedetniemi, and Prins [1967] and by Harary and Hedetniemi [1970]. The problem was posed by Farber, Hahn, Hell, and Miller [1986]. An achromatic number $\psi'(G)$ for edge coloring was defined by Bouchet [1978] and also studied by Jamison [1989]. Both considered the problem of determining $\psi'(G)$ for complete graphs G.

■ 17.7. SUBCHROMATIC NUMBER

Let G be a graph. The subchromatic number $\chi_S(G)$ is the smallest number k such that the vertices of G can be partitioned into k disjoint sets $X_1, X_2, \ldots, X_k$, where each induced subgraph $G[X_i]$ is the disjoint union of complete graphs. Given $n > 0$, what is the maximum subchromatic number $S(n)$ of all graphs on n vertices? ■

Subchromatic numbers were introduced by Mynhardt and Broere [1985] as a special case of a more general concept, and studied by Albertson, Jamison, Hedetniemi, and Locke [1989], who asked this question. Albertson, Jamison, Hedetniemi, and Locke [1989] asked for a characterization of graphs G with $\chi_S(G) = 2$ and gave many partial results. These results seem to indicate that a straightforward characterization should not be expected.

■ 17.8. MULTIPLICATIVE GRAPHS

A homomorphism f from a graph G to a graph F is a mapping $f : V(G) \to V(F)$ such that every edge uv of G is mapped to an edge $f(u)f(v)$ of F.

The direct product $G \times H$ of two graphs G and H is the graph with vertex set $V(G \times H) = V(G) \times V(H)$ and edge set $E(G \times H) = \{(x_1, y_1)(x_2, y_2) : x_1x_2 \in E(G) \text{ and } y_1y_2 \in E(H)\}$.

Is it possible to characterize the graphs M that are multiplicative, that is, have the property that for any G and H, if there is a homomorphism $f : G \times H \to M$, then there is also a homomorphism from at least one of the graphs G and H to M? More precisely: Is there a polynomial algorithm for recognizing multiplicative graphs? ■

Multiplicative graphs seem first studied by Nešetřil and Pultr [1978]. The question above was asked in this general setting by Albertson [1987]. The statement that every complete graph K_k is multiplicative is equivalent to a conjecture due to Hedetniemi [1966] (see Problem 11.1), that for all graphs G and H,

$$\left.\begin{array}{l}\chi(G) \geq k+1\\ \chi(H) \geq k+1\end{array}\right\} \Longrightarrow \chi(G \times H) \geq k+1.$$

Note that there is a homomorphism from G to K_k if and only if G can be k-colored. It is easy to see that Hedetniemi's conjecture is true for $k \leq 2$, and it was proved by El-Zahar and Sauer [1985] for $k = 3$. For $k \geq 4$ the conjecture remains unsolved.

The theorem of El-Zahar and Sauer [1985] was extended by Häggkvist, Hell, Miller, and Neumann-Lara [1988], who proved that all cycles are multiplicative.

Zhu [1992] conjectured that if G_k^d is the graph with vertex set $\{0, 1, \ldots, k-1\}$, where i and j are joined by an edge if and only if $\min\{|i-j|, k-|i-j|\} \geq d$, then G_k^d is multiplicative for all $d, k > 0$ with $2d \leq k$. He showed that this would be equivalent to replacing χ by χ^* in the conjecture of Hedetniemi [1966], where χ^* denotes the star chromatic number defined by Vince [1988] (see Problem 17.4).

Albertson [1987] remarked that results by Hedrlin and Pultr [1965] and Hell [1974] imply the existence of classes of graphs that are not multiplicative.

A question asked by Nešetřil and Pultr [1978] appears to stand open: What is the smallest nonmultiplicative graph?

17.9. REDUCIBLE GRAPH PROPERTIES

A property P of graphs is hereditary if, whenever the graph G has property P, and H is a subgraph of G, then H also has property P. $\mathcal{L}$ denotes the lattice of all hereditary properties ordered by inclusion (see also Section 1.10).

For $P_1, P_2 \in \mathcal{L}$ the property $P_1 \cdot P_2$ is defined as follows. $G \in P_1 \cdot P_2$ if and only if $V(G)$ can be partitioned into two sets V_1 and V_2 such that $G[V_i] \in P_i$ for $i = 1, 2$, where $G[V_i]$ denotes the subgraph of G induced by V_i. A property $R = P_1 \cdot P_2$, where $P_1, P_2 \in \mathcal{L}\backslash\{\emptyset\}$, is said to be reducible, and P_1 and P_2 are said to divide R.

Is the factorization of every property in $\mathcal{L}$ into irreducible properties unique? Equivalently, is every irreducible property Q also "prime," that is, whenever Q divides $P_1 \cdot P_2$, does it follow that Q divides P_1 or P_2?

For a given hereditary property P, a reducible property $R = P_1 \cdot P_2$ is called a "minimal reducible bound" for P if $P \subseteq R$ and there is no reducible property $R' \subset R$ satisfying $P \subseteq R'$. What are the minimal reducible bounds for the class of planar graphs? ■

The first question is due to P. Mihók [personal communication in 1993], who remarked that the answer is probably negative. The property of being k-colorable has factorization O^k, where O denotes the property of not having edges, and it is not difficult to see that this factorization is unique for all $k \geq 1$.

The second question was asked by P. Mihók and B. Toft at the International Conference on Combinatorics held in Keszthely, Hungary, 1993, dedicated to P. Erdős on his eightieth birthday. Mihók [1993] proved that the class of outerplanar graphs, where each member has an embedding in the plane such that every vertex lies on the boundary of the infinite face, has exactly two minimal reducible bounds.

17.10. T-COLORINGS

Let G be a graph and let T be a finite set of nonnegative integers with $0 \in T$. For a positive integer n, an n-tuple T-coloring of G is an assignment S of n-sets of nonnegative integers to the vertices of G, so that if $i \in S(x)$ and $j \in S(y)$ for any edge $xy \in E(G)$, then $i - j \notin T$. Define $\mathrm{sp}_T^n(G)$ as the minimum difference, taken over all n-tuple T-colorings S of G, between the largest number and the smallest number in the union $\bigcup_{x \in V(G)} S(x)$ of all the assigned sets. The number $\mathrm{sp}_T^n(G)$ is called the n-tuple T-span of G. For fixed numbers $k > 1$, $n > 0$, $r \geq 0$ and with $T = \{0, 1, \ldots, r\}$, is it true that

$$\{\mathrm{sp}_T^n(G) : \chi(G) = k\} = \{(r + 1)(k - 1) + 2(n - 1), \ldots, (n + r)k - (r + 1)\}$$

always holds? ■

Füredi, Griggs, and Kleitman [1989] conjectured that the answer to the question is affirmative. They proved, and Tesman [1989] did so independently, that the following inclusion holds:

$$\{\mathrm{sp}_T^n(G) : \chi(G) = k\} \subseteq \{(r + 1)(k - 1) + 2(n - 1), \ldots, (n + r)k - (r + 1)\},$$

and they proved that equality holds when $n = 2$.

It was remarked by Füredi, Griggs, and Kleitman that the truth of their conjecture for $r = 0$ (i.e., for $T = \{0\}$) would imply the theorem of Lovász [1978], conjectured by Kneser [1955], on the chromatic numbers of Kneser graphs (see Problem 9.13). An elementary proof of Lovász's theorem is not known; thus it would seem that an easy proof of the conjecture should not be expected. The Kneser graph $G_t(n)$, where $n \geq t > 0$, is defined as the graph with vertex set $V(G_t(n))$ equal to the set of t-subsets of the set $\{1, 2, \ldots, n\}$, with two t-subsets A and B adjacent if and only if $A \cap B = \emptyset$. The more general graph $G_t(n, d)$, where $n \geq t > 0$ and $d > 0$, has the same vertex set as $G_t(n)$, but A and B are adjacent in $G_t(n, d)$ if and only if $\min\{|a - b| : a \in A, b \in B\} \geq d$. In particular, $G_t(n, 1) = G_t(n)$. Füredi, Griggs, and Kleitman conjectured

$$\chi(G_t(d(k - 1) + 2t - 1, d)) \geq k$$

for all $t, d, k > 0$, and they proved that this would follow from their conjecture on n-tuple T-colorings. For $d = 1$ this last conjecture is the theorem of Lovász.

Bollobás and Thomason [1979] introduced the problem of coloring the vertices of a graph with equally sized sets, subject to the condition that the sets assigned to any two adjacent vertices must be disjoint, and they observed the connection with Lovász's theorem on Kneser graphs.

In the special case $(k,n,r) = (3,2,1)$, Füredi, Griggs, and Kleitman found a complete characterization in terms of homomorphisms. They proved that if G is 3-chromatic, then $\mathrm{sp}^2_{\{0,1\}}(G) = 6$ if and only if G is homomorphic to a cycle of length 5; otherwise, $\mathrm{sp}^2_{\{0,1\}}(G) = 7$. A homomorphism from a graph G to a graph H is a map $f : V(G) \to V(H)$ so that if $xy \in E(G)$ then $f(x)f(y) \in E(H)$. If such an f exists, G is said to be homomorphic to H. Note that any usual k-coloring of G is equivalent to a homomorphism from G to the complete k-graph. See also Problem 10.8.

An n-tuple T-coloring with $n = 1$ is called a T-coloring. In the simplest case $T = \{0\}$, a T-coloring of G is the same as a usual coloring of G. The concept was introduced by Hale [1980], when studying the practical problem of assigning a transmitting frequency to each of a number of radio and television stations $X_1, X_2, \ldots, X_m$. Such an assignment must be done subject to the condition that any pair of stations geographically close enough to interfere with each other should transmit signals that differ sufficiently in frequency. A graph G is defined with vertex set $V(G) = \{X_1, X_2, \ldots, X_m\}$ and with an edge between any pair of stations that are mutually close. T corresponds to the set of so-called "disallowed channel separations," and any T-coloring of G corresponds to an assignment of a frequency to each station as desired.

For further results and references, see the comprehensive survey on T-colorings and their generalizations by Roberts [1991]. Results on T-colorings were also obtained by Liu [1992]. A generalization called "list T-coloring" has been introduced and studied by Tesman [1989, 1993].

■ 17.11. GAME CHROMATIC NUMBER

Let G be a graph and let k be a positive integer. The following two-person game is played on G: Alice and Bob alternate turns, with Alice having the first move. Both players know G and k from the beginning. A move consists in choosing a vertex v from G that has not yet been assigned a color and assigning to v a color from $\{1, 2, \ldots, k\}$ distinct from the colors assigned previously (by any of the two players) to the vertices in G adjacent to v. The game stops with Alice the winner if after $|V(G)|$ moves G has been colored. Bob is declared the winner if at some point of the game there is a vertex that cannot be assigned a color; that is, the vertex is already adjacent to vertices of all the k different colors.

The game chromatic number $\chi_g(G)$ is defined as the smallest k for which Alice has a winning strategy. Clearly,

$$\chi(G) \leq \chi_g(G) \leq |V(G)|.$$

What is the best possible upper bound for the game chromatic number of a planar graph? ■

The game chromatic number was introduced by Bodlaender [1991], who proved that the maximum game chromatic number of all trees is either 4 or 5. Faigle, Kern, Kierstead, and Trotter [1993] proved that the game chromatic number of any tree is at most 4. The question was asked by Kierstead and Trotter [1993], who proved that

$$7 \leq \max_{G \text{ planar}} \chi_g(G) \leq 33,$$

where the upper bound was proved using the four-color theorem (see Section 1.2). The same arguments except using Heawood's weaker five-color theorem would give an upper bound of 41. Moreover, Kierstead and Trotter remarked that the lower bound can be improved to 8. For outerplanar graphs (i.e., graphs with an embedding in the plane such that all vertices lie on the boundary of a single face) Kierstead and Trotter gave the bounds

$$6 \leq \max_{G \text{ outerplanar}} \chi_g(G) \leq 8.$$

Faigle, Kern, Kierstead, and Trotter [1993] have shown that there is no upper bound for $\chi_g(G)$ as a function of $\chi(G)$. In fact, there exist bipartite graphs of arbitrarily high game chromatic numbers. However, Kierstead and Trotter [1993] proved that there exists a function f such that if G does not contain a subdivision of the complete n-graph K_n as a subgraph, then $\chi_g(G) \leq f(n)$. In particular, for any surface $\mathbf{S}$, there is a smallest number $\chi_g(\mathbf{S})$ such that every graph embeddable on $\mathbf{S}$ has game chromatic number at most $\chi_g(\mathbf{S})$.

■ 17.12. HARARY AND TUZA'S COLORING GAMES

Let G be a graph and let k be a positive integer. The two following two-person games, Achievement and Avoidance, can be played on G: Alice and Bob alternate turns, with Alice having the first move. Both players know G and k from the beginning. A move consists in choosing a vertex v from G that has not yet been assigned a color and assigning to v a color from $\{1, 2, \ldots, k\}$ distinct from the colors assigned previously (by any of the two players) to the vertices in G adjacent to v. The player who makes the last move is the winner in the Achievement game, but loses in the Avoidance game.

Characterize those graphs G and integers k for which Alice has a winning strategy in Achievement, and those G and k for which she has a winning strategy in Avoidance. ■

The Avoidance and Achievement coloring games were studied by Harary and Tuza [1993]. They proved for the Petersen graph and $1 \leq k \leq 3$ that Alice has a winning strategy in Achievement if and only if $k = 1$ or $k = 2$, and that she has a winning strategy in Avoidance if and only if $k = 1$ or $k = 3$. They also analyzed the games for paths and cycles in the case $k = 2$.

If the number of colors is greater than the maximum degree $\Delta(G)$, then obviously the winner is determined only by the parity of the number of vertices in G. Harary and Tuza [1993] drew attention to three particular nontrivial cases: $k = \Delta(G)$, $k = \chi(G)$, and $k = 1$, and they remarked that it remains an open problem to analyze the Achievement and Avoidance games on paths and cycles when $k = 1$.

■ 17.13. COLORING EXTENSION GAME

Let m, n, and k be integers with $3 \leq k < m < 2k$ and $k \leq n$. Two players, the "mapmaker" and the "explorer," play the following game. First the explorer presents n independent vertices H and the mapmaker responds with an m-coloring of H. Then the explorer presents a k-colorable extension H_1 of H (i.e., H_1 contains H as a subgraph), and the mapmaker responds with an m-coloring of H that can be extended to an m-coloring of H_1. Then the explorer presents a k-colorable extension H_2 of H_1, and the mapmaker responds with an m-coloring of H that can be extended to an m-coloring of H_2. And so on. The explorer wins if the mapmaker is forced to use all possible m-colorings of H in this process (not considering a renaming of colors as a different coloring). For which values of m, n, and k does the explorer have a winning strategy? ■

This problem is due to Beigel and Gasarch [1991]. They proved that for $3 \leq k = m$ the explorer always has a winning strategy. This also follows from a result of Müller [1975, 1979], even if the explorer may only use graphs where all cycles are long. The remarkable result of Müller is the following.

Theorem (Müller [1975, 1979]). *Let H be an independent set of vertices and let $P_1, P_2, \ldots, P_r$ denote different partitions of H into at most k classes ($k \geq 3$). Then there exists a k-chromatic graph G with $H \subseteq V(G)$ such that G has precisely r different k-colorings $\varphi_1, \varphi_2, \ldots, \varphi_r$ (where colorings obtained by renaming or permuting the colors are considered the same), and for all i the restriction of φ_i to H partitions H into color classes like P_i. Moreover for any given ℓ, there exists such a G in which all cycles have length at least ℓ and any two vertices of H have distance at least ℓ.*

On the other hand, Beigel and Gasarch [1991] remarked that it follows from techniques of Bean [1976] that for $m = 2k$, the mapmaker has a winning strategy. It seems unknown where the situation changes between k and $2k$.

■ 17.14. WINNING HEX

Let $G(m, n)$ be a planar graph consisting of an $m \times n$ grid with a complete set of parallel diagonals added, say all the diagonals directed northeast. More formally, $G(m, n)$ has mn vertices (a, b), where a and b are integers satisfying $1 \le a \le m$ and $1 \le b \le n$. The vertices (a, b) and (c, d) are joined by an edge in $G(m, n)$ if and only if either ($a = c$ and $b - d = \pm 1$) or ($a - c = \pm 1$ and $b = d$) or ($a - c = b - d = \pm 1$). The graph $G(m, n)$ may be thought of as the planar dual of an $m \times n$ diamond-shaped board of hexagons (disregarding, when forming the dual, the infinite face surrounding the board).

The positional game Hex is played on the "board" $G(m, n)$. Two players alternately select a previously unselected vertex of $G(m, n)$. The goal of the first player (respectively, the second player) is to complete a path in $G(m, n)$ joining the two horizontal sides (respectively, the two vertical sides). Find a winning strategy for the first player when playing Hex on $G(m, n)$, at least for small values of n. The case $G(11, 11)$ of the standard Hex board would be of particular interest. ■

Hex was invented in 1942 by Piet Hein while contemplating the four-color problem, and independently in 1948 by John F. Nash (see Gardner [1957, 1959]). Gardner [1957] wrote *"It is something of an occasion these days when someone invents a mathematical game that is both new and interesting. Such a game is Hex, introduced 15 years ago at Niels Bohr's Institute of Theoretical Physics in Copenhagen. It may become one of the most widely played and thoughtfully analyzed mathematical games of the century."*

Hex may be thought of as a variation of the positional games played on a hypergraph (see Problem 15.7). The hypergraph $H(m, n)$ in question has the same vertex set as $G(m, n)$ and two sets of edges $\mathcal{A}$ and $\mathcal{B}$, where the edges of $\mathcal{A}$ (respectively, $\mathcal{B}$) correspond to the vertex sets of paths joining the horizontal sides (respectively, vertical sides). (Note that the vertex sets of paths joining opposite corners are in both $\mathcal{A}$ and $\mathcal{B}$.) The complete bipartite graph $K_{|\mathcal{A}|,|\mathcal{B}|}$ is not vertex list-colorable when associating with the vertices in one side the sets in $\mathcal{A}$ as lists, and in the other side the sets in $\mathcal{B}$. This coloring statement is equivalent to the Ramsey-type statement that if the vertices of $G(m, n)$ are partitioned into two classes, then either there is an edge in $\mathcal{A}$ with all vertices in the first class or there is an edge in $\mathcal{B}$ with all vertices in the second class. In particular, this implies that Hex cannot end in a draw. Interestingly, the fact that Hex cannot end in a draw has been shown by Gale [1979] equivalent to Brouwer's fixed-point theorem.

Since Hex is finite there is always a winning strategy for the first or for the second player. In the symmetric cases when $m = n$ it cannot be a disadvantage to start, hence there is a winning strategy for the first player. It seems that explicit winning strategies have been found only for $n \le 5$ (see Gardner [1959]). It may, of course, be possible that winning strategies can be described in quite simple terms without the strategy being practical. For example, we do not know if the following strategy always works: Select a vertex contained in the largest number of "good" paths not containing any of the vertices already selected by the opponent.

Algorithmic aspects of Hex type games on graphs in general were considered by Even and Tarjan [1975]. They proved that the problem to determine which player has a winning strategy in a given position of the game on a given graph is ***PSPACE***-complete. Reisch [1981] subsequently proved that in fact usual $n \times n$ Hex is ***PSPACE***-complete. Thus if the problem to determine which player has a winning strategy in a given position on an $n \times n$ Hex board is solvable in polynomial time, then any problem solvable in polynomial space is solvable in polynomial time as well. This is considered unlikely (see Garey and Johnson [1979] for an introduction to the class ***PSPACE***, and to ***PSPACE***-complete decision problems).

A similar game played on the edge set of a graph is due to C.E. Shannon and is called Shannon's Switching Game. Given a graph G and two distinguished vertices u and v of G, two players alternately select an edge, one player with the goal of forming a path in G from u to v, the other player with the goal of collecting enough edges that their removal would destroy all such paths. Lehman [1964] generalized this game from graphs to matroids and gave a complete characterization of the instances of the general game that are won for either player, including an efficient method for determining the best move in any possible situation. A solution of Lehman's version of Shannon's Switching Game based on the matroid intersection theorem of Edmonds [1965a] was presented in Chapter 19 of the book on matroid theory by Welsh [1976]. A graph-theoretical solution of Shannon's Switching Game was given by Bruno and Weinberg [1970].

Further variations of Hex have been described by Gardner [1975], Evans [1974, 1976], and Alpern and Beck [1991].

BIBLIOGRAPHY

[1969] Abbott H.L. and D. Hanson. On a combinatorial problem of Erdős. *Canad. Math. Bull.* **12**, 823–829, 1969.

[1993a] Abbott H.L. and B. Zhou. The star chromatic number of a graph. *J. Graph Theory* **17**, 349–360, 1993.

[1994] Abbott H.L. and B. Zhou. Some theorems concerning the star chromatic number of a graph. Manuscript, 1994.

[1987] Albertson M.O. Generalized colorings. In: D.S. Johnson, T. Nishizeki, A. Nozaki, and H.S. Wilf, editors, *Discrete Algorithms and Complexity*, pages 35–49. Academic Press, 1987.

[1989] Albertson M.O., R.E. Jamison, S.T. Hedetniemi, and S.C. Locke. The subchromatic number of a graph. *Discrete Math.* **74**, 33–49, 1989.

[1993] Alon N. Restricted colorings of graphs. In: K. Walker, editor, *Surveys in Combinatorics: Proc. 14th British Combinatorial Conference*, pages 1–33. Cambridge University Press, 1993.

[1991] Alpern S. and A. Beck. Hex games and twist maps on the annulus. *Amer. Math. Monthly* **98**, 803–811, 1991.

[1976] Bean D.R. Effective coloration. *J. Symbolic Logic* **41**, 469–480, 1976.

[1991] Beigel R. and W.I. Gasarch. The mapmakers dilemma. *Discrete Appl. Math.* **34**, 37–48, 1991.

[1991] Bodlaender H.L. On the complexity of some coloring games. In: R.H. Möhring, editor, *Graph-Theoretic Concepts in Computer Science*, volume 484 of *Lecture Notes in Computer Science*, pages 30–40. Springer-Verlag, 1991.

[1979] Bollobás B. and A.G. Thomason. Set colourings of graphs. *Discrete Math.* **25**, 21–26, 1979.

[1990] Bondy J.A. and P. Hell. A note on the star chromatic number. *J. Graph Theory* **14**, 479–482, 1990.

[1978] Bouchet A. Indice achromatique des graphes multiparti complets et réguliers. *Cahiers Centre Études Rech. Opér.* **20**, 331–340, 1978.

[1986] Broere I. and M. Burger. Uniquely cocolorable graphs. *Graphs Combin.* **2**, 201–208, 1986.

[1989] Broere I. and M. Burger. Critically cochromatic graphs. *J. Graph Theory* **13**, 23–28, 1989.

[1970] Bruno J. and L. Weinberg. A constructive graph-theoretic solution of the Shannon switching game. *IEEE Trans. Circuit Theory* **CT-17**, 74–81, 1970.

[1965a] Edmonds J. Minimum partition of a matroid into independent subsets. *J. Res. Natl. Bur. Stand.* **69B**, 67–72, 1965.

[1985] El-Zahar M. and N.W. Sauer. The chromatic number of the product of two 4-chromatic graphs is 4. *Combinatorica* **5**, 121–126, 1985.

[1993] Erdős P. and J. Gimbel. Some problems and results in cochromatic theory. In: J. Gimbel, J.W. Kennedy, and L.V. Quintas, editors, *Quo Vadis, Graph Theory?* volume 55 of *Annals of Discrete Mathematics*, pages 261–264. North-Holland, 1993.

[1992] Erdős P., J. Gimbel, and D. Kratsch. Some extremal results in cochromatic and dichromatic theory. *J. Graph Theory* **15**, 579–585, 1992.

[1990] Erdős P., J. Gimbel, and H.J. Straight. Chromatic number versus cochromatic number in graphs with bounded clique size. *European J. Combin.* **11**, 235–240, 1990.

[1979] Erdős P., A.L. Rubin, and H. Taylor. Choosability in graphs. In: *Proc. West Coast Conference on Combinatorics, Graph Theory and Computing, Arcata, 1979, Congr. Num., 26*, pages 125–157, 1979.

[1974] Evans R. A winning opening in reverse Hex. *J. Recreational Math.* **7**, 189–192, Summer 1974.

[1976] Evans R. Some variants of Hex. *J. Recreational Math.* **8**, 120–122, 1975–1976.

[1975] Even S. and R.E. Tarjan. A combinatorial problem which is complete in polynomial space. In: *Proc. 7th ACM Symposium on Theory of Computing*, pages 66–71. ACM, 1975. *J. Assoc. Comput. Mach.* **23**, 710–719, 1976.

[1993] Faigle U., W. Kern, H.A. Kierstead, and W.T. Trotter. On the game chromatic number of some classes of graphs. *Ars Combin.* **35**, 143–150, 1993.

[1986] Farber M., G. Hahn, P. Hell, and D.J. Miller. Concerning the achromatic number of graphs. *J. Combin. Theory Ser. B* **40**, 21–39, 1986.

[1982] Frank O., F. Harary, and M. Plantholt. The line-distinguishing chromatic number of a graph. *Ars Combin.* **14**, 241–252, 1982.

[1989] Füredi Z., J.R. Griggs, and D.J. Kleitman. Pair labellings with given distance. *SIAM J. Discrete Math.* **2**, 491–499, 1989.

[1979] Gale D. The game of Hex and the Brouwer fix point theorem. *Amer. Math. Monthly* **86**, 818–827, 1979.

[1957] Gardner M. Mathematical games. *Sci. Amer.* **197**, 1957. July issue pp. 145–150; August issue, pp. 120–127; October issue, pp. 130–138.

[1959] Gardner M. Ticktacktoe. In: *The Scientific American Book of Mathematical Puzzles and Diversions*, pages 37–46. Simon and Schuster, 1959. Republished as *Hexaflexagons and Other Mathematical Diversions* (with a new afterword and a new bibliography), The University of Chicago Press, 1988.

[1975] Gardner M. Mathematical games. *Sci. Amer.* **232** and **233**, 1975. June issue, pp. 106–111; December issue, pp. 116–119.

[1979] Garey M.R. and D.S. Johnson. *Computers and Intractability: A Guide to the Theory of NP-Completeness.* W.H. Freeman and Company, 1979.

[1984] Gimbel J. *The chromatic and cochromatic number of a graph.* Ph.D. thesis, Western Michigan University, 1984.

[1986] Gimbel J. Three extremal problems in cochromatic theory. *Rostock. Math. Kolloq.* **30**, 73–78, 1986.

[1992] Gimbel J. Problem. *Discrete Math.* **101**, 353–354, 1992.

[1987] Gimbel J. and H.J. Straight. Some topics in cochromatic theory. *Graphs Combin.* **3**, 255–265, 1987.

[1993] Guichard D.R. Acyclic graph coloring and the complexity of the star chromatic number. *J. Graph Theory* **17**, 129–134, 1993.

[1988] Häggkvist R., P. Hell, D.J. Miller, and V. Neumann-Lara. On multiplicative graphs and the product conjecture. *Combinatorica* **8**, 63–74, 1988.

[1980] Hale W.K. Frequency assignment: theory and applications. *Proc. IEEE* **68**, 1497–1514, 1980.

[1994] Hanson D., G. MacGillivray, and B. Toft. Vertex list-colouring of bipartite graphs. Technical report, University of Regina, 1994. To appear in *Ars Combin.*

[1970] Harary F. and S.T. Hedetniemi. The achromatic number of a graph. *J. Combin. Theory* **8**, 154–161, 1970.

[1967] Harary F., S.T. Hedetniemi, and G. Prins. An interpolation theorem for graphical homomorphisms. *Portugal. Math.* **26**, 453–462, 1967.

[1993] Harary F. and Z. Tuza. Two graph-colouring games. *Bull. Austral. Math. Soc.* **48**, 141–149, 1993.

[1966] Hedetniemi S.T. Homomorphisms of graphs and automata. Technical Report 03105–44–T, University of Michigan, 1966.

[1965] Hedrlin Z. and A. Pultr. Symmetric relations with given semigroup. *Monatsh. Math.* **68**, 318–322, 1965.

[1974] Hell P. On some strongly rigid families of graphs and the full embeddings they induce. *Algebra Universalis* **4**, 108–126, 1974.

[1993] Hoffman D.G. and P.D. Johnson Jr. On the choice number of $K_{m,n}$. *Congr. Num.* **98**, 105–111, 1993.

[1983] Hopcroft J. and M.S. Krishnamoorthy. On the harmonious colorings of graphs. *SIAM J. Algebraic Discrete Methods* **4**, 306–311, 1983.

[1989] Jamison R.E. On the edge achromatic numbers of complete graphs. *Discrete Math.* **74**, 99–115, 1989.

[1993] Kierstead H.A. and W.T. Trotter. Planar graph coloring with an uncooperative partner. In: W.T. Trotter, editor, *Planar Graphs*, volume 9 of *DIMACS Series in Discrete Mathematics and Theoretical Computer Science*, pages 85–93. American Mathematical Society, 1993.

[1955] Kneser M. Aufgabe 360. *Jahresber. Deutsch. Math.-Verein.* **58**, 2. Abteilung, 27, 1955.

[1992] Kundrík A. The harmonious chromatic number of a graph. In: J. Nešetřil and M. Fiedler, editors, *Fourth Czechoslovakian Symposium on Combinatorics, Graphs and Complexity*, volume 51 of *Annals of Discrete Mathematics*, pages 167–170. North-Holland, 1992.

[1987] Lee S.M. and J. Mitchem. An upper bound for the harmonious chromatic number. *J. Graph Theory* **11**, 565–567, 1987.

[1964] Lehman A. A solution of the Shannon switching game. *J. Soc. Indust. Appl. Math.* **12**, 687–725, 1964.

[1977] Lesniak L. and H.J. Straight. The cochromatic number of a graph. *Ars Combin.* **3**, 34–46, 1977.

[1992] Liu D.D.-F. T-colorings of graphs. *Discrete Math.* **101**, 203–212, 1992.

[1978] Lovász L. Kneser's conjecture, chromatic number, and homotopy. *J. Combin. Theory Ser. A* **25**, 319–324, 1978.

[1993] Mihók P. On the minimal reducible bound for outerplanar and planar graphs. Manuscript, 1993. Submitted to *Proc. International Conference on Combinatorics, Keszthely, Hungary, 1993*, dedicated to P. Erdős on his eightieth birthday.

[1991] Miller Z. and D. Pritikin. The harmonious coloring number of a graph. *Discrete Math.* **93**, 211–228, 1991.

[1989] Mitchem J. On the harmonious chromatic number of a graph. *Discrete Math.* **74**, 151–157, 1989.

[1975] Müller V. On colorable critical and uniquely colorable critical graphs. In: M. Fiedler, editor, *Recent Advances in Graph Theory: Proc. Symposium, Prague, June 1974*, pages 385–386. Academia Praha, 1975.

[1979] Müller V. On colorings of graphs without short cycles. *Discrete Math.* **26**, 165–176, 1979.

[1985] Mynhardt C. and I. Broere. Generalized colorings of graphs. In: Y. Alavi, G. Chartrand, L. Lesniak, D.R. Lick, and C.E. Wall, editors, *Graph Theory with Applications to Algorithms and Computer Science*, pages 583–594. Wiley, 1985.

[1978] Nešetřil J. and A. Pultr. On classes of relations and graphs determined by subobjects and factorobjects. *Discrete Math.* **22**, 287–300, 1978.

[1981] Reisch S. Hex ist PSPACE-vollständig. *Acta Inform.* **15**, 167–191, 1981.

[1991] Roberts F.S. T-colorings of graphs: Recent results and open problems. *Discrete Math.* **93**, 229–245, 1991.

[1974a] Seymour P.D. A note on a combinatorial problem of Erdős and Hajnal. *J. London Math. Soc. (2)* **8**, 681–682, 1974.

[1979] Straight H.J. Cochromatic number and genus of a graph. *J. Graph Theory* **3**, 43–51, 1979.

[1980] Straight H.J. Note on the cochromatic number of several surfaces. *J. Graph Theory* **4**, 115–117, 1980.

[1989] Tesman B. *T-colorings, list T-colorings, and set T-colorings of graphs*. Ph.D. thesis, Rutgers University, 1989.

[1993] Tesman B. List T-colorings of graphs. *Discrete Appl. Math.* **45**, 277–289, 1993.

[1975a] Toft B. On colour-critical hypergraphs. In: A. Hajnal, R. Rado, and V.T. Sós, editors, *Infinite and Finite Sets*, volume 10 of *Colloquia Mathematica Societatis János Bolyai*, pages 1445–1457. North-Holland, 1975.

[1988] Vince A. Star chromatic number. *J. Graph Theory* **12**, 551–559, 1988.

[1976] Welsh D.J.A. *Matroid Theory*. Academic Press, 1976.

[1992] Zhu X. Star chromatic numbers and products of graphs. *J. Graph Theory* **16**, 557–569, 1992.

Author Index

Subject Index

WILEY-INTERSCIENCE
SERIES IN DISCRETE MATHEMATICS AND OPTIMIZATION

AARTS AND KORST
Simulated Annealing and Boltzmann Machines: A Stochastic Approach to Combinatorial Optimization and Neural Computing

ALON, SPENCER, AND ERDÖS
The Probabilistic Method

ANDERSON AND NASH
Linear Programming in Infinite-Dimensional Spaces: Theory and Application

BARTHÉLEMY AND GUÉNOCHE
Trees and Proximity Representations

BAZARRA, JARVIS, AND SHERALI
Linear Programming and Network Flows

COFFMAN AND LUEKER
Probabilistic Analysis of Packing and Partitioning Algorithms

DINITZ AND STINSON
Contemporary Design Theory: A Collection of Surveys

GLOVER, KLINGHAM, AND PHILLIPS
Network Models in Optimization and Their Practical Problems

GONDRAN AND MINOUX
Graphs and Algorithms
(*Translated by S. Vajda*)

GRAHAM, ROTHSCHILD, AND SPENCER
Ramsey Theory
Second Edition

GROSS AND TUCKER
Topological Graph Theory

HALL
Combinatorial Theory
Second Edition

JENSEN AND TOFT
Graph Coloring Problems

LAWLER, LENSTRA, RINNOOY KAN, AND SHMOYS, EDITORS
The Traveling Salesman Problem: A Guided Tour of Combinatorial Optimization

LEVITIN
Perturbation Theory in Mathematical Programming Applications

MAHMOUD
Evolution of Random Search Trees

MARTELLO AND TOTH
Knapsack Problems: Algorithms and Computer Implementations

MINC
Nonnegative Matrices

MINOUX
Mathematical Programming: Theory and Algorithms
(*Translated by S. Vajda*)

MIRCHANDANI AND FRANCIS, EDITORS
Discrete Location Theory

NEMHAUSER AND WOLSEY
Integer and Combinatorial Optimization

NEMIROVSKY AND YUDIN
Problem Complexity and Method Efficiency in Optimization
(*Translated by E. R. Dawson*)

PACH AND AGARWAL
Combinatorial Geometry

PLESS
Introduction to the Theory of Error-Correcting Codes
Second Edition

SCHRIJVER
Theory of Linear and Integer Programming

TOMESCU
Problems in Combinatorics and Graph Theory
(*Translated by R. A. Melter*)

TUCKER
Applied Combinatorics
Second Edition